1O Cases in engineering design

Edited by H. O. Fuchs
Professor of Mechanical Engineering
Stanford University, Stanford, California

and R. F. Steidel, Jr.
Professor and Chairman
Department of Mechanical Engineering
University of California, Berkeley, California

Longman

LONGMAN GROUP LIMITED
London
*Associated companies, branches and representatives
throughout the world*

First published 1973

ISBN 0 582 44144 7 cased
44145 5 paper

Made and printed in Great Britain by
William Clowes & Sons, Limited, London, Beccles and Colchester

Contents

Preface

What are case studies?

The use of case studies and the case approach in teaching engineering seek to put the engineering student in the position of the professional engineer faced with a real-life problem. The student is confronted with incidents drawn from actual practice and is given an opportunity to review decisions made by the responsible engineers and the outcome of these decisions as well as professional procedure. The case approach is used in Business Administration, Medicine, Law, and to a lesser extent in some fields of engineering, such as Process Engineering. It has never been widely adopted in engineering, but its use is increasing.

The case approach is the antithesis of the conventional lecture method of teaching engineering. In such lectures, physical principles, material properties or mathematical logic are isolated from reality and examined in essentially a pure environment. The use of case studies in teaching leaves these aspects of the problem in their natural context. One examines them but keeps the examination subordinate to the total real-life problem. In many ways, case studies promote a dialogue between the real world of the engineering profession and the world of engineering education, to the benefit of all. Finally, the case study technique expands the student's ability to communicate technical information verbally and in written form and emphasizes the importance of this skill as exercised by the practising engineer.

The majority of the students participate with enthusiasm in the verbal dissection of the cases, and are strongly interested in suggesting alternate design approaches. The students naturally make a critical examination of each case study looking for a logical design process. In some it is very difficult to find, and in one case included in this collection, they commonly reach the conclusion that it is totally absent.

The most general reaction of the students to the cases is one of consternation. It is difficult for many of them to see how the practising engineer can be so inept in his groping for design solutions. He appears to spend extravagant amounts of money and time for modest progress. This reaction appears natural and healthy when one reflects on the prior conditioning of the students via the 'classroom problem' and the 20/20 hindsight vision with which all of us are blessed. The absence of high quality analysis and the substitution of search, hunch, and hearsay described in several of the cases is a continuing source of frustration to the students.

Although it is invigorating and realistic, the use of case studies is not meant to replace lectures in teaching specific subject matter, since the lecture mode of teaching is still more 'efficient' in the use of time and resources. If measurable quantity of absorbed and reproducible facts is the object of the teacher, he can do better by drill and recitation. Nor is the use of case studies meant to replace projects and project experience. If the instructor seeks to develop flexibility of mind and desire for understanding, the case method can be recommended as a supplement, to enrich classroom experiences, and it does serve as the logical interface with professional engineering.

How are they used?

The case method has been very successfully used for about a century in teaching law and for about half a century in teaching business administration. James B. Conant, former president of Harvard and High Commissioner in Germany, claims that the greater flexibility of American as compared with German administrators is a result of the differences in legal education; by cases in the US, deductively in Germany.

In using the case method, the instructor abandons the illusion that he can teach all that one can know about the subject, or that he can define the fundamentals of the subject. He admits that he can only let the student study examples and prepare the students to learn more by themselves when the need arises or when time permits. Engineering cases can serve as the examples in this method of instruction.

There are many ways of using cases for instruction. Each teacher uses cases in a way suited to his own objectives, his environment and his personal style. Many engineering professors with industrial experience or a consulting practice use their own experiences to enrich the courses they teach. The late Professor John Arnold of Stanford University used case studies frequently and with skill both at the Massachusetts Institute of Technology and later at Stanford University. He was most interested in the creative aspects of mechanical engineering and used fictional cases as a vehicle for innovation and invention.

The Design Division of the Department of Mechanical Engineering at Stanford University uses the case history method extensively in teaching design engineering. It is the custodian of an extensive library of over 150 cases. The Division continually develops new cases each year, keeping the list both current and appealing. These case studies are available to schools and are used in many ways. They are used as reading assignments, background for specific problems, to illustrate theory, as material for problem formulation, as research material and as examination problems.

At the graduate level, a case study program appears to offer some valuable benefits to the students who are lacking in exposure to industrial experience. In the Division of Mechanical Design at the University of California, Berkeley, a design project recently completed outside the University is selected for review and study by a team of graduate students who write a case study of their own. They assemble all related information, weigh and sort facts, and try to establish the basis on which engineering decisions were made in their assigned project. In writing a case study, they begin to appreciate that the degree of investigation and analysis must be consistent with the real circumstances that surround a problem if the engineer is to be successful in design.

Ten case histories

Ten cases have been selected as specific examples of engineering design for use in education. Four of the cases have been developed for the last three years in a course on Case Studies in Engineering at Berkeley. Four were developed at Stanford. Two are from other sources. As a collection, they represent a variety of problems and many fields.

In choosing these cases care has been exercised in the selection to provide a cross-section of engineering problems. Eight of the ten cases involve private industry; the other two are from national laboratories. One is concerned with a product that must sell for $1 and another with a program costing more than $10 000 000. Eight are in between. One involves mass production, and four involve the design of a single item. The fields of nuclear engineering, housing, aerospace, cryogenics, petroleum production, instrumentation, photography, and machine design are all represented. Time constraints, choice of materials or components, and costs seldom have been subjects of classroom problems. In each of the cases at least one of these considerations appears; the interrelations of two or three appear in most of the cases. Decision making, where two or more design solutions are viable alternatives, is the main subject of at least five of the cases. Personnel problems invade three, and the age old problem of whom to believe among a bevy of experts is the main problem of the introductory case.

The collection

The first case is the dilemma of a project engineer whose design has failed. Corrective measures make the problem even worse. In seeking advice, he finds as many opinions as there are advisers. The project engineer has made errors and some of these affect the design. Some do not. Many points of view can be and are presented. The engineer is faced with several critical decisions. The decision making process is thrust upon the student immediately, and he is eager to give his opinion of what went wrong. Invariably, everyone is smarter than the project engineer. The case also shows severe time constraints.

The second case is more conventional. The project engineer in this did not fully define his problem, and because of this, he commits his design prematurely. Because they are different people, two engineers propose to solve the problem in two different ways. The result is a personal conflict between the young project engineer and his older supervisor. This is an excellent case for the discussion of engineering ethics on whether the project engineer should have proceeded without the approval of his boss. It also involves material selection for large nonlinear deformations, a subject not covered in a normal curriculum. This case develops the difference between a case and an engineering report.

The third case is the design of tanks for holding large quantities of liquid oxygen. It is a structural problem involving prestressed concrete. Many people, several companies and several departments of the same company had to work together in order for the problem to be solved.

The fourth case involves pieces of equipment for the emerging technology of the solid state computer industry. It is the first where a qualitative decision of making or buying the equipment must be made. It is an excellent case from which sophisticated design problems can be developed. It is also a case in which management is a constraint on the project engineer.

The make or buy decision is also a part of the fifth case, where two successive project engineers come to different conclusions about a hydraulic control problem. One recommends buying and the other recommends making all the parts within the company. The case study involves a large corporation involving many plants and a central engineering laboratory.

The sixth case is also a control problem, but it is a government rather than a commercial project, and Canadian. It involves a different technology (photography) and requires many iterations before a successful design is reached.

The seventh case is a study of decision making in a big corporation. It is obvious from the start that the problem could be solved if it could be defined. But, who defines the problem, and who says that it has been defined? The case study shows the interaction between a private power company and a controlling government agency. Economics, heat transfer and construction are several of the problems covered in detail.

The eighth case is more sophisticated than any of the previous seven. It involves elastic body mechanics and a thorough understanding of current manufacturing technology. It shows the reward for meticulous, thorough attention to the details of the design. It is excellent to develop the genealogy of a solution.

The ninth case involves a low cost, high production item, invention and patents. It brings out the large effort expended to produce a modest item if it is produced in large quantities.

The tenth and last case is an example of the most sophisticated modern design techniques. In this case study, time and weight are constraints, but money is not. Anything less than an unqualified success is a catastrophic failure, so all effort is unleashed to assure success. Periodic design reviews and failure mode analyses are used. The problem is solved by eliminating the device.

The cases are arranged in order of increasing complexity and sophistication. Cases with similar problems have been placed together, other things being equal. For example, cases VII and VIII are concerned with nuclear technology, cases IV and V contain the make or buy decision, V and VI are control problems and IV and VI involve photography.

How to use this collection

A teacher interested in case studies will probably want to experiment with several modes of using them.

Each case is followed by a short section called 'suggested study' which lists some problems, assignments, or topics for discussion. They are intended as examples of the type of study which the teacher might assign to his students. For those who have never used the case approach we might mention that class discussion of the cases, preferably in sections of fewer than fifteen students, is one of the best ways to bring out the content of the cases and to exercise the student's skills in qualitative analysis and in verbal communication. The discussions are easily started by asking a leading, preferably controversial, question. The teacher must then be on guard however to keep the discussion focused, to prevent it from becoming a 'bull session'.

To balance the qualitative and free-floating class discussion it may be advisable to assign more tightly structured homework before the discussion. This might require sketching alternative solutions or calculating some numbers which would have been useful to the engineer whose work is described in the case, or sketching a diagram of the morphology of the design work described in the case.

We think that you will find this collection a stimulating addition to your educational experience, whether it be as an instructor, student, or casual reader.

Acknowledgements

A large part of the work presented in this volume was made possible by grants from the US National Science Foundation to Stanford University and from the Ford Foundation to the University of California, Berkeley and Los Angeles. Mr Peter Booker and the Institution of Engineering Designers contributed Case VIII. Professor Geza Kardos of Carleton University prepared Case VI and took a very active part in the editorial work for this volume. It is a pleasure to express our thanks to these people, to the authors of the cases, the engineers who talked with the authors, and the companies who released the information.

H. O. Fuchs
R. F. Steidel, Jr.

October 1971

Case 1 Failure of a ball bearing

by Karl H. Vesper

Prepared with support from the National Science Foundation for the Stanford Engineering Case Program. Assistance from Jack Wireman, Elmer F. Ward, Dino Morelli and Thomas Barish is gratefully acknowledged.

This case is available in pamphlet form as ECL 14 from the Engineering Case Library, Room 500, Stanford University, Stanford, California 94305.

PART 1

In September 1963, Mr Jack Wireman, an engineer at Task Corporation was faced with the question of what should be done about the failure of an electric motor shaft bearing. The motor was a new design to power an aircraft hydraulic pump. A contract for 300 such motors had been placed with Task, and the terms of this contract required that the motor be capable of operating for 2 500 hours without failure. When the motor in test began to draw excessive current after only 1 800 hours operation, it was shut down and disassembled. Examination showed severe pitting and wear of the balls and races in the front shaft bearing. Pictures of the bearing after failure appear in Exhibit 1. Cross-section drawings of the motor and the hydraulic pump to which it attached appear in Exhibits 2 and 3.

Task Corporation was at this time a company of 140 employees in Anaheim, California, which manufactured a variety of products. These included electric motors, pumps, fans, blowers, refrigeration systems, and measuring instruments for wind-tunnel testing. The electric motors ranged in size from fractions to hundreds of horsepower and were for special applications where high performance such as great speed, low weight or small volume per horsepower were required. Confidence in Task's ability to design high performance motors was reflected in a comment by Mr Elmer Ward, the company's president: 'I once told a man that he could have anyone he wanted design him an electric motor, and if we couldn't come up with a better one in 24 hours, I'd pay one hundred dollars. He never collected, and the offer is still open.'

In April 1962, Task was invited by the Geyser Pump Company* to bid on the job of designing and producing an electric motor to be used for driving an aircraft hydraulic pump. Geyser was, in turn, preparing its own bid to make the pump, attach the motor to the pump, and sell the completed units to the Thunder Aircraft Company* for installation in Thunder 99* aircraft. Two such units were required per aircraft. Each pump would operate continuously during flight. A specification described the pump as: 'The primary source for the flight power control units as well as an alternate source of hydraulic pressure. It is also used as a hydraulic pressure source for use during pre-flight testing and routine aircraft maintenance operations.' In addition to these electrically powered hydraulic pumps, each 99 had its main pumps which were driven mechanically by the turbojet engines.

Operating requirements of the electric pump units included the following:

Hydraulic fluid	Skydrol 500A (See fluid specification, Exhibit 4)
Fluid temperature	65°F (18°C) to 180°F (82°C) ±5
Inlet pressure	45–50 psi
Outlet pressure	3 000 psi ±50
Flow rate	1 to 6 gal/min ±0·1
Input power	400 cycle A.C. at 200 volts
Output power (motor only)	11·75 horsepower
Weight (motor only)	18·75 lb
Efficiency (motor only)	65%
Maximum motor case temp.	225°F (107°C)
Explosion proof	
Life	2 500 hours with 'no failure of parts or excessive wear'

In addition to these operating conditions, the motor was required to comply with eight US Government specification documents, eleven non-government specification documents, and various other requirements relating to torque at different speeds, rotational acceleration at 65°F (18·0°C), flammability, corrosion and fungus resistance, electrical bonding, radio interference, noise level, fume propagation, safety wiring, dielectric arc-over resistance, and fluid leakage under pressure.

Testing of the performance and life was to be done by Geyser Pump on two

* disguised name

'qualification units'. A schematic diagram of the test rig is shown in Exhibit 5. In the life test, each motor was to be run through several different cycles as described in the schedule of Exhibit 6.

After some discussion between engineers of Task and Geyser it was decided that a 6 000 rpm synchronous unit would allow the best compromises of weight and size. Task engineers planned on the motor being liquid cooled by the hydraulic fluid enroute to the pump. The fluid would, after passing through a ten micron filter, enter one side of the motor at the front, flow back through annular slots around the stator, circulate forward through the rotor, then pass through a centrifugal impeller at the front of the motor and into the Geyser pump for boost up to 3 000 psi. Both front and rear bearings of the motor would be kept fully immersed by the hydraulic fluid as it circulated through to the pump under a pressure within the motor of around 50 psi. A schematic of this flow pattern appears in Fig. A. Task offered to design and build the motors at a price between $600 and $1 000 each, depending on the quantity ordered.

In May 1963 Geyser Pump Company won a contract for the motor-pump units from Thunder and gave Task a contract to provide 300 motors in total over a 12-month period. The first two units would be for qualification tests to assure that the motors performed as required. All remaining units would be for installation in Thunder aircraft. It was expected that contracts for more of the motors would be placed with Task in the future since Task was the only company selected to make them. If, however, the motors did not meet performance specifications of the contract, Geyser could refuse to accept them from Task and would not have to pay for them. Similarly, by the contract between Thunder and Geyser, Thunder could refuse to accept the complete pump units if either the motor or the pump did not measure up to specified performance.

When the Task proposal was accepted, engineers at Task proceeded with detailed design of the motor. Electrical components, numbers of turns, amount of iron and other aspects of the electromagnetic circuitry were determined. Dimensions of the path for circulating the hydraulic fluid and the shape of the centrifugal impeller were designed. The shape and dimensions of the aluminium motor housing had to be arranged to fit the electrical and hydraulic requirements and also to mate with the Geyser pump. Power transmission between motor and pump was to be through a spline, the dimensions of which were given by Geyser. Shaft dimensions of the motor were picked by Task engineers to transmit expected loads with abundant safety and to blend with the required spline.

Three types of loading were expected on the motor shaft, torsional, radial, and axial. Torsional loading would be primarily due to the torque output of the motor. Radial loading would be mainly caused by the weight of the rotor and impeller (5·3 lb), and by rotor dynamic imbalance, which by careful manufacture, testing and balancing would be limited to five pounds. The main axial load on the motor shaft was expected to be imposed by the pump shaft which was axially pre-loaded by a spring in the pump. The pump had a male spline which was pushed into the mating motor shaft, compressing the spring, according to Geyser engineers, to a force of 50 to 70 lb.

Ball bearings made by the Barden Corporation were picked for the shaft. The use of Barden bearings had become something of a standard practice at Task after problems had been encountered with bearings of several other manufacturers. Barden specialized in bearings of high precision. Although relatively expensive, such bearings were considered highly reliable.

The front bearing chosen for the motor was a Barden No. 204SST5 conrad, with an outer race centered* cage and a contact angle† of 14°. This bearing was chosen

---

* 'Outer race centered' refers to the manner in which the ball cage or retainer of the bearing is centered. In most commonly used bearings, the ball cage is centered by contact against the inner race. At high rotational speeds, centrifugal force imparted by the cage can make it difficult for lubri-

† See next page.

because it apparently provided a sufficient margin of load capacity and its inner race was large enough to slip easily over the internally splined shaft, and its outer race would fit conveniently into the motor housing. Radial allowances between 0·0001 and 0·0003 in were provided between inner race and shaft and between outer race and housing. To absorb axial loading on the shaft, the inner race of this bearing was rigidly fixed to the shaft by a nut and lockwasher, and the outer race was fastened into the aluminium motor housing by pieces of the housing being bolted together.

A small bearing, Barden No. 203SS5, was picked for the opposite end of the shaft. This rear bearing was not rigidly fixed to either shaft or housing, so it could move axially to allow for dimensional variations of manufacture and for expansion and contraction of motor parts with changes in temperature. A radial allowance of 0·0004 in up to 0·001 in was left between outer race and housing, and an allowance of zero up to 0·0003 in between inner race and shaft on the rear bearings. To insure proper bearing geometry and to prevent the bearing from having its outer race turn in the housing, a pair of belleville springs were compressed with a force of 4 to 26 lbs between the rear of the housing and the outer race.

Although no detailed life calculations were made on the bearings, catalogue data and design experience indicated to Task engineers that they would last well beyond the 2 500 hours required life, and that if failure occurred, it would first happen elsewhere in the motor. It was assumed that the greatest loads on the front bearing would be axial and less than 100 lbs, since the maximum pump preload spring force predicted by Geyser was 70 lb. The Barden 204, it was felt, should be more than adequate for such loading. For the tail of the shaft, a smaller bearing was considered sufficient. There would be no need for the rear bearing to be large enough to slip over the spline, and the axial loading on it was expected to be very small, equivalent to the belleville spring load. Both bearings cost around $10, but by using the smaller bearing at the rear, there would be a few cents saved on each motor. Another benefit of the smaller bearing would be lighter weight. Excerpts from a Barden Company catalogue giving specifications on the chosen bearings appear in Exhibit 7.

The first qualification test motor was made in the Task shop and tested on the Task dynamometer, without symptoms of difficulty. After this test, the motor was shipped, in late August 1963, to the Geyser Pump Company where it was mated to a pump, installed in a test rig, and started on life tests according to the schedule of Exhibit 6. After 1 800 hours, Geyser technicians monitoring the test, noticed the motor was drawing excessive current. They shut the motor down and removed it from the pump. Turning the shaft over by hand, they could feel roughness in the bearings. (It wasn't possible to hear the bearings during test because the pump made too much noise.)

At 10 a.m. the next morning, several Geyser engineers appeared at Task with the motor asking for an answer to the problem. The motor was immediately disassembled in the Task shop. The front bearing had become severely galled, and the grooves of the races had been widened about 0·040 in by wear all the way round. Wear was so extensive on both balls and races that the engineers could see no clues in the bearing as to how it had failed.

Meanwhile, the second qualification test motor had been finished and was ready for testing. Tooling had been completed and production was under way, with over a

---

cant to reach the inner race. Also at high speeds the cage may become centrifugally expanded and thereby lose centering contact. By centering through contact with the outer race, such loss of centering contact is better controlled for many applications.

† 'Contact angle' refers to the angle between a line perpendicular to the shaft axis of rotation and a line through both the centre of a ball and the point at which that ball contacts with the outer race. The initial contact angle is defined by assuming contact with no deflections and is fixed by radial looseness.

dozen motors complete and others in various stages of manufacture. Geyser Pump was progressing similarly in production of pumps to match the motors. Task top management had asked Mr Jack Wireman, a mechanical engineer, to prescribe action for curing the problem.

(See page 62, Suggested Study, question 1.)

EXHIBIT 1   204SST5 bearing after 1 800 hours operation

EXHIBIT 2   Pump motor

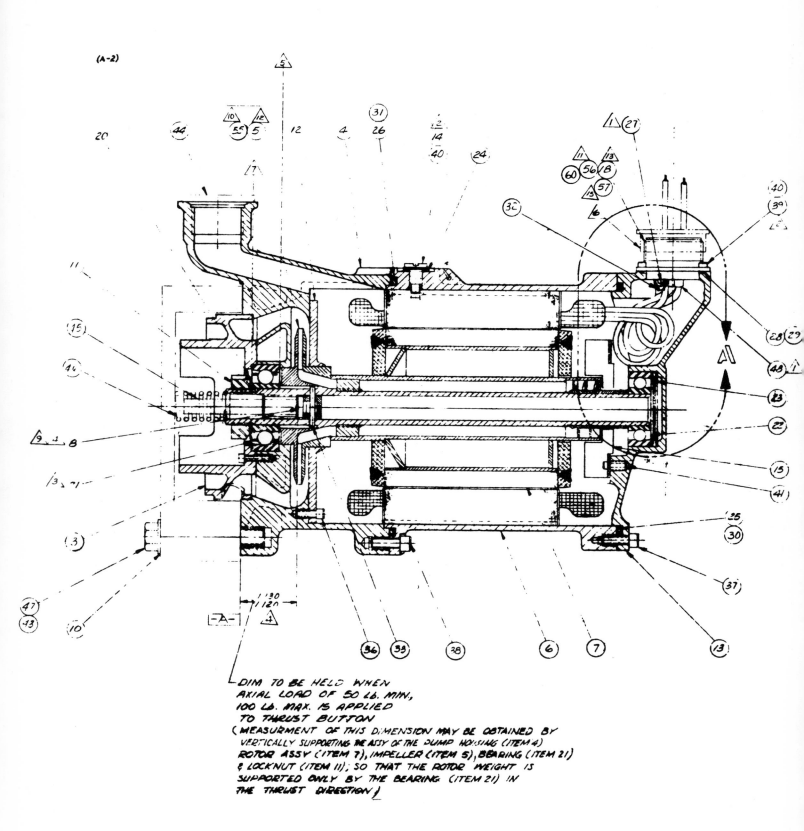

DIM TO BE HELD WHEN
AXIAL LOAD OF 50 LB. MIN,
100 LB. MAX. IS APPLIED
TO THRUST BUTTON
(MEASUREMENT OF THIS DIMENSION MAY BE OBTAINED BY
VERTICALLY SUPPORTING THE ASSY OF THE PUMP HOUSING (ITEM 4)
ROTOR ASSY (ITEM 7), IMPELLER (ITEM 5), BEARING (ITEM 21)
& LOCKNUT (ITEM 11); SO THAT THE ROTOR WEIGHT IS
SUPPORTED ONLY BY THE BEARING (ITEM 21) IN
THE THRUST DIRECTION)

EXHIBIT 3   Pump assembly

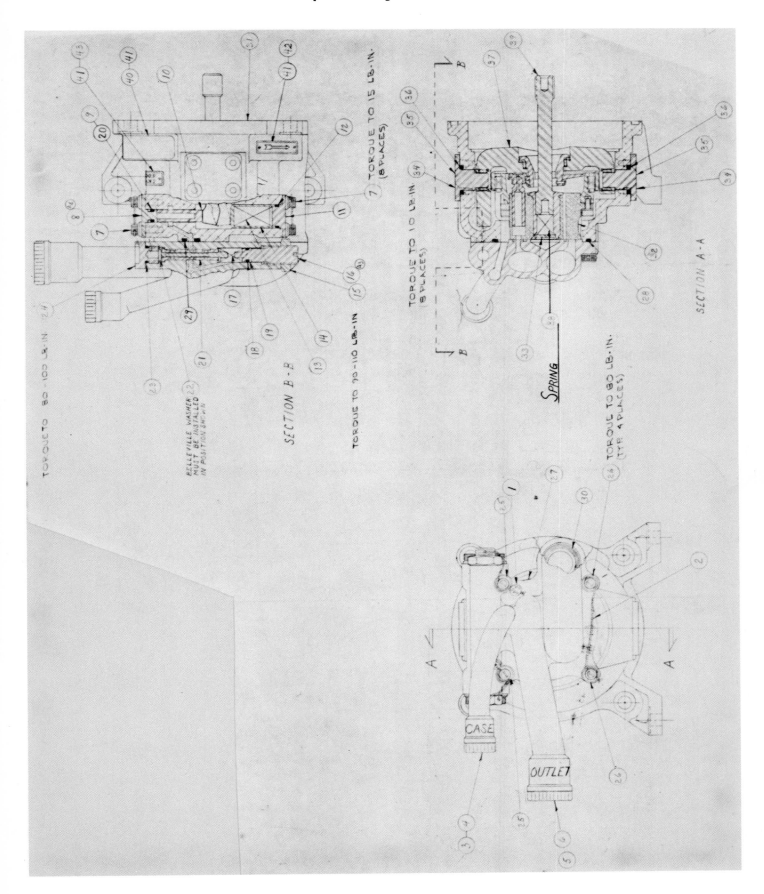

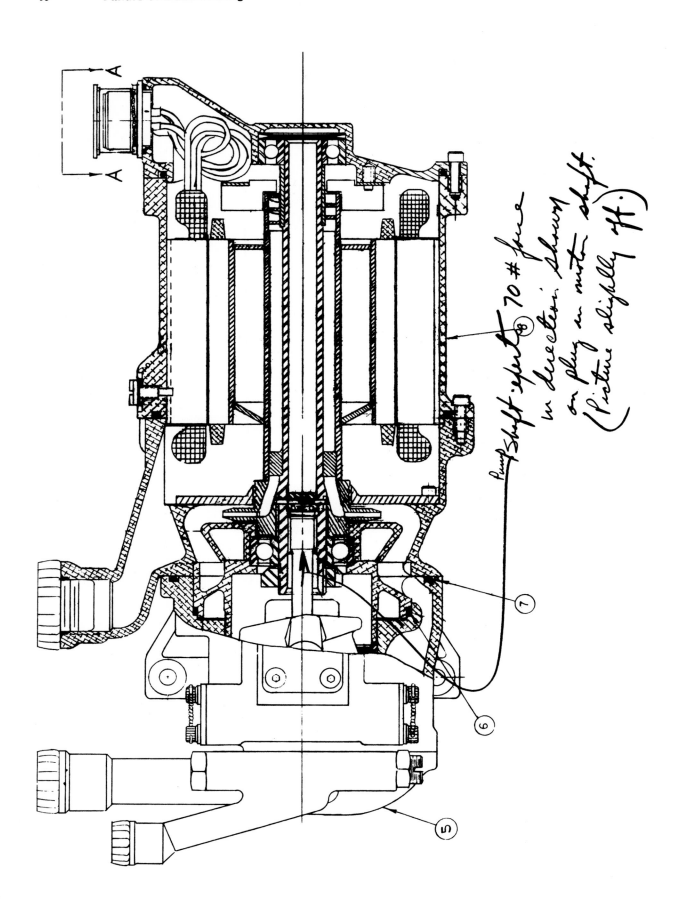

Pump shaft exerts 70 # force
in direction shown
on plug in motor shaft.
(Picture slightly off.)

EXHIBIT 4   Properties

Properties

SKYDROL SPECIFICATIONS FOR NEW FLUIDS

Property	Skydrol 500A	Skydrol 7000
Appearance	Clear, purple liquid	Clear, green liquid
Neutralization number, mg. KOH/gm.	0.10 maximum	0.20 maximum
Specific gravity 25°C./25°C.	1.060 to 1.066	1.080 to 1.086
Viscosity, CS		
@ 210°F. (99°C.)	3.90 to 4.00	3.95 to 4.05
@ 100°F. (38°C.)	11.30 to 12.10	15.00 to 16.00
Moisture, per cent	0.20	0.25

TYPICAL PHYSICAL AND CHEMICAL PROPERTIES

Property	Skydrol 500A	Skydrol 7000
Appearance	Clear, purple liquid	Clear, green fluid
Odor	Mild, pleasant	Mild, pleasant
Autogenous ignition Temperature, (ASTM D 286-58T)	Above 1100° F. (593°C.)	1060°F. (571°C.)
Pour Point, (ASTM D97)	Below −85° F. (−65°C.)	Below −70°F. (−57°C.)
Neutralization number, (ASTM D-974-58T) mg. KOH/gm.	0.01	0.01
Specific gravity, (ASTM D941) (See Graph, page ..) 77°F. (25°C.)	1.065	1.086
Viscosity index	+238	+160
Viscosity, CS, (ASTM D445) (See Graph, page 14)		
210°F. (99°C.)	3.95	4.00
100°F. (38°C.)	11.70	15.50
Pressure Viscosity, CS 100°F., @ 4000 psi	20.8	Not Available
Thermal conductivity, 82°F. (28°C.)	0.0777 Btu/Hr./Ft.²°F./Ft. (32.2 X 10⁻⁵Cal/Sec. CM.² °C. CM)	0.0723 Btu/Hr./Ft.²°F./Ft. (29.9 X 10⁻⁵Cal/Sec. CM.² °C. CM)
Specific heat, @ 100°F., Btu/lb.,°F. or Cal/gm.°C.	0.39	0.42
Isothermal secant Bulk modulus 77°F., (See Graph, page 15)	340,000 psi	328,000 psi
Foaming (ASTM D892-46T)	Essentially nonfoaming	Essentially nonfoaming
Shear stability	Comparable to MIL-H-5606A	Exceeds MIL-H-5606A
Surface Tension, 77°F. (25°C.)	30.8 dynes/cm.	28.9 dynes/cm.
Hydrolytic stability	**Skydrol** 500A and **Skydrol** 7000 are not seriously affected by low concentrations of water (less than 2%), but water contamination should be avoided.	
Refractive index, nD 25°C.	1.470 to 1.475	1.5067 to 1.507
Heat of Combustion	12,900 Btu/lb.	Not available
Thermal coefficient of expansion	0.000452/°F. 0.000813/°C.	0.000418/°F. 0.000753/°C.
Moisture %, (Karl Fischer)	0.05	0.05
Dielectric Strength (ASTM D877-49) 25°C.	12 KV	36 KV
Dielectric constant 25°C., 1 KC.	8.81	8.87
Volume resistivity (ASTM D-1169) OHM-CM 25°C.	43 X 10⁶	500 X 10⁶

PHYSICAL PROPERTIES

Compressibility

Skydrol 500A and *Skydrol* 7000 both have a better bulk modulus than MIL-H-5606A, resulting in a more positive response of the hydraulic system. Table I shows the bulk modulus of *Skydrol* 500A, *Skydrol* 7000, water and MIL-H-5606A.

Table I   ISOTHERMAL BULK MODULUS COMPARISON
77°F. (25°C.), 0 to 500 psi

Fluid	Bulk Modulus psi
Skydrol 500A	340,000
Skydrol 7000	328,000
Water	337,000
MIL-H-5606A	259,000

Volatility

The low volatility of *Skydrol* 500A and *Skydrol* 7000 assure low evaporation rates in the hydraulic system. Both fluids are relatively unaffected by changes in pressure and temperature; *Skydrol* 500A and *Skydrol* 7000 have far lower volatility than MIL-H-5606A or similar products. The Graph on page 16 shows vapor pressure vs. temperature for MIL-H-5606A, *Skydrol* 500A and *Skydrol* 7000.

Table II

Fluid	Viscosity Pressure Coefficient (VPC)*
Skydrol 500A	1.98
MIL-H-5606	2.52
MLO 7557	3.25
2-Ethyl Hexyl Sebacate	2.52

See SKYDROL 500A Viscosity vs. Pressure on page 14.
*Measured as per cent of change in viscosity per 100 psi pressure change.

Pressure Viscosity

The change of fluid viscosity with pressure will alter the fluid flow characteristics which could slow system response.

Skydrol fluids simplify this design engineering problem, for pressure has relatively little effect on these fluids. The measure of the effect of pressure on viscosity, the Viscosity Pressure Coefficient, for *Skydrol* 500A is compared with other commonly used aircraft fluids in Table II.

Table III   FOAMING TENDENCY OF SKYDROL 500A AND SKYDROL 7000 (ASTM D 892-46T)

	SKYDROL 500A Tendency	SKYDROL 500A Stability	SKYDROL 7000 Tendency	SKYDROL 7000 Stability
75°F. (24°C.)	40 ml.	22 sec.	5 ml.	2 sec.
200°F. (93°C.)	20 ml.	5 sec.	5 ml.	2 sec.
75°F. (24°C.)	40 ml.	19 sec.	5 ml.	2 sec.

Foaming Tendency

Both *Skydrol* 500A and *Skydrol* 7000 have a very low foaming tendency. The ability of these fluids to resist air "pick-up" and reject entrained air add greater reliability of the hydraulic system. Table III gives the sequence and results of foaming tests for both fluids.

THERMAL AND CHEMICAL PROPERTIES

Chemical Stability

Skydrol 500A and *Skydrol* 7000 are heat stable and resist oxidation at temperatures beyond those encountered in actual service. In hydraulic systems, the upper temperature limit for continuous operation for *Skydrol* 500A and *Skydrol* 7000 is approximately 225°F., (107°C.). Portions of the system can operate for a short time at higher temperatures without excessive deterioration of the fluids. At operational pressures of 3000 psi, *Skydrol* fluids have shown no tendency to thicken or form sludge over thousands of service hours.

Non-corrosiveness

Skydrol 500A and *Skydrol* 7000 are phosphate ester-based fluids which have little corrosive effects on the metallic parts of the hydraulic system—a fact proven by years of flight experience. In general, the *Skydrol* fluids act as metal passivaters, which means they may be used as the preservative fluid in hydraulic components.

In mineral oil systems, a special preservative grade fluid is usually used to protect the component during storage. Systems equipped with the *Skydrol* fluids require no such special preservative grade fluid, the *Skydrol* fluids providing sufficient protection.

Table IV shows the low corrosive effects of both *Skydrol* fluids to common metals.

Table IV   CORROSION AND OXIDATION TEST RESULTS
(per MIL H-5606, 168 hrs. at 250°F. (121°C.)

Property	Skydrol 500A Initial	Skydrol 500A Final	Skydrol 7000 Initial	Skydrol 7000 Final
Viscosity at 130°F. (54°C.) (cs.)	8.37	8.55	9.60	9.81
Neutralization number, mg KOH/gm.	0.02	0.03	0.13	0.28

Metal	Effect on Metals (weight change mg./cm.°) Skydrol 500A	Skydrol 7000
Cu	−0.03	−0.92
Fe	−0.01	0.15
Al	0.00	0.02
Mg	−0.01	−0.12
Cd/Fe	−0.01	0.00
Fluid evaporation	<1.0%	2.05%
Fluid Separation	None	None
Color change	Fluid darkens slightly	Changes to darker green

Heat Transfer

Skydrol 500A and *Skydrol* 7000 have good heat transfer properties. Table V lists thermal data.

Table V   THERMAL DATA OF SKYDROL 500A AND
SKYDROL 7000

Property	Skydrol 500A	Skydrol 7000
Specific Heat		
90 to 120°F. (32 to 49°C.)	— —	0.45 Btu/lb./°F.
−40°F. (−40°C.)	0.34 Btu/lb./°F.	— —
75°F. (24°C.)	0.38 Btu/lb./°F.	— —
145°F. (63°C.)	0.41 Btu/lb./°F.	— —
212°F. (100°C.)	0.44 Btu/lb./°F.	— —
Thermal Conductivity		
82°F., Btu/Hr.Ft.^{2}°F.Ft.	0.0777	0.0723
28°C., Cal/Sec.Cm.^{2}°C.CM	29.9 X 10^{-5}	32.2 X 10^{-5}
178°F., Btu/Hr.Ft.^{2}°F.Ft.	0.0779	0.0716
81°C., Cal/Sec. Cm^2°C.CM	32.2 X 10^{-5}	29.6 X 10^{-5}

Table VI

Dielectric strength	Skydrol 500A	Skydrol 7000
25°C., 0.01" Gap	12 KV	36 KV
Dielectric Constant		
25°C., 100KC	8.81	8.87 (1 KC)
100°C., 100KC	6.95	
Power Factor		
25°C., 10KC	67%	0.49 (1 KC)
25°C., 100KC	6.7%	
100°C., 10KC	100%	
100°C., 100KC	39%	
Volume Resistivity, ohm-cm		
25°C., 500V., D.C., 0.1"Gap	43 x 10^6	500 x 10^6
100°C., 500V., D.C., 0.1"Gap	5.7 x 10^6	

Table VII.   SHELL 4-BALL LUBRICITY TEST FOR
SKYDROL 500A, SKYDROL 7000 and MIL-H-5606A
(600 RPM, 167°F., 1-hr., Scar Diameters in Millimeters)

Fluid	Steel On Steel		Steel On Bronze	
	1 kg.	40 kg.	1 kg.	40 kg.
Skydrol 500A	0.154	0.809	0.327	0.894
Skydrol 7000	0.251	0.598	0.424	1.113
MIL-H-5606A	0.365	0.849	0.726	2.181

Electrical Properties

Both *Skydrol* 500A and *Skydrol* 7000 possess good dielectric properties. If hydraulic system leaks develop, the *Skydrol* fluids do not cause short circuits, nor will they cause electrolytic corrosion in hydraulic systems. With the hundreds of miles of electrical wiring and the hundreds of circuits in today's aircraft, this lack of conductivity is another safety feature of *Skydrol* 500A and *Skydrol* 7000, which adds to their importance for hydraulic systems. Table VI shows the dielectric properties of the *Skydrol* fluids.

Performance Properties

Fire Resistance

Skydrol 500A and *Skydrol* 7000 are much less susceptible to ignition than MIL-H-5606A, as shown by tests conducted according to AMS 3150. These tests were designed to evaluate the fluids under actual operating conditions to realistically pinpoint their fire-resistance value.

In specific tests, high-pressure sprays of *Skydrol* fluids through the white heat of a welder's torch (often above 6000°F.) does not cause burning, but only occasional flashing. In the same test, MIL-H-5606A ignites instantly and continues burning. On a red hot manifold at 1300°F., *Skydrol* fluids do not burn. In other tests simulating hot manifolds, sparks, exhaust flames or electrical arcing, *Skydrol* fluids do not support fire. Even though they might flash at exceedingly high temperatures, *Skydrol* fluids could not spread a fire because burning is localized at the source of heat. Once the heat source is removed or the fluid flows away from the source, no further flashing or burning can occur because of the self-extinguishing features of the *Skydrol* fluids. To those who know ignition sources these facts explain why *Skydrol* 500A and *Skydrol* 7000 protect the material investment in an aircraft and the lives of its passengers and crew.

Lubricity

The experience of millions of flight hours on virtually every type of aircraft hydraulic pump have proven that *Skydrol* 7000 generally increases the service life of most pumps. *Skydrol* 7000 in Douglas cabin superchargers is approved for 4000 hours of operation as compared to only 250 hours for petroleum oil. Experience with *Skydrol* 500A since its introduction in 1956 offers ample evidence that it too is an excellent lubricant. Tables VII and VIII show lubricity tests for both fluids.

Compatibility of Fluid Materials

Mineral Oil

Although *Skydrol* fluids are miscible with mineral oil, the mixing of *Skydrol* with mineral oil must be avoided to maintain *Skydrol* fire-resistant performance. Mineral oil will seriously degrade the ability of *Skydrol* to resist combustion and fire, the vital justification of equipping systems with *Skydrol*. Similarly, mineral oil, mixed with a *Skydrol* fluid, will degrade seals and packings used in *Skydrol*-equipped hydraulic systems.

Additives used in mineral oils can also damage *Skydrol* systems components; for example, some viscosity index (VI) improvers used with mineral oils may not be soluble in *Skydrol*. In this case, the VI improver may precipitate from the hydraulic fluid mixture, leaving a gum-like residue which may interfere with proper operation of valves and filters.

Silicone and Silicate Fluids

Mixture of these fluids with a *Skydrol* fluid should be avoided for the same reasons mentioned for mineral oil.

Turbo Oil

Both *Skydrol* 500A and *Skydrol* 7000 are miscible with Turbo Oil 15 and 35 in concentrations up to 50 per cent; however, mixing of the fluids is not recommended, since such mixtures can degrade the fire-resistant properties of *Skydrol* and the packings and seals used in *Skydrol* equipped systems.

Miscibility of Skydrol 500A and Skydrol 7000

Skydrol 500A and *Skydrol* 7000 are completely miscible, and the major effect of mixing the two fluids is to increase the viscosity of *Skydrol* 500A at low temperatures, as shown in Table X.

Table VIII   VICKERS PF-3911 PISTON PUMP TEST

	Skydrol 500A		Skydrol 7000	
Duration Time:	224.2 hours		450 hours	
Lubricity:	Excellent. Superior to MIL-H-5606A		Excellent. Superior to MIL-H-5606A	
Packings:	Inspection of the Butyl elastomeric materials, including accumulator bladders, showed no degradation and practically no swelling.			
Viscosity (cs);	Initial	Final	Initial	Final
210°F. (99°C.)	3.37	2.26	3.92	2.95
100°F. (38°C.)	10.09	6.97	14.9	12.09
−40°F.(−40°C.)	576	407	6750	6130
Shear Loss:	20% in first 100 hours. Leveled off at approximately 33%.		17% in 400 hours	
Neutralization Number Change:	0.20 to 0.37		0.20 to 1.00	

Table IX COMPARISON OF SHEAR STABILITY SKYDROL 500A, SKYDROL 7000 AND MIL-H-5606A

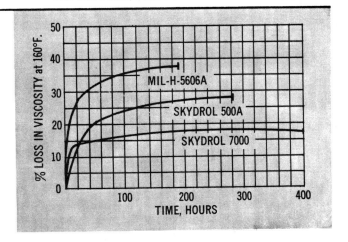

EXHIBIT 5   Test schematic

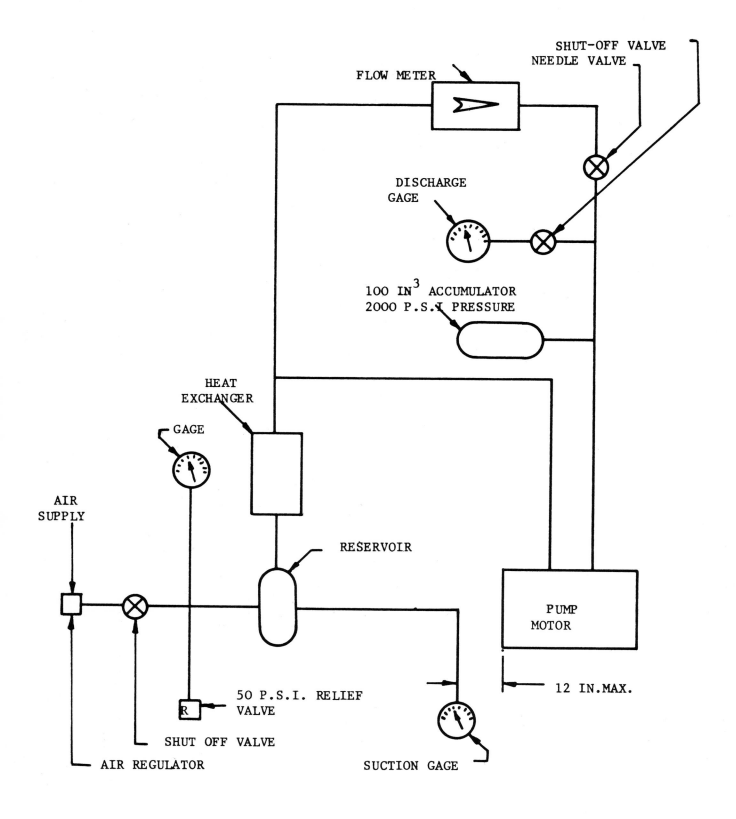

EXHIBIT 6   Endurance test schedule

HOURS	OUTPUT	FLUID INLET PRESSURE (MIN.)	FLUID INLET TEMP. ± 5°F
225	Cycle A	5 psig	150°F
325	Cycle B	45 psig	150°F
110	1/2 gpm	45 psig	150°F
1820	1 gpm	45 psig	150°F
20	1 gpm	5 psig	Room Temp.

CYCLE A

Motor	Flow	Time
On	5.7 gpm min.*	15 min.
On	0	30 min.
Off	-	15 sec.

CYCLE B

Motor	Flow	Time
On	5.7 gpm min.*	6.25 min.
On	2 gpm	5 min.
On	1/2 gpm	5 min.

*Flow which produces a motor current draw of 38.5 amperes or 5.7 gpm min., whichever occurs first.

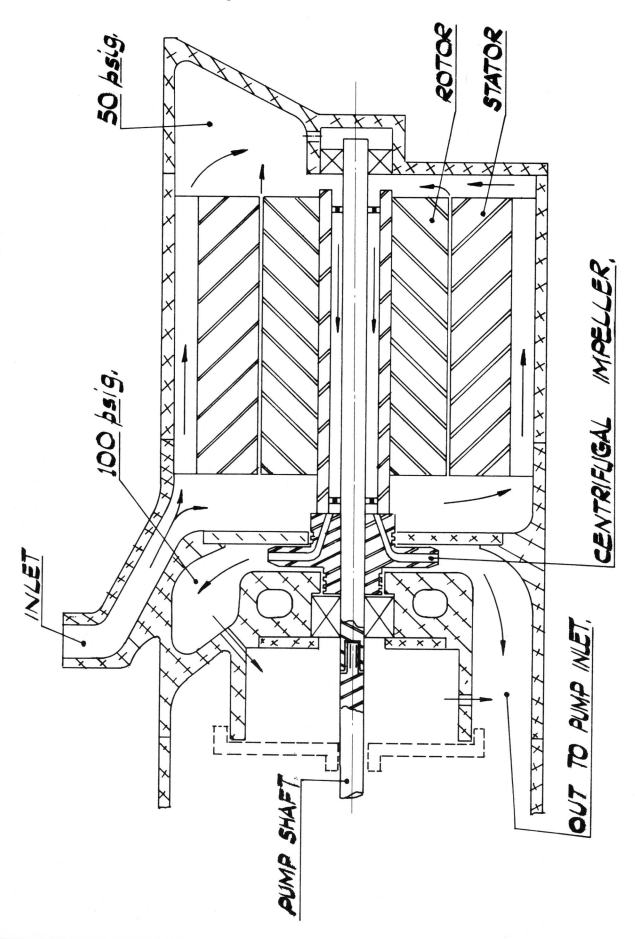

BARDEN STANDARD PRECISION BEARING TYPES

MATERIALS

Stainless steel (AISI 440C) rings and balls are used in most miniature and instrument bearings listed in this catalog. As heat treated and processed by Barden, these bearings have high corrosion resistance and good dimensional stability for high temperature operation. Average running torque values are the same as for chrome steel bearings.

Chrome steel (SAE 52100) rings and balls are used in most spindle and turbine bearings and special design bearings listed here. Chrome steel is preferred for many applications where corrosion resistance is not the governing factor.

Either material may be readily obtained in many standard bearings. Other materials for high temperature use, such as AISI M50 steel, can be supplied on special order.

CONSTRUCTION

Deep groove bearings have equal raceway shoulders on both sides of inner and outer rings, will support radial loads and thrust in either direction, and have lower torque at very low speeds than angular contact bearings.

Extra wide deep groove bearings, normally furnished with two shields, have increased lubricant space for long life at high speeds with the initial charge of lubricant.

Flanged deep groove bearings provide accurate positioning surfaces for good alignment; they permit through-boring, eliminating the need for housing shoulders or shoulder rings.

Angular contact bearings have one shoulder cut away on either inner or outer ring, will support radial loads and thrust in one direction only, and require light axial loading; they are generally preferred for the higher speed applications.

Separable angular contact bearings have one shoulder eliminated on the inner ring. This permits removal and mounting of ring on shaft while the outer ring assembly is mounted separately in housing, an aid to assembly and dynamic balancing.

Non-separable angular contact bearings have one shoulder only partially cut away on the outer ring to prevent separation of bearing parts. Construction allows use of one-piece retainer and more balls for greater static and dynamic load capacity than deep groove bearings.

BALL RETAINERS deep groove bearings

Pressed steel standard retainers, used for moderate to high speeds, are one-piece in smaller sizes, two-piece in larger. Two-piece "W" retainers, also available in most sizes, have "antiwindup" design to prevent retainer lock, reduce torque peaks and increase bearing life at speed. Retainers are relatively unaffected by temperature; lubricant limitations govern operating temperatures.

Phenolic retainers, used for high speeds, are one-piece "TA" in smaller sizes, two-piece "T" in larger; both are outer ring piloted for increased lubricant penetration and circulation, less wear and longer life. "T" retainers have aluminum reinforcement and positive riveting to strengthen lightweight, long wearing phenolic. Retainers operate up to about 300°F (150°C) and for short periods to about 350°F (175°C).

BALL RETAINERS angular contact bearings

Phenolic retainers for separable bearings are one-piece, outer ring piloted for best endurance at high speeds. Will operate to about 300°F (150°C) and for short periods to about 350°F (175°C).

Phenolic retainers for non-separable bearings are one-piece, lightweight Barden "halo" type, outer ring piloted for better lubricant penetration and circulation, longer life at high speeds. Will operate to about 300°F (150°C) and for short periods to about 350°F (175°C).

Bronze retainers for non-separable bearings, used for high speeds, have high conductivity to dissipate heat, are one-piece, thin section and outer ring piloted for increased lubricant penetration, circulation and cooling, resulting in longer bearing life at high speeds and temperatures. Bronze retainers are relatively unaffected by temperatures up to 550°F (285°C); range of operating temperatures is largely governed by lubricant limitations.

SHIELDS AND SEALS

Stainless steel shields, single or double, are available in most sizes of deep groove bearings. Barden close-clearance shield design protects against contamination and lubricant loss; precision snap wire shield retention avoids outer ring distortion. Single-shield flanged bearings have shields on flange side. Shields and snap wires are removable.

Flexeal seals, exclusive with Barden, are low friction, wear resistant fiber bonded to aluminum; in light contact with ground surfaces on inner rings, Flexeals have positive but free sliding action that effectively seals in lubricant, seals out contaminants and minimizes air or fluid flow through bearings. Precision snap wire retention avoids outer ring distortion. Available in many deep groove sizes.

DUPLEX MATCHED PAIRS

Barden-matched duplex pairs are available in most deep groove and angular contact non-separable bearings listed in this catalog. For further data see page 52.

Matched preloaded pairs, either DB (back to back) or DF (face to face), support radial loads and thrust in either direction; they provide exact axial positioning, increased rigidity, controlled axial and radial yield rates, and increased radial load capacity.

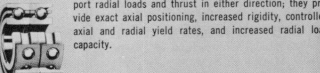

Matched tandem pairs, DT (back to face), support radial loads and thrust in one direction only; they should be mounted opposed with thrust loading between pairs or between a pair and a single bearing. Used mainly for increased load capacity and rigidity.

EXHIBIT 7   Bearing data

NOMENCLATURE

STANDARD BEARING TYPES. Standard types of Barden Precision bearings can be identified by use of the following nomenclature which is used on order acknowledgments, packages, invoices and correspondence. For new designs or initial orders, Barden should be furnished complete environmental and performance requirements so that bearings best suited for the application may be specified. Well established bearing requirements may be specified precisely by using the complete nomenclature shown here.

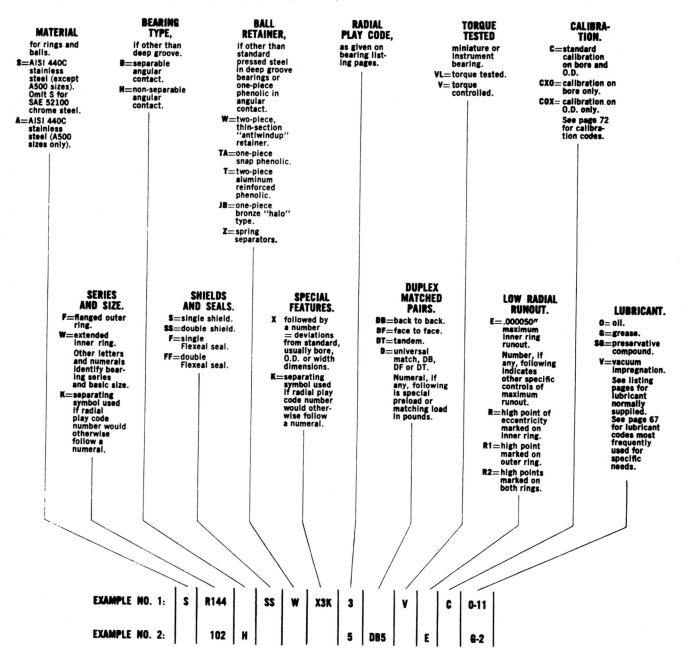

MATERIAL
for rings and balls.
S=AISI 440C stainless steel (except A500 sizes). Omit S for SAE 52100 chrome steel.
A=AISI 440C stainless steel (A500 sizes only).

BEARING TYPE,
if other than deep groove.
B=separable angular contact.
H=non-separable angular contact.

BALL RETAINER,
if other than standard pressed steel in deep groove bearings or one-piece phenolic in angular contact.
W=two-piece, thin-section "antiwindup" retainer.
TA=one-piece snap phenolic.
T=two-piece aluminum reinforced phenolic.
JB=one-piece bronze "halo" type.
Z=spring separators.

RADIAL PLAY CODE,
as given on bearing listing pages.

TORQUE TESTED
miniature or instrument bearing.
VL=torque tested.
V=torque controlled.

CALIBRATION.
C=standard calibration on bore and O.D.
CXO=calibration on bore only.
COX=calibration on O.D. only.
See page 72 for calibration codes.

SERIES AND SIZE.
F=flanged outer ring.
W=extended inner ring.
Other letters and numerals identify bearing series and basic size.
K=separating symbol used if radial play code number would otherwise follow a numeral.

SHIELDS AND SEALS.
S=single shield.
SS=double shield.
F=single Flexeal seal.
FF=double Flexeal seal.

SPECIAL FEATURES.
X followed by a number = deviations from standard, usually bore, O.D. or width dimensions.
K=separating symbol used if radial play code number would otherwise follow a numeral.

DUPLEX MATCHED PAIRS.
DB=back to back.
DF=face to face.
DT=tandem.
D=universal match, DB, DF or DT.
Numeral, if any, following is special preload or matching load in pounds.

LOW RADIAL RUNOUT.
E=.000050" maximum inner ring runout.
Number, if any, following indicates other specific controls of maximum runout.
R=high point of eccentricity marked on inner ring.
R1=high point marked on outer ring.
R2=high points marked on both rings.

LUBRICANT.
O=oil.
G=grease.
SB=preservative compound.
V=vacuum impregnation.
See listing pages for lubricant normally supplied. See page 67 for lubricant codes most frequently used for specific needs.

EXAMPLE NO. 1:	S	R144		SS	W	X3K	3		V	C	O-11
EXAMPLE NO. 2:		102	H				5	DB5	E		G-2

SPECIAL DESIGN BEARINGS AND ASSEMBLIES. Barden bearings or assemblies designed specifically to customer requirements are identified by Y or Z followed by a number, such as Z155. This basic nomenclature generally includes all specifications normally spelled out by additional nomenclature in the case of standard bearing types. Occasionally, later modifications will be reflected in the bearing nomenclature by the use of X followed by a number, such as the Z202X1.

BARDEN PRECISION SPINDLE AND TURBINE BEARINGS

LOW TO ULTRA HIGH SPEED

DEEP GROOVE STEEL OR PHENOLIC RETAINERS

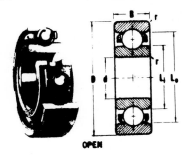

OPEN

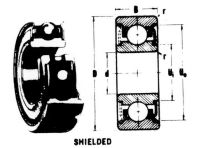

SHIELDED

AVAILABILITY. All sizes listed are normally available from stock in chrome steel.

DIMENSIONS

BORE, O.D., WIDTH						MAX. FILLET RADIUS inches	BASIC ORDERING NUMBER						OTHER DIMENSIONS inches				APPROX. WEIGHT pounds	DATA REFERENCE NUMBER
							STEEL RETAINER			PHENOLIC RETAINER								
d		D		B		r	Open	Single shield	Double shield	Open	Single shield	Double shield	L_i	L_o	U_i	U_o		
mm	inches	mm	inches	mm	inches													
15	.5906	32	1.2598	9	.3543	.012				102T			.798	1.053			.07	102
15	.5906	35	1.3780	11	.4331	.025	202K	202S	202SS	202T	202ST	202SST	.815	1.153	.755	1.223	.10	202
15	.5906	35	1.3780	12.70	.5000	.025					202STX1	202SSTX1			.755	1.223	.20	202
17	.6693	35	1.3780	10	.3937	.012	103K	103S	103SS	103T	103ST	103SST	.895	1.153	.835	1.215	.09	103
17	.6693	40	1.5748	12	.4724	.025	203K	203S	203SS	203T	203ST	203SST	.952	1.292	.890	1.372	.20	203
20	.7874	42	1.6535	12	.4724	.025				104T	104ST	104SST	1.050	1.390	.989	1.458	.18	104
20	.7874	47	1.8504	14	.5512	.040				204T	204ST	204SST	1.130	1.530	1.060	1.610	.20	204
25	.9843	47	1.8504	12	.4724	.025				105T			1.247	1.587			.21	105
25	.9843	52	2.0472	15	.5906	.040	205K	205S	205SS	205T	205ST	205SST	1.320	1.720	1.250	1.800	.30	205
30	1.1811	55	2.1654	13	.5118	.040				106T	106ST	106SST	1.511	1.869	1.451	1.949	.30	106
30	1.1811	62	2.4409	16	.6299	.040				206T	206ST	206SST	1.580	2.060	1.500	2.200	.50	206
35	1.3780	62	2.4409	14	.5512	.040	107K	107S	107SS	107T	107ST	107SST	1.710	2.110	1.620	2.190	.35	107
35	1.3780	72	2.8346	17	.6693	.040				207T	207ST	207SST	1.857	2.382	1.777	2.523	.70	207
40	1.5748	80	3.1496	18	.7087	.040				208T			2.081	2.643			.80	208
45	1.7717	85	3.3465	19	.7480	.040				209T			2.289	2.850			.90	209

APPLICATIONS. Motors, generators, aircraft accessories, gear drives, pumps, power tools, compressors, machine tool spindles, stable platforms, magnetic recording devices and other low to ultra high speed applications.

DESCRIPTION. Metric series deep groove bearings for smooth, quiet operation with minimum vibration under moderate to heavy loads; support radial loads and thrust in either direction. Some sizes (suffix X) are extra wide to hold more lubricant and provide wider mounting surfaces for accurate alignment.

Ball retainers for low to high speeds are two-piece pressed steel; for ultra high speeds, two-piece aluminum-clad phenolic laminate, outer ring piloted for increased lubricant penetration and circulation, less wear and longer life. Aluminum reinforcement and positive riveting add strength to light weight and long wearing qualities of phenolic.

Close clearance shields protect against contamination and lubricant loss; precision snap wire shield retention avoids outer ring distortion. Shields and snap wires are removable.

MATERIAL. Rings and balls: SAE 52100 chrome bearing steel. **Retainers:** pressed stainless steel or aluminum-reinforced phenolic. **Shields and snap wires:** stainless steel. Further data on page 48.

LUBRICANTS NORMALLY SUPPLIED. Oil: MIL-L-6085A (Barden code 0-11) or MIL-L-7808C (Barden code 0-14) for open bearings only; both require subsequent lubrication wth continuous oil mist or spray application. **Grease:** Andok C (Barden code G-6) or UniTemp 500 (Barden code G-18). **Preservative compound:** MIL-C-11796A (Barden code SG-1) for open bearings only; requires subsequent lubrication with mineral oil. Further data on page 66.

PRELOADED PAIRS. Most sizes listed may be obtained in DB or DF duplex pairs with controlled axial preload. Further data on page 52.

FURTHER DATA. Tolerances, page 46. Radial and axial play and yield, page 49. Shaft and housing shoulders and fits, page 68.

LOW TO
ULTRA HIGH SPEED
BORE .5906"-1.7717"
O.D. 1.2598"-3.3465"

PERFORMANCE DATA

DYNAMIC LOAD RATINGS. Radial load ratings below are for 500-hour design life, or 2500-hour average life, with proper mounting, lubrication and protection of bearings. Use value C_S for fatigue life computation procedure on page 62.

○ Normal speed limit for open or shielded bearings with steel retainers.

□ Normal speed limit for open or shielded bearings with phenolic retainers, grease lubrication.

△ Normal speed limit for open bearings with phenolic retainers, oil mist or spray lubrication.

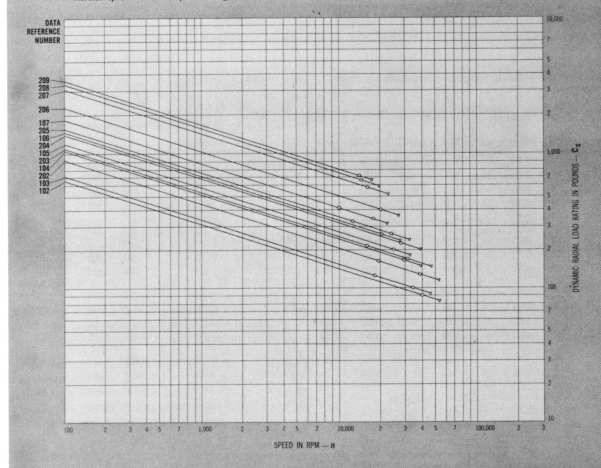

STATIC LOAD RATINGS. Static ratings are shown at left for radial play ranges normally supplied. These values indicate peak loads that can be sustained by bearings without permanent effect on smoothness of operation. Further data on page 60.

AVERAGE RUNNING TORQUE. These bearings are used mainly at speeds where the predominant factor in torque is lubricant drag, which varies with lubricant type and quantity, speed and operating temperatures. When needed for estimate of power requirement, approximate torque values will be supplied.

STATIC LOAD RATINGS—pounds

DATA REFERENCE NUMBER	THRUST LOAD—T_0			RADIAL LOAD C_0
	RADIAL PLAY RANGE			
	Code 3 .0002"-.0004"	Code 5 .0005"-.0008"	Code 6 .0008"-.0011"	
102	2030	2120	1800	650
103	2300	2350	1700	730
	.0002"-.0005"	.0005"-.0009"	.0009"-.0014"	
202	2100	2240	2380	770
203	2720	2880	3000	1000
104	2160	2210	2340	900
204	3900	4000	4160	1400
105	2000	2300	2400	950
205	4400	4550	4680	1600
106	3300	3520	3630	1380
107	4070	4290	4400	1700
		.0007"-.0012"	.0012"-.0017"	
206	6210	6300	6750	2280
207	8100	8640	9000	3100
208	9450	9900	10400	3600
209	10500	11000	11500	4040

LOAD CAPACITY

For many applications final bearing selection can be made from dimensional and performance data shown on the bearing listing pages of this catalog. Where an application runs continuously at speed, or where static load capacity is a major factor, bearing selection can be finalized only by computation of life and load factors.

Fatigue life of ball bearings that run continuously at speed under appreciable loading is considered to be the number of hours or revolutions that the bearings run before the first evidence of ball or raceway spalling develops. Fatigue is directly related to bearing load and speed. Calculation of fatigue life is of primary importance when selecting bearings for use in power-driven devices sush as motors, generators, gear drives, turbines and similar equipment.

Bearing life may be limited by factors other than fatigue life, such as lubricant exhaustion, contamination, misalignment or improper fitting, and thermal constraint. In some cases, static capacity may be more important than fatigue life in selection of a bearing, as in lightly loaded components such as gyro gimbals, synchros and computer gear trains where little sustained speed is involved.

STATIC LOAD RATINGS

The static load capacity of a bearing is the peak load that can be sustained without appreciable permanent effect on smoothness of operation. Static load ratings shown on the listing pages of this catalog are based on load rating evaluation methods developed by the Anti-Friction Bearing Manufacturers Association. These ratings may be exceeded for some applications, but users anticipating heavier loadings should consult Barden before finalizing selection.

In a lightly loaded ball bearing at rest, small elastic deformations are developed at ball-to-raceway contacts. As loads increase, a portion of these deformations become permanent and, if sufficiently large, impair subsequent performance by causing high torque or rough operation. For most precision applications, a reasonable limit value for permanent indentations is one ten-thousandth of the ball diameter.

Under pure radial loading, the maximum depth of indentations approaches the center of the raceway. Under thrust loading, the indentations approach the edge of the raceway. With increased radial play, the contacts approach still closer to the edge of the raceway. Under heavy thrust loads and large values of radial play, the contact areas may run over the edge of the raceway, resulting in increased stress and deeper indentation. These effects, along with raceway depth limitations, have been considered in establishing the static thrust load ratings in this catalog.

In general, less damage is done to bearings if loads are imposed while the bearings are rotating than when at rest. Even in high speed operation, however, if the load duration is so short that there is no rotational overlap of the most heavily stressed ball contact areas, separate permanent indentations may result. Bearings used in high speed applications where smooth operation is important, such as gyro rotors, should be selected so that maximum peak or shock loads are within static ratings.

In many cases, a simple statement of g loading is not sufficient for use in bearing selection. Transient and vibratory loads must also be considered, but they are frequently difficult to determine, since resonance and damping characteristics of the complete rotating assembly and supports will greatly affect peak loadings imposed on the bearings. Complete vibration analysis or qualification tests under actual vibratory conditions may be needed.

BEARING SELECTION FOR STATIC CAPACITY

Pure radial loads on the bearing tentatively selected can be checked directly against the radial load rating in column C_o of the static load table on the appropriate listing page. Pure thrust loads can be checked against the thrust load columns T_o under the radial play range selected. Combined radial and thrust loads must be computed in terms of their radial and thrust load components, with bearing selection made so that each of the component loads is within its respective static radial or thrust rating. If either radial or thrust loading rating is lower than the peak load expected, a bearing with higher static rating should be chosen.

Static load ratings for DT (tandem) duplex pairs are double the thrust and radial ratings of single bearings. With usual preload, the thrust rating of a DB (back to back) or DF (face to face) pair is equal to that of a single bearing but radial capacity is double that of the single bearing.

SPEED CAPABILITIES

Speed capability of bearings is almost impossible to determine exactly because of the wide variety of environmental, design and life requirements. However, the approximate speed limits given in the dynamic load rating charts on the listing pages of this catalog will serve for general guidance. On request, Barden engineers will assist you in making detailed studies of specific speed and life requirements as an aid in bearing selection.

Bearings with phenolic or bronze ball retainers have highest speed capability; bearings with pressed steel retainers have lowest speed capability. Speeds for bearings with phenolic or bronze retainers are limited by type of lubrication or by outer ring centrifugal ball loading; speeds for bearings with steel retainers are limited by retainer performance.

Bearings with phenolic or bronze retainers have highest speed-life capability when lubrication is supplied during operation by continuous oil spray or mist. Prelubrication with grease gives lower speed-life capability. Lowest speed-life capability results when bearings are prelubricated with instrument oil and not supplied with additional lubricant during operation.

FATIGUE LIFE

For bearings that are properly mounted, lubricated and protected, fatigue life may be estimated as the number of hours or revolutions at a given speed and load that a group of similar bearings will operate before the first evidence of spalling or flaking of raceways or balls becomes apparent.

Since some variability in fatigue life is inevitable, it is necessary to establish a probability factor for the percentage of a group of bearings that will endure the given conditions of load and speed. In accordance with bearing industry practice, a probability factor of 90% is employed for load-life computations in this catalog.

Design fatigue life is considered to be the estimated life which will be exceeded by 90% of a group of identical bearings operating under identical load conditions, assuming proper mounting, lubrication and protection against foreign substances. Average fatigue life is approximately five times this figure.

Moderate changes in the load applied to the bearing have a pronounced effect on fatigue life. Halving the bearing load increases life eight times, but doubling the load reduces life to one-eighth. For this reason, an accurate determination of actual operating loads is most important for life computations.

BEARING SELECTION FOR DYNAMIC CAPACITY
Equivalent radial load computations

Fatigue life computations are based on two assumptions: a constant direction radial load and a condition of stationary housing (outer ring) and rotating shaft (inner ring). Therefore, combined radial and thrust loads must be converted into an equivalent radial load. Also, even where there is zero thrust load, an equivalent radial load must be calculated to take account of any radial load that tends to concentrate on a small portion of the inner ring raceway. Such an effect is developed by dynamic unbalance loading on a rotating inner ring or by a dead weight or constant direction load on a stationary inner ring. The formulas below for equivalent radial loads take these factors into consideration.

Equivalent radial loads for individually mounted bearings are found by solving both Formulas 1 and 2 below. The larger of the two values derived is then used in subsequent formulas to find fatigue life.

FORMULA 1 $P = R_h + 1.2R_s$

FORMULA 2 $P = X (R_h + 1.2R_s) + YT$

where: P = equivalent radial load, pounds

R_h = radial load fixed in relation to outer ring, pounds

R_s = radial load fixed in relation to inner ring, pounds

X = radial load factor

Y = thrust load factor

T = thrust load, pounds

Identification of radial loads R_h and R_s is found in Table 7.

Factor X is found by using the DATA REFERENCE NUMBER of the bearing tentatively selected to enter Table 8 (pages 27 and 28) to find the contact angle for the radial play range chosen. This contact angle is then entered in Chart A (page 25) to determine factor X.

When appreciable thrust loads are involved, select the highest possible radial play range for the highest resulting contact angle, since both factors X and Y decrease as radial play is increased.

To find factor Y first note the appropriate value ZD² in Table 8 and compute value $\frac{T}{ZD^2}$. Select Chart B or Chart C (page 25) as shown by Table 8 and enter with $\frac{T}{ZD^2}$ and the contact angle of the bearing to determine factor Y.

When Y has been established, Formulas 1 and 2 may be computed and the larger resulting value taken to Formula 3 to find the design fatigue life of the bearing selected.

TABLE 7—RADIAL LOAD SYMBOLS R_h AND R_s

Component rotation	Nature of radial load	Symbol
Housing (outer ring) stationary, shaft (inner ring) rotating	Dead weight or constant direction load fixed in relation to housing (outer ring)	R_h
	Dynamic unbalance or rotational load fixed in relation to shaft (inner ring)	R_s
Shaft (inner ring) stationary, housing (outer ring) rotating	Dead weight or constant direction load fixed in relation to shaft (inner ring)	R_s
	Dynamic unbalance or rotational load fixed in relation to housing (outer ring)	R_h

EXAMPLE—EQUIVALENT RADIAL LOAD COMPUTATION

Application	high speed turbine
Operating speed	80,000 rpm
Rotating member	shaft (inner ring)
Lubrication	oil spray or mist
Dead weight radial load	4.3 pounds
Dynamic unbalance at operating speed	2.3 pounds
Thrust from turbine	15.0 pounds
Thrust from preload spring	8.0 pounds
Bearing tentatively chosen	38H
DATA REFERENCE NUMBER	38H

From Table 7:

    Dead weight radial load of 4.3 pounds = R_h

    Dynamic unbalance of 2.3 pounds = R_s

Total thrust T = 15.0 + 8.0 = 23.0 pounds

Using Formula 1: $P = R_h + 1.2R_s = 4.3 + 1.2(2.3) = 7.1$ pounds

To obtain values X and Y for Formula 2, enter Table 8 with DATA REFERENCE NUMBER 38H to find the contact angle of the bearing. The code 6 radial play range shows a contact angle of 17° as compared with 14° for code 5. Selecting code 6 for the higher contact angle and entering Chart A with 17°, factor X = .43. Entering Table 8 with reference 38H, ZD² = .220

Value $\frac{T}{ZD^2} = \frac{23}{.220} = 105$

LOAD CAPACITY

CHART A—RADIAL LOAD FACTOR X

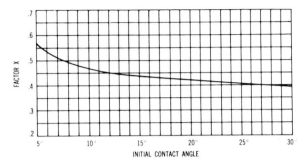

CHART B—THRUST LOAD FACTOR Y OR Y_d (SEE TABLE 8)

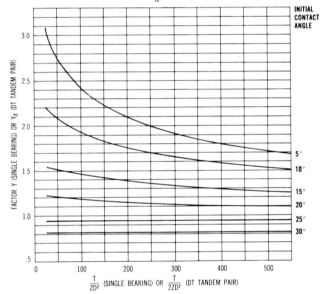

CHART C—THRUST LOAD FACTOR Y OR Y_d (SEE TABLE 8)

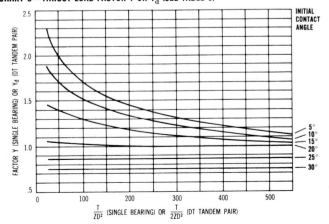

Entering Table 8 with 38H, select Chart C as shown. Entering Chart C with 105 for $\frac{T}{ZD^2}$ and 17° for contact angle, factor Y = 1.2

Using Formula 2:
$$P = X(R_h + 1.2R_s) + YT$$
$$= .43(4.3 + 2.8) + 1.2(23)$$
$$= 3.1 + 28 = 31.1 \text{ pounds}$$

Formula 2 results in a greater load than Formula 1; in the computation below P = 31.1 pounds, the higher value.

FATIGUE LIFE COMPUTATIONS—HOURS

Since fatigue life is usually computed in terms of hours at a given speed, the dynamic load ratings C_s are given in the dynamic load charts on the listing pages of this catalog for speeds of 100 rpm and higher. The design life basis for C_s values is 500 hours for 90% survival, as developed from load rating evaluation methods standardized by the Anti-Friction Bearing Manufacturers Association.

Formula 3 gives the simplified computation for fatigue life in hours using equivalent radial load value P, as derived above, and dynamic load rating C_s from the listing pages to find the life ratio.

FORMULA 3 (simplified computation)

$$L_{ra} = \frac{C_s}{P}$$

where:
L_{ra} = life ratio
C_s = dynamic radial load rating, pounds
P = equivalent radial load, pounds

Dynamic load rating C_s is found in the dynamic load chart on the appropriate listing page in the front of this catalog. This chart is entered with the speed of rotation and the DATA REFERENCE NUMBER for the bearing selected.

Equivalent radial load P is the larger of the two values resulting from solution of Formulas 1 and 2 (page 24).

Life ratio L_{ra} is used to enter Chart 6 to determine design life or average life.

EXAMPLE—SIMPLIFIED LIFE COMPUTATION

Entering the dynamic load rating chart with 80,000 rpm and DATA REFERENCE NUMBER 38H, C_s = 43 pounds.

This chart shows that the 38H bearing has a speed limit below 80,000 rpm for grease lubrication; it should not be used in this application without oil spray or mist.

Using Formula 3: $L_{ra} = \frac{C_s}{P} = \frac{43}{31.1} = 1.38$

Entering Chart 6 with L_{ra} of 1.38, design life L_{10} is approximately 1300 hours and average life is approximately 6500 hours.

Formula 4 gives the full computation for fatigue life in hours, using value P and ratings C_s to find life in hours directly. This will give approximately the same value derived from Formula 3, but with greater accuracy.

FORMULA 4 (full computation)

$$L_{10} = 500 \left(\frac{C_s}{P}\right)^3$$

where: L_{10} = life in hours

C_s = dynamic radial load rating

P = equivalent radial load, pounds

Factors C_s and P are identical to those used in Formula 3.

EXAMPLE—FULL LIFE COMPUTATION

$$L_{10} = 500 \left(\frac{C_s}{P}\right)^3 = 500 \left(\frac{43}{31.1}\right)^3$$
$$= 500(1.38)^3 = 1314 \text{ hours}$$

CHART 6—DESIGN OR AVERAGE FATIGUE LIFE—HOURS

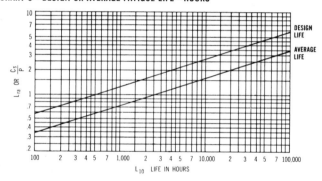

FATIGUE LIFE COMPUTATION—REVOLUTIONS

Life computation in revolutions uses basic dynamic radial load ratings C, shown in Table 8, which are based on a life of 1,000,000 revolutions with 90% survival being the equivalent of 500 hours design life at 33-1/3 rpm only.

The C rating for a given bearing is greater than the C_o, rating, the static radial load rating given on the listing page. It would therefore be unwise to use only rating C for life computation in applications operating at low speeds where shock or vibratory loads may be involved. Another limitation of using C ratings is that they do not give consideration to speed limits in terms of lubrication, ball retainer type and centrifugal ball loading effects on capacity at very high speeds.

Formula 5 may be used when life in revolutions is needed for reference or for purposes of comparison.

FORMULA 5 $$L_{10r} = \left(\frac{C}{P}\right)^3 \times 10^6 \text{ revolutions}$$

where: L_{10r} = design life in revolutions

C = basic dynamic radial load rating, pounds

P = equivalent radial load, pounds

DUPLEX BEARING LIFE COMPUTATION

For DT tandem duplex pairs, equivalent radial load ratings are found by solving Formulas 6 and 7. The larger of the two values derived is used in Formula 3 or 4 to find the design fatigue life of the bearing pair selected.

FORMULA 6 $P = .62 (R_h + 1.2R_s)$

FORMULA 7 $P = .62X (R_h + 1.2R_s) + .62Y_dT$

where: P = equivalent radial load of tandem pair, pounds

R_h = radial load fixed in relation to outer rings, pounds

R_s = radial load fixed in relation to inner rings, pounds

X = radial load factor

Y_d = thrust load factor of tandem pair

T = thrust load, pounds

The factor .62 takes account of slight errors in load division between the two bearings and the probability that one bearing of the pair may fail before the other, making advisable the replacement of both bearings of the pair for maximum reliability in later operation.

Factors R_h, R_s and X are identical to those used in Formulas 1 and 2 for individually mounted bearings.

To obtain factor Y_d, first compute value $\frac{T}{2ZD^2}$. ZD^2 is found in Table 8 under the DATA REFERENCE NUMBER of the single bearing to be used in a duplex DT pair. Select Chart B or Chart C as shown by Table 8 and enter with $\frac{T}{2ZD^2}$ and the contact angle of the bearing to determine factor Y_d.

Formulas are not shown for computing life of DB or DF preloaded pairs since preload requirements, rather than bearing capacity, often limit the amount of external loading that may be applied safely. This is particularly true for miniature and instrument bearings, less true for spindle and turbine bearings. However, speed capability of spindle and turbine preloaded pairs is generally limited by energy loss, heating or by type of lubrication. Users of duplex preloaded pairs for new applications are urged to consult Barden for complete analysis and bearing recommendations.

INCREASED LIFE RELIABILITY

If desired, factors for use with dynamic ratings will be supplied by Barden for computation of fatigue life on the basis of reliability greater than 90% survival. However, at high speeds and high temperatures many other factors must be considered, such as accuracy of mounting, degree of dynamic balance, effectiveness of lubrication system and thermal constraining influences. A complete review of requirements by Barden engineers is recommended to attain the increased life reliability desired.

LOAD CAPACITY

TABLE 8—VALUES FOR FATIGUE LIFE COMPUTATION

MINIATURE AND INSTRUMENT BEARINGS

| DATA REFERENCE NUMBER | Chart used for factor Y (page 62) | INITIAL CONTACT ANGLE—degrees Radial play range | | | | | | Ball complement | | Value ZD^2 | Basic dynamic load rating C pounds |
		Code 2	Code 3	Code 4	Code 5	Code 6	Std.	Number Z	Diameter D		
R0	B	12	17	18	25	27		6	1/32″	.0059	21
R1	B	11	14	16	22	26		6	1 mm	.0093	30
R1-5B	B						16	6	1/16″	.0234	60
R2	B	9	11	12	16	20		7	1/16″	.0273	69
R2B	B						16	7	1/16″	.0273	69
R2H	C						16	8	1/16″	.0312	92
R2-5	B	9	11	12	16	20		6	1/16″	.0234	60
R2-5B	B						20	6	1/16″	.0234	60
R2-5H	B						12	7	1/16″	.0273	66
R2-6	B	9	11	12	16	20		7	1/16″	.0273	80
R3	B	7	9	10	13	16		7	3/32″	.0615	140
R3B	B						16	7	3/32″	.0615	140
R3H	B						10	8	3/32″	.0703	153
R3W	B	7	9	10	13	16		6	3/32″	.0528	127
R4	B	7	9	10	13	16		8	3/32″	.0703	159
R4B	B						16	8	3/32″	.0703	159
R4H	B						10	9	3/32″	.0791	172
R8	C	12	15	17	22	27		10	5/32″	.244	748
R133	B	12	17	18	25	27		7	1/32″	.0068	23
R133W	B	12	17	18	25	27		8	1/32″	.0078	26
R144	B	11	14	16	22	26		8	1 mm	.0124	39
R156	B	11	14	16	22	26		9	1 mm	.0140	40
R156W	B	11	14	16	22	26		11	1 mm	.0171	46
R166	B	9	11	12	16	20		8	1/16″	.0312	89
R168	B	11	14	16	22	26		11	1 mm	.0171	40
R188	B	9	11	12	16	20		11	1/16″	.0430	108
Z114	B		9		13	16		12	3/32″	.105	204
34	B	6	7	9	10	13		6	1/8″	.0938	202
34BX4	B						12	6	1/8″	.0938	164
34H	C			13				8	1/8″	.125	336
34-5B	B						15	6	1/8″	.0938	202
36	B	6	7	8	10	13		6	9/64″	.119	258
36BX1	B						12	6	9/64″	.119	209
36H	C				14	17		8	9/64″	.159	424
38	B	6	7	8	10	13		7	5/32″	.171	357
38BX2	B						14	7	5/32″	.171	355
38H	C				14	17		9	5/32″	.220	570
103	C	11	12	16	20	24		10	3/16″	.352	1040

LARGE BORE, EXTRA THIN BEARINGS

| BASIC BEARING NUMBER | Chart used for factor Y (page 62) | INITIAL CONTACT ANGLE—degrees Radial play range | | Ball complement | | Value ZD^2 | Basic dynamic load rating C pounds |
		Code 3	Code 5	Number Z	Diameter D		
A538T	C	15	22	10	1/8″	.156	495
A539T	C	15	22	12	1/8″	.188	548
A540T	C	15	22	14	1/8″	.219	596
A541T	C	15	22	16	1/8″	.250	629
A542T	C	15	22	18	1/8″	.281	646
A543T	C	15	22	22	1/8″	.344	721

TABLE 8—VALUES FOR FATIGUE LIFE COMPUTATION (Continued)

SPINDLE AND TURBINE BEARINGS

DATA REFERENCE NUMBER	Chart used for factor Y (page 62)	INITIAL CONTACT ANGLE—degrees Radial play range				Ball complement		Value ZD²	Basic dynamic load rating C pounds
		Code 3	Code 4	Code 5	Code 6	Number Z	Diameter D		
34	B	7		10	13	6	1/8″	.0938	202
34H	C		13			8	1/8″	.125	336
36	B	7		10	13	6	9/64″	.119	257
36H	C			14	17	8	9/64″	.159	424
36J	C			14	17	8	9/64″	.159	424
38	B	7		10	13	7	5/32″	.171	357
38H	C			14	17	9	5/32″	.220	570
38J	C			14	17	9	5/32″	.220	570
38X2	C	14		20	25	7	5/32″	.171	570
100	C	12		20	24	7	3/16″	.246	793
100H	C			13	16	9	3/16″	.316	795
100J	C			13	16	10	3/16″	.352	851
100X1	C	10		13	16	7	3/16″	.246	734
101	C	12		20	24	8	3/16″	.281	884
101H	C			13	16	10	3/16″	.352	868
101J	C			13	16	11	3/16″	.386	926
102	C	12		20	24	9	3/16″	.316	970
102H	C			13	16	11	3/16″	.386	936
102J	C			13	16	12	3/16″	.422	990
103	C	12		20	24	10	3/16″	.352	1040
103H	C			13	16	13	3/16″	.457	1050
103J	C			13	16	13	3/16″	.457	1050
104	C	9		13	16	10	1/4″	.625	1620
104H	C			12	15	11	1/4″	.688	1570
104J	C			12	15	13	1/4″	.813	1750
105	C	9		13	16	10	1/4″	.625	1850
105H	C			12	15	13	1/4″	.813	1750
105J	C			12	15	14	1/4″	.875	1830
106	C	9		13	16	11	9/32″	.870	2280
106H	C			12	15	14	9/32″	1.11	2260
106J	C			12	15	14	9/32″	1.11	2260
107	C	8		12	15	11	5/16″	1.07	2750
107H	C			11	14	15	5/16″	1.46	2850
107J	C			11	14	15	5/16″	1.46	2850
200	C	11		16	21	7	7/32″	.335	1040
200H	C			13	16	9	7/32″	.431	1230
200J	C			13	16	9	7/32″	.431	1230
201	C	11		16	21	7	15/64″	.385	1180
201H	C			13	16	9	15/64″	.495	1400
201J	C			13	16	10	15/64″	.549	1500
202	C	10		15	20	7	1/4″	.438	1340
202H	C			13	16	10	1/4″	.625	1700
202J	C			13	16	10	1/4″	.625	1700
203	C	10		15	20	8	17/64″	.565	1650
203H	C			12	15	10	17/64″	.706	1920
203J	C			12	15	11	17/64″	.776	2050
204	C	10		14	18	8	5/16″	.781	2210
204H	C			12	15	10	5/16″	.976	2570
204J	C			12	15	11	5/16″	1.07	2740
205	C	10		14	18	9	5/16″	.879	2420
205H	C			12	15	11	5/16″	1.07	2760
205J	C			12	15	13	5/16″	1.27	3090
206	C	9		14	18	9	3/8″	1.26	3360
206H	C			12	15	12	3/8″	1.69	4070
206J	C			12	15	12	3/8″	1.69	4070
207	C	8		13	16	9	7/16″	1.72	4420
207H	C			11	14	12	7/16″	2.29	5350
207J	C			11	14	13	7/16″	2.49	5650
208	C	8		13	16	9	15/32″	1.98	5020
208H	C			11	14	12	15/32″	2.64	6080
209	C	8		13	16	10	15/32″	2.20	5370
209H	C			11	14	13	15/32″	2.86	6400

PART 2

Following the failure of the 204SST5 bearing, Mr Wireman used Barden Company catalogue data to calculate the manufacturer's design life of the bearing. According to his calculations, the bearing had a 'B-10' life of 6 500 hours. He then sent the failed bearing to Barden for examination. Barden replied with recommendation that a bearing with higher load capacity but identical external dimensions, either number M204BJHX2 or 204HJB1519 be used. Mr Wireman decided to use the 204HJB1519.

The conclusion of the Barden Corporation upon examination of the 204SST5 bearings from the first motor was that failure in the front bearing had been due to poor lubricity between balls and races. This opinion was expressed in a letter from Barden of 11 September 1963, which is attached as Exhibit 8. The answer, Barden asserted, was to use a bearing of higher load capacity and, therefore, longer design life. Of the two bearings suggested by Barden, Mr Wireman chose the 204HJB1519 because it was much the cheaper, costing only a few cents more than the previous 204SST5 bearing. The 204HJB1519 had a higher initial contact angle* (18° instead of 14°) and more balls (11 balls instead of 8) than the previous front bearing, and a design life of 19 000 hours, according to Mr Wireman's calculations.† A Barden catalogue page giving specifications for the HJB1519 appears as Exhibit 9.

Mr Wireman was puzzled as to why it should be necessary to use a bearing design life so much higher than the required operational life, but somehow the idea of using a heavier bearing seemed reasonable. Experience with another motor also lent support to this approach. The other motor was an air-cooled version which attached in the same way to run the same pump. It differed in that although its front shaft bearing was fully immersed, the oil flow path was around, not through the bearing, and there was a seal which kept the hydraulic fluid out of the case, so the rotor and rear shaft bearing were not immersed. The frame also differed by having cooling fins and by allowing room for a larger front bearing, a Barden 205 double-shielded steel retainer bearing which had, according to Mr Wireman's calculations, a lower contact angle but a design life of approximately 200 000 hours under the same loading conditions as the liquid-cooled pump motor. Another difference of the air-cooled motor was that it was used with MIL-5606 hydraulic fluid (a specification for which appears in Exhibit 10), instead of Skydrol. No bearing failures had occurred on endurance tests of 2 500 hours with the air-cooled motor.

Since the use of a heavier-duty bearing seemed reasonable, and since the 204HJB1519 could be installed in place of the 240SST5 without otherwise altering the motor (the 205 bearing could not) and without much affecting costs, the decision was made to install it in all the liquid-cooled pump motors for Geyser Pump. Both qualification test motors were fitted with the 204HJB1519 and twenty production motors which had been shipped were called back, torn down and reassembled with the higher capacity bearing, a process which consumed about two man-hours per motor. Drawings were modified to require the 204HJB1519 on all subsequent units.

Consequently, there was considerable surprise and dismay when it was found after 700 hours running on the reassembled first qualification unit that the new bearing was already starting to fail. This motor had been put back on test to complete the original 2 500 hours. Since the first bearing had failed at 1 800 hours, there were 700 hours remaining of the test. After running the 700 uneventful hours with the new bearing, the motor was stopped and disassembled to inspect those parts which had now completed 2 500 hours. The front bearing, which had as yet shown no symptoms of difficulty while running, was found to be just slightly spalled, signifying

---

* The higher angular contact was produced by increasing the radial clearance in the bearing.

† The 1519 in the bearing number means diametral allowance between balls was held between 15 and 19 ten-thousandths of an inch.

the onset of failure. Pictures of this bearing appear in Exhibit 11. A written description of the failed bearings is given in Exhibit 12. This description, dated 9 December 1963, was prepared by an independent consultant, Mr Thomas Barish, a noted mechanical engineering consultant and author of over twenty published articles on bearings.

EXHIBIT 8   Letter from the Barden Corporation (i)

THE BARDEN CORPORATION
Barden Precision Ball Bearings
Danbury, Connecticut

September 11, 1963

Mr. Elmer Ward, Chief Engineer
Task Corporation
1009 East Vermont
Anaheim, California

Dear Mr. Ward:

This will summarize our recent telephone conversations and letters regarding the failure of a Barden Precision 204SST5 bearing in a
            pump motor qualification unit and our recommendations to prevent recurrence of such failures. The pump runs at 6000 rpm in Skydrol at 150°F with a thrust load of 75 lbs on the 204SST5 bearing and 20 lbs on the opposed bearing, a 203SS5. The failure in the 204 size bearing was experienced at 1800 hours.

Examination of the 203SS5 bearing showed that it operated with normal contact angle in the presence of considerable contamination. There is also a very light running band where the bearing apparently operated under a reverse thrust condition. In spite of the evidence of contamination this bearing could have continued operating for a considerable time.

The 204SST5 bearing is a typical fatigue failure due to poor lubricity of the hydraulic fluid. In this type of failure a very fine surface spalling occurs on the inner ring raceway and erodes the metal until increased axial play causes failure by rubbing of the rotor and housing or increased loads due to the roughened raceway cause normal fatigue failure of the balls and outer ring. In this case both of these apparently happened. There is no evidence of inadequate alignment or poor mounting to aggravate the failure. It is doubtful that the contamination seen in the 203 size bearing had any significant effect. It is, however, possible that the use of a double shielded bearing helped keep debris in the ball path and accelerate the failure.

In order to increase the life of the bearing with Skydrol as lubricant and with the loads stated, it would be necessary to increase the capacity of the bearing. Because the failure at 1800 hours was a progressive type failure it would be necessary to make a substantial increase in capacity in order to prevent the application from being continually marginal in operation. The two alternates are first, a duplex pair of bearings which, as you say, would require considerable machine modification and, therefore, were not quoted; or a single bearing of different design for greater capacity. This greater

EXHIBIT 8 *(contd.)*

capacity could be obtained by using the Barden Precision M204BJHX2 or
the 204HJB1519. Both of these bearings have the same envelope dimension
as the 204SST5 originally used. The M204BJHX2 was developed particularly
for use with heavy loads and low lubricity fluids. It has M50 tool steel
rings and balls as shown on drawing SA-3711 which has been forwarded to
you. The open, unshielded construction which permits an added flow of
fluid which would help wash away the contamination or wear particles
which might otherwise collect on the bearings. The oversize balls,
high radial play, and high shoulders give the bearing a capacity
slightly higher than that of the 205J bearing shown in the Barden Catalogue. Based on the experience with the 204SST5 bearing there should
be no problem meeting the desired life of 6000 hours with this bearing.

If, for economic reasons, the M204BJHX2 bearing cannot be used, the
next recommendation is the Barden Precision 204HJB1519 bearing. This
bearing, because of its larger ball complement than the 204SST5, open
construction, and high radial play, should give significantly longer
life than the 204SST5.

The design life of the 204SST5 bearing is 26,500 hours, for the
204HJB1519 it is 98,000 hours, and for the M204BJHX2 it is 235,000
hours. Assuming the failure at 1800 hours to be typical, this would
give an expected life of 6600 hours for the 204HJB1519 and 16,000
hours for the M204BJHX2. This figure will be increased or the chances
of attaining it will be much better with the use of the open, unshielded
204HJB1519 but it would be impossible to put a quantitative figure on
this relationship without a considerable amount of testing. If the
filtering system can be improved to effectively filter the fluid to a
10 micron level this should also have a beneficial effect on life.

In this discussion we have recommended use of bronze retainers based
on your experience with this material in Skydrol. We have no field
data on this and would normally have recommended comparative checks
with phenolic and bronze. The bronze retainer has the added advantage
of an extra ball in the complement, giving increased design life. The
material used in these retainers is continuous cast Asarcon (80% copper,
10% tin, 10% lead) bronze. We have found it to be superior to all other
types of bronze for use in high performance precision ball bearings.

Summing up, the failure of the 204SST5 bearing is a typical low lubricity fluid failure which we feel can be corrected to give good
reliability at 6000 hours by use of the Barden Precision 204HJB1519
bearing. It is further suggested that, in the qualification unit, the
machine be disassembled at 3000 hours for bearing analysis at Barden,
reassembled, and continued for the full life of the bearing, with
periodic checks.

EXHIBIT 9   Bearing data

BARDEN PRECISION SPINDLE AND TURBINE BEARINGS

HIGH TEMPERATURE
MODERATE TO
ULTRA HIGH SPEED

ANGULAR CONTACT
BRONZE RETAINER

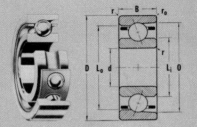

AVAILABILITY.
All sizes listed are normally available from stock in chrome steel.

■ Sizes also available in stainless steel; specify by adding prefix S, as S38HJB, etc.

DIMENSIONS

BORE, O.D., WIDTH						MAX. FILLET RADIUS—inches		BASIC ORDERING NUMBER	OTHER DIMENSIONS inches			APPROX. WEIGHT pounds	DATA REFERENCE NUMBER
d		D		B									
mm	inches	mm	inches	mm	inches	r	r_o		L_i	L_o	0		
6	.2362	19	.7480	6	.2362	.012	.010	36HJB	.383	.596	.636	.027	36J
7	.2756	22	.8661	7	.2756	.012	.010	37HJB	.463	.692	.739	.035	38J
8	.3150	22	.8661	7	.2756	.012	.010	■ 38HJB	.463	.692	.739	.034	38J
10	.3937	26	1.0236	8	.3150	.012	.010	■ 100HJB	.583	.837	.902	.05	100J
10	.3937	30	1.1811	9	.3543	.025	.015	■ 200HJB	.656	.953	1.028	.08	200J
12	.4724	28	1.1024	8	.3150	.012	.010	101HJB	.670	.924	.989	.06	101J
12	.4724	32	1.2598	10	.3937	.025	.015	201HJB	.721	1.040	1.122	.10	201J
15	.5906	32	1.2598	9	.3543	.012	.010	102HJB	.798	1.053	1.117	.08	102J
15	.5906	35	1.3780	11	.4331	.025	.015	■ 202HJB	.815	1.153	1.240	.12	202J
17	.6693	35	1.3780	10	.3937	.012	.010	■ 103HJB	.895	1.153	1.217	.09	103J
17	.6693	40	1.5748	12	.4724	.025	.015	203HJB	.952	1.292	1.394	.18	203J
20	.7874	42	1.6535	12	.4724	.025	.015	■ 104HJB	1.050	1.390	1.474	.17	104J
20	.7874	47	1.8504	14	.5512	.040	.020	204HJB	1.130	1.530	1.649	.30	204J
25	.9843	47	1.8504	12	.4724	.025	.015	105HJB	1.247	1.587	1.673	.20	105J
25	.9843	52	2.0472	15	.5906	.040	.020	205HJB	1.320	1.720	1.840	.35	205J
30	1.1811	55	2.1654	13	.5118	.040	.020	106HJB	1.511	1.869	1.978	.30	106J
30	1.1811	62	2.4409	16	.6299	.040	.020	206HJB	1.580	2.060	2.203	.55	206J
35	1.3780	62	2.4409	14	.5512	.040	.020	107HJB	1.710	2.110	2.229	.40	107J
35	1.3780	72	2.8346	17	.6693	.040	.020	207HJB	1.857	2.382	2.565	.75	207J

APPLICATIONS. Aircraft accessories, turbines, compressors, pumps and other high speed components operating at high temperatures.

DESCRIPTION. Metric series non-separable angular contact high temperature bearings for smooth, quiet operation with minimum vibration at high speed under moderate to heavy loads; support radial loads and thrust in one direction only; generally need light thrust loading.

Angular contact construction permits larger number of balls and greater load capacity than deep groove bearings of same size.

Extremely thin section but sturdy one-piece machined bronze ball retainer has high conductivity to dissipate heat. Outer ring piloted "halo" design gives increased lubricant penetration, circulation and cooling, resulting in longer bearing life at high temperatures.

MATERIAL. Rings and balls: SAE 52100 chrome bearing steel; some sizes also available in corrosion resistant AISI 440C stainless steel. High temperature materials, such as M-50 tool steel, also available on special order. **Retainers:** SAE 64 bronze. Further data on page 48.

LUBRICANT NORMALLY SUPPLIED. Oil: MIL-L-6085A (Barden code 0-11); requires subsequent lubrication with continuous oil mist or spray application. Further data on page 66.

TANDEM PAIRS. All sizes listed may be obtained in DT duplex pairs for additional load capacity. Further data on page 54.

FURTHER DATA. Tolerances, page 46. Radial and axial play and yield, page 49. Shaft and housing shoulders and fits, page 68.

EXHIBIT 9 *(contd.)*

PERFORMANCE DATA

**HIGH TEMPERATURE
MODERATE TO
ULTRA HIGH SPEED**
BORE .2362″-1.3780″
O.D. .7480″-2.8346″

DYNAMIC LOAD RATINGS. Radial load ratings below are for 500-hour design life, or 2500-hour average life, with proper mounting, lubrication and protection of bearings. Use value C_s for fatigue life computation procedure on page 62.

Normal speed limits are shown for oil mist or spray lubrication.

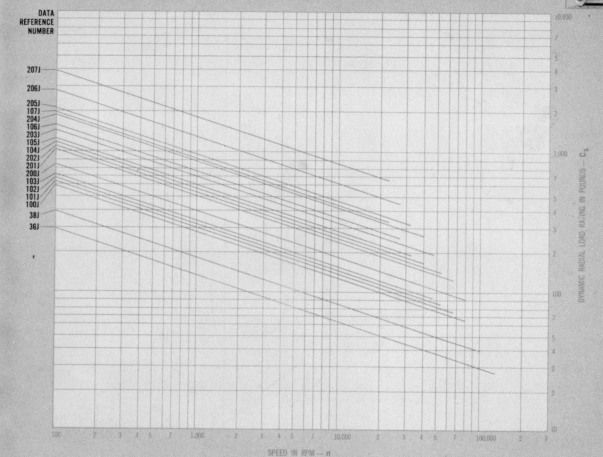

STATIC LOAD RATINGS. Static ratings are shown at left for radial play ranges normally supplied. These values indicate peak loads that can be sustained by bearings without permanent effect on smoothness of operation. Further data on page 60.

AVERAGE RUNNING TORQUE. These bearings are used mainly at speeds where the predominant factor in torque is lubricant drag, which varies with lubricant type and quantity, speed and operating temperatures. When needed for estimate of power requirement, approximate torque values will be supplied.

DATA REFERENCE NUMBER	STATIC LOAD RATINGS—pounds		RADIAL LOAD C_9
	THRUST LOAD—T_0		
	RADIAL PLAY RANGE		
	Code 5 .0005″-.0008″	Code 6 .0008″-.0011″	
36J	528	560	222
38J	720	765	310
100J	1200	1250	505
101J	1320	1375	567
102J	1450	1500	620
103J	1630	1690	695
	.0005″-.0009″	.0009″-.0011″	
200J	1760	1890	715
201J	2150	2300	900
202J	2400	2600	1030
203J	2970	3190	1260
104J	2600	2800	1200
204J	4180	4400	1760
105J	2940	3080	1330
205J	4940	5200	2080
106J	3780	4060	1680
107J	4950	5100	2230
	.0007″-.0012″	.0012″-.0017″	
206J	8280	8400	3100
207J	12100	12200	4580

EXHIBIT 10   Specification excerpts – MIL 5606 hydraulic fluid

in the formulation of an approved product shall require requalification.

3.2 Materials. The fluid shall be clear and transparent consisting of petroleum products with additive materials to improve the viscosity-temperature characteristics, resistance to oxidation, and antiwear properties of the finished product.

REQUIREMENTS

3.1 Qualification. The fluid furnished under this specification shall be a product which has been tested and passed the qualification inspection specified herein, and has been listed on or approved for listing on the applicable qualified products list. Any change

3.3 Petroleum base stock requirements. The properties of the petroleum base stock used in compounding the finished fluid, before the addition of any other ingredients required herein, shall be as designated in Table I when tested as specified in 4.7.2.

TABLE I. *Properties of Petroleum Base Stock*

Property	Value
Pour Point (max) [1]	—59.4° C. (—75.0° F.)
Flash Point (min)	93.3° C. (200.0° F.)
Acid or Base No. (max)	0.10
Color, ASTM Std (max)	No. 1

[1] Pour point depressant materials shall not be used.

3.3.1 *Specific gravity.* The specific gravity of the base stock shall be determined as specified in 4.7.2 but shall not be limited. Samples of base stock submitted for acceptance tests shall not vary by more than $\pm$ 0.008 at 15.6/15.6°C (60.0°F) from the specific gravity of the original sample submitted for qualification tests.

3.4 Additive materials.

3.4.1 *Viscosity-temperature coefficient improvers.* Polymeric materials may be added to the base petroleum oil in quantities not to exceed 20 percent by weight of active ingredient in order to adjust the viscosity

of the finished fluid to the values specified in 3.5.

3.4.2 *Oxidation inhibitors.* Oxidation inhibitors shall be added to the base oil in quantities not to exceed 2 percent by weight.

3.4.3 *Antiwear agent.* The hydraulic fluid shall contain 0.5 $\pm$ 0.1 percent by weight of tricresyl phosphate, conforming to Specification TT–T–656.

3.5 Finished fluid. The properties of the finished fluid shall be as specified in Table II and 3.5.1 through 3.5.11.

EXHIBIT 10 *(contd.)*

MIL–H–5606B

TABLE II. *Properties of Finished Fluid*

Property	Value
Viscosity in centistokes at 54.4° C. (130° F.) (min).	10.0
Viscosity in centistokes at —40° C. (—40° F.) (max).	500
Viscosity in centistokes at —54° C. (—65° F.) (max).	3000
Pour point (max) [1]	—59.4° C. (—75.0° F.)
Flash point (min)	93.3° C. (200.0° F.)
Acid or base No. (max)	0.20

[1] Pour point depressant materials shall not be used.

3.5.1 *Color.* The fluid shall contain red dye in concentration not greater than 1 part of dye per 10 000 parts of oil by weight. There shall be no readily discernible difference in the color of the finished fluid and the standard color when tested as set forth in 4.7.3.

3.5.2 *Corrosiveness and oxidation stability.*

3.5.2.1 *Corrosiveness.* When tested as specified in 4.7.2, the change in weight of steel, aluminum alloy, magnesium alloy, and cadmium-plated steel subjected to the action of the hydraulic fluid shall be not greater than ± 0.2 milligrams per square centimeter of surface. The change in weight of copper under the same conditions shall be no greater than ± 0.6 milligram per square centimeter of surface. There shall be no pitting, etching, nor visible corrosion on the surface of the metals when viewed under magnification of 20 diameters. Any corrosion produced on the surface of the copper shall be not greater than No. 3 of the ASTM copper corrosion standards. A slight discoloration of the cadmium shall also be permitted.

3.5.2.2 *Resistance to oxidation.* When tested as specified in 4.7.2, the fluid shall not have changed more than —5 or +20 percent from the original viscosity in centistokes at 54.4°C (130.0°F) after the oxidation-corrosion test. The acid or base number shall not have increased by more than 0.20

over the acid or base number of the original sample. There shall be no evidence of separation of insoluble materials nor gumming of the fluid.

3.5.3 *Low temperature stability.* When tested as specified in 4.7.2 for 72 hours at a temperature of —54 ± 1°C (—65 ± 2°F), the fluid shall show no evidence of gelling, crystallization, solidification, or separation of ingredients. Any turbidity shall be not greater than that shown by the turbidity standard.

3.5.4 *Shear stability.* When tested as specified in 4.7.4 the percent viscosity decrease of the hydraulic fluid, measured in centistokes at 54.4°C (130.0°F) and at —40°C (—40°F), shall be no greater than the percentage viscosity decrease of the shear stability reference fluid nor shall the acid or base number have increased by more than 0.20 over the original acid or base number.

3.5.5 *Swelling of synthetic rubber.* When tested as specified in 4.7.2, the volume increase of the standard synthetic rubber **L** by the fluid shall be within the range of 19.0 to 28.0 percent.

3.5.6 *Evaporation.* The residue after evaporation for 4 hours at 65.6 ± 3°C (150 ± 5°F) shall be oily and neither hard nor tacky when tested as specified in 4.7.2.

EXHIBIT 11   204HJB1519 bearing after 700 hours operation

EXHIBIT 12   Report of Thomas Barish (i)

Task Corporation   Pump-Motor 56383

Examination of Failed Bearings:

(1) <u>First Set</u>: 1800 hrs. Pump End Brg. Barden 204SSTX5

   <u>Inner Race</u>: completely failed all around circle, from over center
   at bottom to shoulder and shoulder rolled. Pitting rather shallow,
   no deep spots, and over all area. No signs of heating.
        Race also showed one shallow band of contact near opposite
   shoulder, and narrow: may be rubbing of failed parts.
        Too far gone to tell initial failure point. No difficulty in
   bore or clamped shoulders.

   <u>Outer Ring</u>: Outer same but smeared over pitting, probably foreign
   matter pits from inner as surface not nearly so rough as inner.

   <u>Balls</u> also badly pitted, shallow type over nearly half of surface,
   with a few deeper pits or breaks from riding edge of inner after
   inner failed.
        Measured one ball where not failed: .3125" indicating no wear.
        (Shield rubbed on booster pump, after inner failed)
   <u>Cage</u>: 2 piece riveted bakelite compound with alum. side plates.
   Broken to disassemble brg. Heavy rubbing on bore of outer, one
   side only and severe pocket stress with many imbedded small steel
   flakes. Secondary failure: inner went first.

   <u>Conclusion</u>: Heavy load, mostly thrust with possible contribution
   from lubricant of poor lubricity: (indicated by peculiar type of
   shallow pitting.)

(2) Closed End Bearing: Barden 203SSTX5;
        No failure: only surfaces badly spread with fine hard foreign
   matter pits. May have come from front bearing failure, but difficult
   to see how.

   <u>Inner Ring</u>: Inner edge of contact fuzzy and measurements questionable.
   May have been two separate contact overlapped.
        Bore had turned on shaft, rubbing mostly on both sides and not
   middle as the shaft were hollowed. Also shoulder rub and wear.

   <u>Outer Ring</u>: also pitted. O.D. rubbed and turned with most of
   contact at one end (outer end).

   Balls and pressed steel cage very good.

                                        Thomas Barish
                                        December 9, 1963

EXHIBIT 12 *(contd.)*

(3)   Second Set: 700 hours Pump End: Barden 204HJB1519

<u>Inner Ring</u>: shallow pitting load failure as above but caught much earlier. Extended only about 180° with tapering off at each end, and only for width of contact at max.

<u>Outer Ring</u> shows only small pitting from hard particles off inner. Otherwise in very good condition. No difficulty on bore, O.D. or shoulders of either race.

<u>Cage</u> (halo bronze type) extremely good in view of inner failure. No deterioration at all. Likewise balls good. No wear and only very faintly banded.

Closed End; Second Set: Barden 203SSTX5 but shields removed.

Entire bearing in very good condition. No wear or loss of initial polish except slight greying of contact areas.

Inner showed fairly heavy load, per calculation on page 4. However contact may have been two contacts overlapped but still high and indicates more than 3 to 7 lbs. expected.

Complete absence of any pitting at all indicates that back bearing does not need shielding normally.

Thomas Barish
December 9, 1963

Further recommendations on the bearing problem were submitted in writing after galling appeared on the second bearing (204HJB1519) by both the Barden Company and the bearing expert, Mr Barish. Both recommended that bearings of still higher capacity be used. A number of possible causes of failure were suggested, but there was some conflict between the opinions of Mr Barish and those of Barden as to the reasons for failure. Mr Wireman wondered which of the suggested causes of failure he should consider most likely and what should be done to cure it. He was especially puzzled about why the change from the first bearing, No. 203SST5 to one of higher capacity, No. 204JHB1519 had not cured the failure, since both the bearing company and Mr Barish advocated higher capacity as the answer.

Opinions of the Barden Bearing Company

After receiving the second set of failing bearings from the first motor, the Barden Company returned its analysis in a letter dated 4 December 1963. Barden repeated its earlier injunction that a bearing of heavier capacity was needed. The failure, Barden explained, was due to the poor lubricity (their opinion) of Skydrol. As a cure, Barden recommended that custom bearings, costing from $60 to $175, in small quantities, be used. A copy of the Barden letter appears as Exhibit 13.

Opinions of the Geyser Pump Company

Geyser Pump Company engineers said they had in the past experienced somewhat similar wear problems in bearings of pumps like that being used with the Task motor. Their answer had been to use a custom bearing of higher angular contact which had been designed by another bearing company. At the suggestion of Geyser Pump, Task contacted the company which had made the special bearing. The bearing company suggested for Task a custom bearing of higher contact angle similar to the custom bearing proposed by Barden and costing abound $1 000.

Opinions of Thomas Barish

The first two bearings which had failed, together with prints of the motor and pump as shown in Part 1 were sent to Mr Barish. His analysis of these bearings, which was dated 14 December 1963 and appears as Exhibit 14, was more extensive than that of Barden. He, too, suggested a number of possible causes of failure. Rather than lubricity, Mr Barish saw the main cause of failure as being fatigue, but he also suggested several other possible sources of difficulty. Among the suggestions of his report were modifications to the design of the pump as well as the use of heavier bearings.

Mr Barish recommended that special bearings be avoided and standard bearings used. With standard bearings, he said, there was less likelihood of unforeseen 'bugs'.

Meanwhile, more motors were being made and tested. The second qualification test motor was shut down and disassembled after 630 hours when metal particles began showing up in the discharge fluid. Already the front bearing, a Barden No. 204HJB1519 was failing seriously. Pictures of this bearing after failure appear in Exhibit 15. With failure of this bearing at 630 hours, Mr Wireman thought the evidence more strongly than ever suggested that heavier load capacity was not the answer, but he did not know how to explain the phenomenon or what to do about it.

PART 3

EXHIBIT 13   Letter from the Barden Corporation (ii)

THE BARDEN CORPORATION
Precision Bearings
Danbury, Connecticut

December 4, 1963

Mr. Jack Wireman
Task Corporation
1009 East Vermont
Anaheim, California

Reference: My letter of September 11, 1963 to Mr. Elmer Ward

Dear Jack:

The bearings from the hydraulic pump which you sent for analysis
after 700 hours of test running have been examined. The 240
size bearing is still in serviceable condition in spite of some dirt
denting including a few rather large depressions caused by hard contamination in the .--2/.005 size range. The general quantity of dirt
denting, however, is less than that in the first bearing we reported
on last September.

The 204HJB1519 bearing shows evidence of advanced fatigue. There are
two types of fatigue patterns, the more prominent, which is fairly
deep spalling, extends for approximately one half of the circumference
of the inner ring contact area. There are also, in the remaining
contact area, several small spots of surface fatigue. In addition,
there is appreciable dirt denting which most probably resulted from
the material plucked out of the raceway. There is no evidence of excessive radial unbalance or thrust loading or of improper mounting.

In comparing this bearing which failed at 700 hours with the original
204SST5 bearing which failed at 1800 hours, we find that this bearing
has light traces of surface fatigue and moderately heavy spalling in
the raceway while the earlier bearing showed extensive metal erosion
from surface fatigue and resultant heavy wear in the bearing. In
other words, in this bearing the distress that has occurred happened
quite swiftly while that in the earlier bearing was the result of a
much slower process over a very long time. It is possible, indeed
probable, that the surface fatigue in the 204SST5 bearing had begun even
before the 700 hour point was reached. We cannot explain why the
surface fatigue in the 204HJB1519 bearing generated the heavy spalling
while that in the other bearing continued as a surface fatigue with
general wear resulting. It is however, quite apparent that the heavy
fatigue does not have the same pattern as a typical, heavy load, sub-
surface fatigue. The fatigued ring was checked for hardness with
readings of 60.75 to 61 Re with a specification tolerance of 58.5 to
62 Rc.

EXHIBIT 13 *(contd.)*

The corrective action to take to overcome the failure condition is
not changed -- to increase the capacity of the bearing so that the
unit stress will be decreased to a point where the lubricating qualities of the hydraulic fluid will be adequate to sustain life for the
required 3000 hours. In accordance with our recent telephone conversations we have designed a bearing within the envelope dimensions of
the 204 bearing which we feel should be adequate for the application.
This is the 204HJBX31 bearing. In this bearing the ball complement
has been increased to ten 11/32 balls and the radial play has been
increased to give a nominal contact angle of 35°. Inasmuch as the
retainers in both of the above bearings were in practically new
condition, there has been no change made in the retainer. The one
being used is, again, 80-10-10 bronze. The increased ball complement
and high radial play result in a design life of 500,000 hours at 75 lbs.
thrust load. Bearings can be modified to this design in four weeks
at a price of $175.00 each for six bearings or $60.00 each for 25
bearings. Prices for production quantities will be forwarded as soon
as an inquiry can be processed.

The bearings are being returned herewith. Should you have any
question on the above or if we can be of further assistance please
don't hesitate to contact us.

                              Yours very truly,

                              THE BARDEN CORPORATION

                              Herbert D. Williams
                              Senior Product Engineer

HDW:etk

CC: Carl Berg, Purchasing Agent

EXHIBIT 14   Report of Thomas Barish (ii)

For Task Corporation, 1009 East Vermont Ave., Anaheim, California   Page 1
                    Attention: Jack Wireman, Project Engineer

Pump-Motor 56383: Failure of 204 Ball Bearing.

(1) General Conclusions: Bearings failed because of excessive thrust
load: plus minor contribution from fluid not being the best lubricant;

(2) Detail Bearing examination. First pump end bearing
too far gone to draw conclusions, except primarily inner race fatigue
failure: cage behaved very well. No heating from rub or bad lubrication.
     Second pump bearing showed clear fatigue failure, primarily thrust
load, on inner. May have been appreciable radial load.
     Both small end bearings showed much larger thrust load than expected
3 to 7 lb. from spring. Areas showed 92 lb. thrust but this is
misleading as it resulted from 2 or more load paths overlapping which
looked like one large load. Nevertheless the actual load was 5 to 10
times the expected: indicating outer race did not slide in aluminum
housing in operation (or spring was bottomed in error).

(3) Expected Loads: page 3; Pump spring, 40 to 70 (on test, 47)
                      Booster Pump   36
                      Spline Friction   58 to 64

                Total Thrust   134 to 190 lb.

If the spline imbeds, the thrust can be very much higher. With the
present design, imbedding with a sharp edge is almost certain because

(a) the unit pressure is considerable, 2190 psi projected area,
(b) the spring takes out all initial shake, and holds spline in
    one place with all motion at pump, (none in motor).
(c) female spline at pump too soft.

In addition, there may be considerable axial expansions due to
(a) alum. housing and steel shaft at lower of higher (same temp.),
(b) shaft temporarily much hotter than housing, and
(c) any lift-off at valve plate in pump, when operating.

     Radial loads can come from misalignment, since long heavily
loaded splines will have very little rocking effect. If assume spline
tight, and max. runout of 0.004", will give 31 lb. on motor bearing.
     Also magnetic loads may be appreciable. However failures indicate
primarily thrust loads.

(4) Measured Loads from contact areas on the bearings, page 4 show:

          169 to 197 lb. thrust on second 204 bearing

but these figures are not exact: estimate -0 to + 50%.

                                        Thomas Barish
                                        December 14, 1963

EXHIBIT 14 *(contd.)*   Exhibit 14 (contd.)

Task Corp.   Pump-Motor 56383   Page 2

(5) <u>Estimated Life</u>: Using latest ABEC formulas, second bearing
(Barden 204 HJB) has a rating of 473 lb. radial at 6000 rpm. (B10 life
of 500 hours.) At 183 lb. thrust load, this gives <u>4200 hours</u> B10 life.
    Apparently, it took appreciably more thrust load, + some radial,
+ inferior lubricity to bring this down to 700 hrs.

(6) Even a loose bearing would not give appreciable radial motion to
permit large magnetic loads. With thrust down to 80 lbs., max. radial
motion is under 0.0002" at 120 lb. radial. Curves, page 5.

<u>Recommendations</u>:

(1) Check unit for possible <u>axial bind</u>. The second bearing, 204HJB
showed a fairly constant heavy load, whereas spline thrust would vary.
    Beck check: thru any available aperture (or make one), push shaft
or rotor against spring. Should move 0.010 to 0.020, after reaching 40
to 70 lbs., without much load increase.

(2) Eliminate possibility of <u>spline imbedding</u>: For changeover that
does not require any major parts change (except maybe truing motor
shaft bore) see design 2, page 6. Need only replace quill by 2 piece
design, and new longer spring. For new motors, recommend design 3.

(3) <u>Increase thrust capacity</u> by changing to 40° or 35° bearing. These
are in regular production by MRC (7204P); SKF (7204B, maybe): New
Departure (30204): and Fafnir (7204W). Be sure to obtain either bakelite-
compound or bronze one-piece cages: one with lots of space to allow oil
(and foreign matter) to pass thru easily.
    It is not necessary to have these made by others as specials, at
greatly increased cost, and with some experimentation involved. The
first two makes above have 10 to 15% more ball capacity, (34 to 54%
more life.)

(4) On new designs, cut diameter at booster pump next to bearing, and
leave some debris space around bearing. See design 3, page 6.

(5) Improve shaft fit for smaller (203) bearing. One bearing returned
showed turning and loose fit. Needs tighter fit when no locknut. Use
mfr standards, and be sure bearing slightly loose to allow for press fit.
Recommend mfrrs "loose fit" standard.

(6) Believe steel liner in housing will be necessary for small bearing.
In spite of many efforts, no one seems to have been able to make this
work without a sleeve for long life and any but very small loads.
    In the meantime, for current units, recommend 0.0003 in loose min.

(7) Review airgap selection; Petroff's equation shows about 3 h.p. loss
in gap even at 2 centrifpoises viscosity and 0.007 in gap. At larger gap,
less fluid loss may offset magnetic losses.

                                        Thomas Barish
                                        December 14, 1963

EXHIBIT 14 *(contd.)*

Task Corp.   Pump-Motor 56383   Page 3

<u>Load Calculations:</u>

<u>Thrust Load:</u> (1) Preload Spring on Quill: 40 to 70 lb.
Actual on Test Units 47 lb.

(2) Friction from Spline:

11.5 H.P. x 63000/6000 rpm = 120.7 in lb.

Tangential force, 120.7 / 0.23 inches radius = 524 lb.

Axial Friction at 0.11 to 0.16 coef. of friction

= <u>58 to 84 lb.</u>

This number may be exceeded if spline tends to imbed.

Spline Loading: Area: 0.46 in p.d. x 0.60 in long = 0.28 sq. in
(tangential projected area).

Unit Loading: 524 / 0.28 = <u>2190 psi:</u>

(inches if equally divided.)

(3) Thrust from Booster Pump:

2.8 in O.D. at 6000 rpm = 73.3 ft./sec.

Hydraulic head at 80% eff. v^2 x 0.8/ 2g = 67 ft. head

= 16 psi

Thrust on seal diam., 1.7 in = <u>36 lb.</u>

<u>Total Thrust:</u> <u>134 lb. to 190 lb.</u> with possible large increase if
spline imbeds.

<u>Radial Load:</u> 1/2 rotor weight + 1 G unbalance = 10 lbs.

Possible load from eccentricity if splines imbed:

(or from pump housing tilt)

For 0.004 inches eccentricity (0.002 inches radius)
on beam 2 1/2 long x 0.34 in diam. (ends fixed) 31 lb.

Thomas Barish
December 9, 1963

EXHIBIT 14 *(contd.)*

Task Corp.   Pump-Motor 56383   Page 4

<u>Load Calculation</u> from Brg Contact <u>Measurements</u>:

	<u>First Set - 1800 Hrs.</u>		<u>Second Set - 700 Hrs.</u>	
Barden Brg. No.	204SS-TX5	203SS	204HJB1519	203SS
Balls	8-5/16	8-17/64	11-5/16	8-17/64

<u>Data from Barden</u>

Curvatures %	52-52	52-52=	52-54	52-52
Inner Groove Rad.	0.1625	0.1382	0.1625	
Groove Diam.	1.0175	0.8564	1.0175	
Land Diam.	1.1300	0.9520	1.1311	
Groove Depth	0.0562	0.0478	0.0568	
degrees	48.5	48.5	49.5	
Radial Looseness	0.0005-9	2.0005-9	0.0015-19	
Initial Contact Angle	11.47-15.85	12.5-16.8	16.25-18.3	
Hardness-Inner			60.75-61 C.	

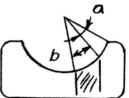

<u>Measurements</u> and <u>Calculations</u> (inner races)

Contact	a	0.050	0.050	0.060	0.050
	b	0.060	0.060	0.050	0.060
Load/Ball (from b)		46	43	55	46
a + 1/2 b		0.08	0.08	0.085	0.08
in degrees		34.	28.6	30.	34.
Contact Angle		14.5	20.9	19.0	14.5
Thrust Load		<u>92 lb.*</u>	<u>169 lb.</u>	<u>197 lb.</u>	<u>92 lb.*</u>

Contact angle, calculated using measured load and initial angle

T/nd^2K	0.00153	0.00148	0.000860	0.00153
angle	18-20.4	18.5-21.2	19.6-21.3	18.5-21.2

(* Both of these may have been two or more paths overlapped)

Thomas Barish
December 9, 1963

EXHIBIT 14 *(contd.)*

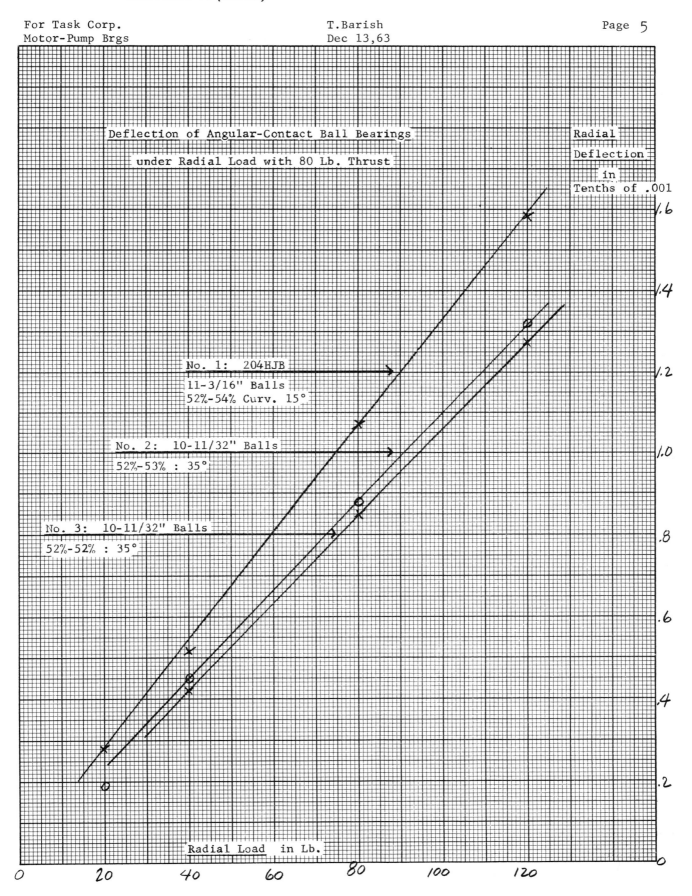

Deflection of Angular-Contact Ball Bearings under Radial Load with 80 Lb. Thrust

Radial Deflection in Tenths of .001

No. 1: 204HJB
11-3/16" Balls
52%-54% Curv. 15°

No. 2: 10-11/32" Balls
52%-53% : 35°

No. 3: 10-11/32" Balls
52%-52% : 35°

Radial Load in Lb.

EXHIBIT 14 *(contd.)*

Sketches for Correction

Design No. 3:

For new motors.

Design 1:

Method for using 1905R as
preload Brg to eliminate
radial play.
    Not needed since
deflection small.

Suggest
smaller
nuts
(aircraft)

Design 2:

To eliminate spline
imbedding. Only
change in cplg.
Rest of unit same.

    Large Diam Spline,
cuts thrust to 1/4
Long spring in
center.

May need
grind
shaft
inside
here.

No spring
here.

| THOMAS BARISH | 3210 Warrensville Center Road Cleveland 22, Ohio | Thomas Barish | Dec 14, 1963 |

EXHIBIT 15   204HJB1519 bearing after 630 hours operation

Following the failure of the third bearing at 630 hours, several further actions were taken by Task to determine the cause and cure. Additional bearings were sent to Mr Barish, who examined them and supplemented his earlier reports with more possible explanations and reiteration of some of his earlier recommendations. A second consultant, Professor Dino Morelli of the California Institute of Technology, was asked to examine the bearings and give his opinions. Professor Morelli said he thought the cause was skidding of the balls due to too light an axial loading for the particular type of bearing running fully immersed in fluid. The recommendations of Professor Morelli differed from those of Mr Barish, leaving Mr Wireman with the question of which of the recommendations should be followed, if any.

Further opinions of Mr Barish

The third set of bearings was sent to Mr Barish for examination. In addition, at his request, additional sets were copper-plated and run for ten hours, one set on a pump and one set on the Task Dynamometer with no pump, after which they were sent to Mr Barish. From impressions on the copper, it could be seen where the balls were riding in the races and how wide was the band of contact, which in turn enabled him to compute axial shaft loadings.

A report on the third failure and an analysis of the copper-plated bearings was submitted by Mr Barish on 26 December. Some possible causes of failure suggested in this report were, (a) binding between housing and rear bearing and (b) axial vibrations due to the hydraulic pump. From the copper-plated bearings, Mr Barish concluded that axial forces on the shaft were within the predicted range, though the pump imposed a somewhat higher thrust than was to be expected from the loading spring alone. This report appears as Exhibit 16.

Another report dated 31 December by Mr Barish, gives his findings for examination of five additional bearing sets which had run in service from 10 to 20 hours. In this analysis, which appears as Exhibit 17, Mr Barish cited binding or misalignment of the rear bearing as the most likely source of difficulty.

Still later in correspondence, Mr Barish made the following comments:

'I apparently lost the needed emphasis on the major problems in trying to cover all possibilities and also because the first reports were made without seeing parts or bearings.

'Major Problems: friction in the spline produces thrust load, estimated at 54 to 68 lb. This always occurs, whether there is imbedding or not. Imbedding can cause much greater loading.

'The spline problem is aggravated because the spring held the "floating" coupling tight one-way, and even small movements would start to build up thrust. Also the small diameter and greater length mean higher loads and little alignment capacity.

'Second: Usual practice uses steel liner in aluminum housings for ball bearing outer rings.

'Temperature changes *and* temperature differentials cause radial binding between outer ring and housing on the small bearing. Hence, there is much more thrust until the bearing is free to slide. Even transients here can be destructive.'

A characteristic of the bearings which was discussed with Mr Barish over the telephone was that of 'ball banding', which had appeared on some of the service units' bearings. On some of the bearings, each ball had acquired a narrow circumferential stripe of wear, indicating that it had rolled in a continuously repetitive pattern. When some of the bearings had been taken for copper plating to the Bearing Inspection Company, a specialized Los Angeles testing laboratory, Mr Wireman had discussed the ball banding with representatives of the laboratory. They told him the banding showed that overloading of the bearing had prevented the balls from moving in a random fashion. Mr Wireman asked Mr Barish about this on the 'phone and was told that ball banding always occurred under operation with continuous thrust and was not an indication of overload.

Bearing Inspection also informed Mr Wireman that bearing problems on pumps such as that being used with the Task motor were not uncommon. In other pumps having the same general design and operational characteristics there had been problems of failure of bearings within the pumps. Mr Wireman understood that the approach taken to solve such problems had been to install bearings of higher load capacity.

Opinions of Professor Morelli

Ball skidding due to a combination of factors was seen by Professor Morelli as the main cause of failure. These factors were as follows:

1. By operating fully immersed, a retarding force of viscous drag was imposed on the balls as they passed around the race. Catalogues on bearings, it was pointed out, specify 'mist lubrications' for ball bearings. (In contrast to Professor Morelli, Mr Barish saw full immersion of the bearings as an advantage. In his opinion, it assured full lubrication and the continuous flow tended to reduce the danger of localized heating.)
2. Being 'outer race centered' there was an additional retarding force on the balls exerted by the retainer as it dragged through viscous shear against the stationary outer race. The point at which the retainer contacted each ball also caused it to apply a force inward on the balls lowering the contact pressure between the balls and the outer race.

Professor Morelli concluded that the answer was to increase ball contact stress, to use an 'inner race centered' bearing, or to modify the design so the bearings would not be running fully immersed. His recommendations were to use an inner race centered bearing of load capacity comparable to that of the bearing originally used.

EXHIBIT 16   Report of Thomas Barish (iii)

for the <u>Task</u> Corporation, 1009 E. Vermont Ave., Anaheim Calif.

Attn. Jack Wireman, Project Engineer

Supplement A to Report of Dec. 14, 1963 on

Motor <u>Bearing Failures</u> in <u>Motor-Pump</u> 56383.

(1)   Examination of the new failure, SN/5, Barden 204HJB (page A3) is helpful because failure had not progressed so far. It showed that failure <u>started</u> with the balls <u>riding on the edge</u> of the groove. The ball surface breaks were typical.

The races did not have any real surface breaks, only severe roughening: except outer did begin some slight surface breaks.

Severe roughening extended over entire grooves.

(2)   The balls could ride the edge of the inner under two conditions: the first a very severe tilt of the outer, of the order of .005 to .010"/". The outer race surface roughening was over all the race except a few long islands in the center (sketch, page A3). This also agrees with bad tilt. Likewise cage deterioration with severe ball-pocket wear and light wear between cage and outer.

(3)   The second possibility (and I believe more likely) is that the outer race of the small brg was bound in the aluminum housing and did not slide: happens if aluminum housing is cooler than shaft and brg. Then also, the outer housing would shrink endwise and bind small brg against large end brg. Practically every small end brg showed much larger thrusts than spring preload. (109 lb. this set, page A3)

On the first failure (conrad brg) this reverse load on the 204S would be less harmful; than on later 204 hjb. Here, the reversing would quickly cause ball to ride on shallow shoulder (about .008" end movement) and this would break up the ball surface rapidly.

The new failure rode on both edges of outer path. Brg would then break loose and bounce back to other side: showing clear islands again. Also cage could show the same distress, mostly from broken ball surfaces. Also inner contact extended about 20° over-center towards wrong side.

(4)   Other possible causes:
<u>Critical frequencies</u>, axial: first using pump spring, = 1700 rpm. Too low to matter. Second using brg spring rate, pages A5 and A6 give critical of 38,000 per min. Still low, since one-per-cylinder forcing function is 54000.

<u>Shock loads</u>, when small brg lets go quickly, only rise to 660 lb.; small compared to shock capacity (brinell load) of over 2000.

Note discussion on plated brgs, page A2.

EXHIBIT 16 *(contd.)*

Task Corp.   Pump-Motor 56383   Page A2

Recommendations:

(1) Same as before: check carefully for errors in handling, chips, binding etc. in assembling pump and motor.

(2) As before: eliminate spline imbedding by two-piece coupling per sketches. This will also isolate pump further from motor for eccentricities, transfer of forcing vibrations, and will permit longer preload Spring.

(3) Increase thrust capacity on 204 as before.

(4) Add much emphasis on eliminating bind in 203S outer race fit. A good steel liner, and looser fits are indicated. In view of trouble, recommend about 0.0004" min.

A good changeover is available with present parts by changing from 203 to 103 size leaving room for 0.10" thick liner.

Plated Brg experiments proved very helpful: The contacts were all where they should be and showed no trouble (measurements and calculations, page A4) except that load on small brg was large as on all previous cases. Also the no. 3 brg (small end) shows a slight misalignment of outer race, and smaller load.

The positive conclusion that can be drawn from this is: these brgs would not have failed if continued as they were. The failures came from conditions introduced by some other operating conditions such as cold starts or unusual temperature differentials: or else by some assembly or machining errors existing on other units.

Thomas Barish Dec. 26, 1963

EXHIBIT 16 *(contd.)*

Task Corp.   Pump-Motor 56383   Page A3

<u>Examination</u> of Additional <u>Failed Brgs</u>: S/N.No. 5 Motor-   600 Hrs.

(1) Brg. Barden 204HJB1519. This brg shed new light on the problem.
Both races were fairly completely covered by very bad roughing up
of the surface with only some parts of outer groove breaking thru
the surface slightly. This existed over all the outer, and on inner
from shoulder to about 20° over center. Both races had the ball run
hard enough on the shoulder edge to raise a positive burr.

In addition the outer race showed some long islands, about
at the middle of the roughened area, and about 1/5 of the width
where the surface was relatively smooth. (sketch below)

Most of the balls were badly broken on the surface and some
of them along a diam, about 180°, and much deeper than the race
surfaces. This means that the balls rode the edge of the groove
and the breaks were started by this edge cutting, (both on inner and
outer).

This also means that the initial failure was on the ball surfaces
from riding this edge and that the race roughening resulted from
the ball failure.

In addition, the outer showed signs of having rotated slightly,
on both O.D. and thrust shoulder.

The above condition, wide path with clear area in center of
outer, is produced by (1)* a badly cocked outer race plus thrust
load. Reference- "Effect of Misalignment on Forces acting on Cage"
by K. Kakuta, ASME Paper 63-Lub-12-.

This was confirmed by the unusual retainer deterioration.
All the ball-pocket surface showed bad spreading, some very severe
with high burrs: whereas the O.D.-Outer-ring contact showed only
a little distress. It was thought that unequal ball size might
account for the variation in pocket wear but enough of the balls
could be measured on unbroken diameters to show no wear (all 0.3125)
and all equal.

Also the cage enlarged slightly
under severe forces and bound slightly
on outer ring.

Normally, such a tilted outer
would show a bad tilt at the ball path,
instead of the wide spread. But in this
case, the outer turned somewhat in the
housing.

There were no signs on inner or
outer of appreciable chips under the
thrust shoulders, or uneven clamping.

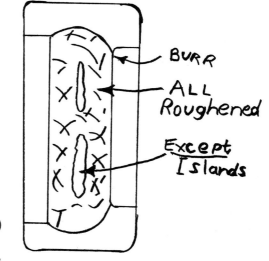

(2) The small Brg. Barden 203S showed
no failure. Contact area measurements,
a = 0.045": b= 0.06" (page 4, Exhibit 14)
which indicates 109 lb. thrust or two
overlapping paths. Thrust was right way.

Bore showed no contact except one corner, and also appreciable
rub at shoulder. Again, too loose a fit on shaft and probably badly
tapered. Outer showed no motion in housing.

*(2) Alternate: rear brg outer sticking in housing, building up
reverse thrust and then letting go. Perhaps building up thrust in
opposite direction.

Thomas Barish Dec. 26, 1963

EXHIBIT 16 *(contd.)*

Task Corp.   Pump-Motor 56383   Page A4

<u>Examination of Copper-Plated Test Brgs.</u>

These brgs all showed just what might be expected (except for larger thrust on small end brg) and nothing like what had shown on failed 204 size. In each case a fairly narrow uniform width of contact. Measurements and calculations follow: (fig., page 4, Exhibit 14)

	<u>With Spring alone</u>		<u>With pump</u>	
Brg. No	no. 2	no. 3	no.1	no.4
Size	204 HJB	203S	204HJB	203S
inner: a, b at 90°		0.045 0.058 0.045 0.055 0.045 0.058 0.045 0.060		0.050 0.050 " " " " " "
average	0.062 0.046	0.045-0.0578	0.060-0.058	0.050-0.050
Outer: a,b		0.030-0.050 0.038-0.050 0.050-0.042 0.045-0.050		0.065-0.035 0.065-0.035 0.057-0.042 0.060-0.038
average	.068-.035	0.0408-0.048	0.060-0.042	0.0618-0.0375

Load/ball
  Inner
  Outer
	19
	20
		40
		26
			38
			33
				30
				12

1/2a+b:inner 30° Outer 29.2		30.6 27	31.3 27.5	31.2 35.4

Contact angle
  Inner 19.5°
  Outer 20.2
		17.9
		21.5
			18.2
			22
				17.4
				13.4

Measured (initial by BII)	18.3	15.0	18.0	15.5

Total Thrust
  Inner 69.5
  Outer 76
		98
		76
			131
			128
				72
				22**

<u>Conclusions:</u> Thrust on main brg about where they should be: (note about 73 lb. spring alone, and 130 with pump).

Thrust on small brg always larger than expected.

**(These readings show small but positive outer tilt).

Thomas Barish   Dec. 26, 1963

EXHIBIT 16 *(contd.)*

Task Corp.   T.Barish 26Dec63   Page A5
Pump-Motor 56383

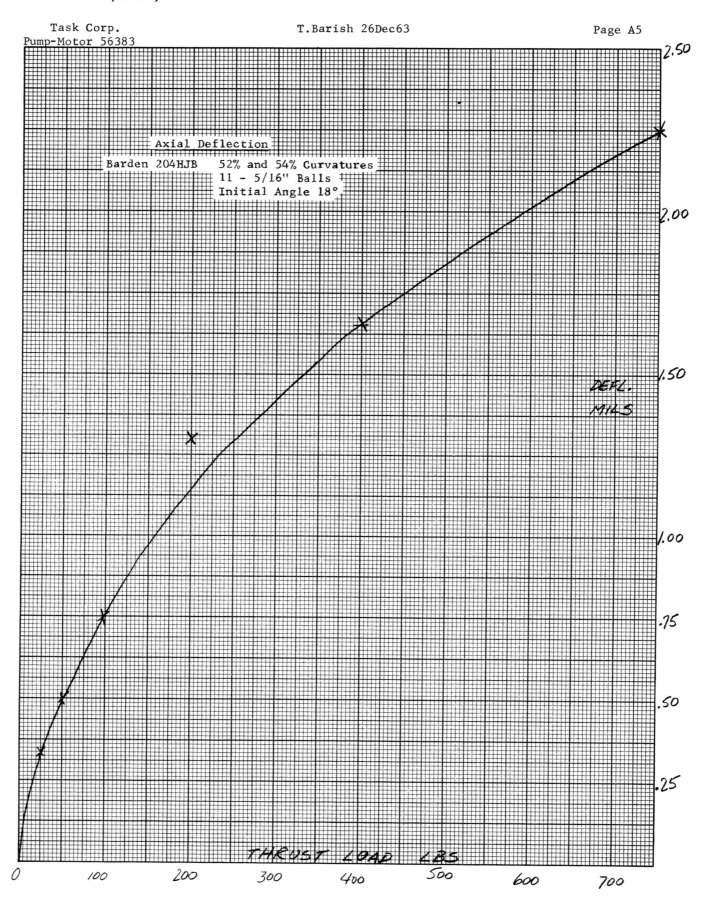

Axial Deflection

Barden 204HJB 52% and 54% Curvatures
11 - 5/16" Balls
Initial Angle 18°

DEFL. MILS

THRUST LOAD LBS

EXHIBIT 16 *(contd.)*

Task Corp.   Pump-Motor 56383   Page A6

Other Possible Causes:

(1) Critical Frequencies: axial: Rotor Weight, 5.5lb.

Spring Constant taken from curve, page A5, for 100 lb. load

   = 187,000 lb. per inch.

   Deflection under own weight = 5.5 / 187,000 = 0.000 0294

   Critical Frequency = $197.7 / (.000\ 0294)^{1/2}$ = <u>38000 per min.</u>**

   Forcing functions: 6000 for rpm and <u>54,000</u> for 9 cylinders.

(2) Shock Effect: if rear brg hangs up and lets go quickly:

   Total Travel: estimated at .012"max. (endplay of 204 brg.)

   Energy build up: 70 lb. spring x .012"

   Using area under deflection curve, page A5, this would
   require brg force to go from 100 lb. to <u>660 lb.</u> max. shock

   This is still not excessive since brinell capacity or
   shock capacity of the brg is over 2000 lb.

** (Note: 35° Brg will increase spring rate about 60 to 70% and will
        raise this critical to 48,000 to 50,000 per min.)

Thomas Barish Dec.26,1963

EXHIBIT 17   Report of Thomas Barish (iv)

Page B1

for the Task Corporation, 1009 East Vermont, Anaheim, California
                    Attention: Jack Wireman, Proj. Engr.

<u>Motor-Pump</u> 56383: Bearing Failures.

<u>Supplement B</u> to Report of Dec. 14, 1963: Review of 5 additional Brgs.
     Barden 204HJB1519 that had seen small amounts of use.

(1) <u>Bearing examination</u> on page B2. No deterioration. But ball paths just distinguishable in 4 cases.

<u>Conclusions</u>:

(2) The thrust loads were about where they should be, perhaps a little high, but nothing to cause any trouble.

(3) There were no signs of any tilt or misalignment of ball path.

(4) However two of the bearings (and slightly in a third) showed that at some time, the small end bearing took most of the thrust and left this large end bearing under pure radial load; in one case leaning slightly towards reverse thrust.

(5) This confirms previous analysis (emphasized in Exhibit 16) that the major cause of trouble is seizing of outer race of the small end bearing preventing axial sliding.
     And that this trouble is aggravated in longer runs by small heavily loaded spline held in one place and tending to imbed:
     And by possible eccentricities at Pump-Motor joint causing larger radial load.

(6) One possible alternate for this lot; if motor was run without pump or spring load, bearings would show radial load as above.

(7) Previous discussion only considered as source for binding of O.D. of small bearing, the possible greater temperature of shaft and bearing over housing. One other possibility now suggested is that unsymmetrical end-bell under fairly large pressure load distorts bearing housing out-of-round.

(8) <u>Recommendations</u> still as in Exhibit 16, p. A2.

Thomas Barish
December 31, 1963

EXHIBIT 17 *(contd.)*

Task Corp.   Motor-Pump 56383   Page B2

<u>Examination of Bearings</u>:

5 <u>Barden 204 HJB1519</u> that had seen small amounts of miscellaneous use.
(numbered 1 to 5 by TB)

None of the bearings showed failure: practically no deterioration.
In fact, they could be reassembled after very light polishing, with
new (high quality) balls and used as new.

The <u>cages</u> were all in practically perfect condition. The <u>balls</u>
all showed very light banding; one set, except 2 sets in bearing 2.
The bands were very faint in bearings 4 and 5.

The <u>races</u> showed no signs of any motion on bores, O.D. or sides.
However, the ball paths could be distinguished and were measured and
checked for load below: (except no. 4 where marks were too faint.)

The major point of interest was that bearings 2 and 3 showed
beside the standard thrust load path, a <u>second lighter area</u>, a small
scraping on the inners, under approximately radial load. In no. 3,
this even edged slightly towards the wrong side thrust. No. 5 also
showed a very faint spot of radial load on the inner. On the outer
races, these were not discernable because the polish was not so high
and it would take more running to show these faint paths.

Ball path measurements and
      load calculations

<u>Bearing No.</u>	1	2	3	5
Inner Race a	0.060	0.060	0.060	0.060
b	0.050	0.050	0.045	0.040
a + 1/2b	0.085	0.085	0.0825	0.080
Contact Angle, degrees	19.5	19.5	20.5	21.3
Load per Ball lb.	22 1/2	22 1/2	17	22 1/2
Total Thrust lb.	83	83	66	90
Outer Race a	0.055	0.065	0.065	can't see
b	0.040	0.035	0.040	
a + 1/2b	0.075	0.0825	0.085	
Contact Angle, degrees	23	20.5	19.5	
Load per ball	28	20	28	
Total Thrust, lb.	120	78	102	

Thomas Barish
December 31, 1963

To test the opinions advanced by the different experts as to why the bearings had failed, Mr Wireman conducted several investigations. In his judgment, the results of these investigations cast doubt on all the opinions advanced and verified none. Some action, however, had to be taken to solve the problem. Mr Wireman rejected the idea of trying several different bearings at once. Such an approach seemed to him 'excessively expensive and sloppy engineering'. His decision was to try another standard bearing, New Departure No. 030203, which had a higher load capacity, a contact angle of 35°, larger balls, and an inner-race centered plastic retainer. The thickness of the retainer was greater enough so its point of contact on the balls was halfway between the inner and outer races. The 030204 was manufactured to only a standard A.B.E.C. class 3 quality, a class compared to A.B.E.C. class 7 for the Barden bearings, and its price consequently was around $3 as compared to around $10 for the Barden bearings.

As of 24 February 1964, a test motor had run with the 030204 bearing for 600 hours, and no failure had yet occurred. The main question in Mr Wireman's mind was what he would do if the bearing did fail before the required 2 500 hours test was completed.

The possibility that the rear bearing was misaligned or binding was eliminated by careful measurement of the parts concerned while operating and computations as to the expansion effects to be expected with higher temperatures. Computations indicated that the aluminum motor housing could be expected to expand from 0·003 to 0·006 in farther axially than the shaft as temperature rose to that of normal operation. Mr Wireman concluded that there was more than enough room for this much axial movement of the rear bearing, and movement would be made easier at higher temperature since the aluminum housing would expand radially more than the outer race.

Careful examination revealed no evidence of spline imbedding. Axial loading of the shaft under operating conditions was checked by measuring how far the shaft moved with the motor running. No excessive deflection resulted, and it was therefore concluded that the maximum load on the shaft was not excessive.

Varying the applied axial loading by varying throttling of the hydraulic fluid outflow, the shaft was forced to move about 0·008 in back and forth with the motor running and it was observed that the rear bearing slid longitudinally in the housing, as it was supposed to, without binding. Observation through a plexiglas cover which was substituted over the rear bearing made it possible to see that the belleville load was sufficient to keep the rear bearing balls from skidding.

The possibility of ball skidding of the front bearing was also studied. The front of the motor was replaced with a transparent plastic through which the bearing could be seen. The motor was then filled with oil, and run. With stroboscopic light, it was possible to tell by observing the retainer whether the balls were revolving at the right speed or going too slow and skidding. By varying the axial load hydraulically, the loading at which skidding occurred could be measured. It was found that the 204HJB1519 bearing would not skid unless the axial load dropped below 60 or 70 lb. With the lighter load 204SST5 bearings, only 20 to 30 lb were needed to prevent skid. With the 030204, around 40 lb were needed. All these figures were less than the loading expected in operation, so Mr Wireman concluded that ball skidding was not the cause of failure.

Mr Wireman also looked for signs of cavitation in the hydraulic fluid, but he could not see any. He surmised that the pressure of 50 psi in the motor was enough to prevent cavitation.

These tests were performed on the Task Dynamometer. It was not possible to conduct them with a pump attached to the motor, but this fact, in Mr Wireman's opinion, did not cast doubt on the validity of the results.

Another possibility which had occurred to Mr Wireman was that there might be some sort of high frequency pressure feedback from the Geyser pumps. He had inquired with Geyser about this and been told that no such feedback was present. He commented, however, 'We've been told that bearing problems have occurred with that type of pump before. The pump wails like a banshee when it's running. Something has to be vibrating quite a bit.'

If the bearings were failing due to excessive loading imposed by the pump, it was presumed that Geyser Pump might be liable for the expenses of determining the cause of the bearing problem and for correcting it. Such liability would have to be determined on a negotiated basis, however, since the original specification to which the motor had been designed had made no reference to axial loadings on the motor shaft due to the pump.

As of mid-March 1964, all the pump motors were being assembled with the New Departure 030204 bearing. A test motor was also being run in the Task shop with this bearing, and after 600 hours no problems had yet occurred. It had been decided that all the bearings should be 'sound tested'* by the Bearing Inspection Company before installation. At first a rejection rate of 50% was experienced with this inspection. The cause was found to be that the supply house from which the bearings were bought had been repackaging them with insufficient care. In repackaging, some of the bearings had apparently become exposed to moisture which caused slight corrosion and consequent roughness of the balls and races.

Aircraft were flying with the pumps installed, some having the Barden 204HJB1519 and some having the New Departure unit. Maintenance instructions on the aircraft were set to require tear down and inspection of the pump motors after 1 000 hours operation, but none had yet been in use this long. A couple of pumps had failed during pre-flight testing due to clogged outlet filters (10 micron filters), but none had yet suffered bearing failures. Some other problems with the motors, such as burning of the electrical connector pins, need for higher flowrate, faster cold starting and lower current drain had been corrected by various design modifications, none of which were expected to affect shaft bearing life. Not all of these changes were Task's responsibility, and the costs of making them had accordingly been divided among Task, Geyser Pump, and Thunder Aircraft by mutual agreements.

Over 120 motors had now been shipped, and it had been agreed among Task, Geyser Pump, and Thunder that Task was responsible for performance of the bearings. There were 180 motors yet to be completed and shipped on the contract.

* In 'sound testing' an unloaded ball bearing is spun next to a microphone. The noise level produced is compared to a standard. Excessive noise indicates roughness. Such testing costs about $1 per bearing, depending on quantity.

SUGGESTED STUDY

The following questions are suggested as examples of problems which may be raised by this case:

1. After reading Part 1, and before reading further, what do you think that Jack Wireman should do and why?
2. State possible causes of the bearing failures. List all causes and then evaluate the credibility of each.
3. Check the theoretical life of the bearing. Explore the importance of parameters and their possible variations.
4. Was there any attempt to eliminate causes of bearing failure on the basis of likelihood?
5. What constraints of time, money and customer relations did Jack Wireman face?
6. With all your hindsight, how would you have attacked this problem to begin with?
7. After reading the case study, what would you do next if the final action had failed?

Case 2   Development of an oil-well stripper rubber

by Paul E. Bickel

Prepared with support from the Ford Foundation for the University of California, Los Angeles. Assistance from the National Supply Division of Armco Steel Corporation is gratefully acknowledged.

This case is available in pamphlet form as ECL 1-13 from the Engineering Case Library, Room 500, Stanford University, Stanford, California 94305.

In April 1961 Paul Bickel applied for registration as a Professional Engineer in Ohio, where he had obtained a BS degree in Mechanical Engineering from the University of Cincinnati in 1955. After graduation he had been employed by the National Supply Company in California. He was still working for the same company, but had meanwhile been transferred to Houston, Texas.

Registration in Ohio requires, among other things, submission of a report which shows that the candidate has been in responsible charge of a professional job. Paul chose to submit a report on the development of an oil-well stripper rubber which he had done from May 1959 to August 1960.

A few years later Paul was offered another job. He became manager of the Chicago plant of Metal Improvement Company. One of Paul's friends in that company was interested in engineering case histories. Paul showed him the registration report. The friend asked Paul to tell him more about the circumstances of this job and Paul obliged by giving him a more personal account.

The job as described for the Ohio State Board of Registration for Professional Engineers is Part 2 of this case. The circumstances as described to the friend are Part 3. Working drawings of the device are reproduced in Part 4.

PART 2

An engineering report submitted by P. Bickel* in connection with his professional registration

Introduction

The stripper rubber is a component member of the well-head assembly common to all oil and gas wells. One function of the well-head equipment is to provide various pressure control means during the drilling, completion, and production of oil and gas wells. This function of well head equipment at times requires the use of a stripper rubber. Figures A(i) and A(ii) show the stripper rubber and its placement in a typical well-head assembly.

The stripper rubber is contained in the body of the assembly immediately below the tubing suspension; it incorporates a pressure seal on the OD; the main body of the rubber is a flexible, resilient material which provides a pressure seal on the OD of the smallest diameter tubular string, generally called tubing.

The function of the stripper rubber lies entirely in this tubing seal. While the stripper rubber is rigidly contained in a body at all times, its twofold functions are, generally speaking, of a static and a dynamic nature. That is to say, the stripper rubber must provide a tubing seal either when the tubing is held stationary or when the tubing is being run into or out of the well. Under static conditions the stripper rubber must be able to contain a maximum fluid or gas pressure of 6 000 psi; under dynamic conditions, 200 to 2 000 psi.

The problem

The principal design problem and the condition which distinguishes this seal design from other common mechanical seals is the considerable variance in diameter that each size stripper rubber must seal. For example, in the common $2\frac{3}{8}$ in and $2\frac{7}{8}$ in tubing sizes, the coupling diameters are $3\frac{1}{16}$ in and $3\frac{43}{64}$ in respectively. This represents a

* National Supply Division, Armco Steel Corporation, Houston, Texas.

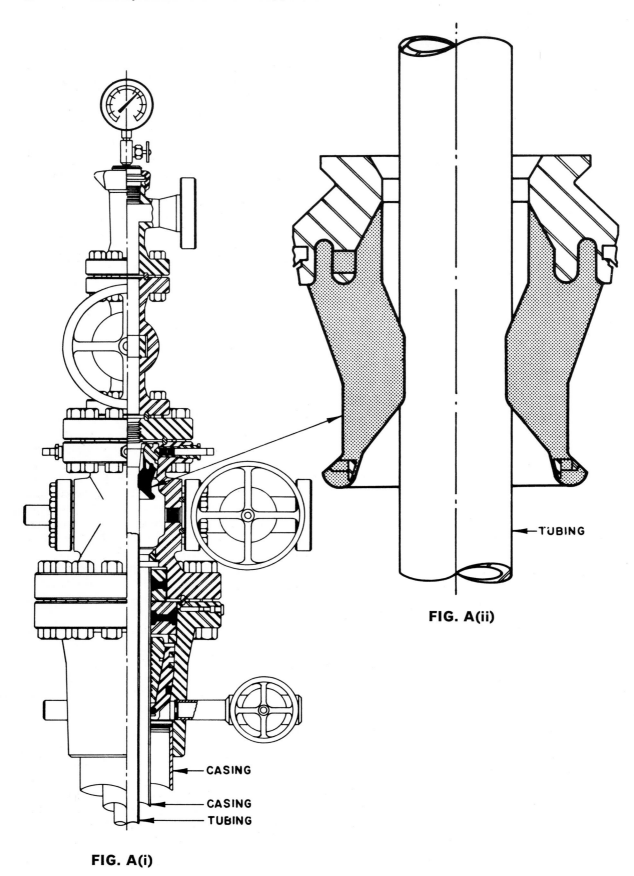

CASING

CASING

TUBING

FIG. A(i)

TUBING

FIG. A(ii)

minimum to maximum diameter difference of $\frac{11}{16}$ in and $\frac{51}{64}$ in; that is, the stripper rubber must seal on the smallest diameter and still be able to pass and seal a diameter $\frac{11}{16}$ in or $\frac{51}{64}$ in larger. See Fig. B.

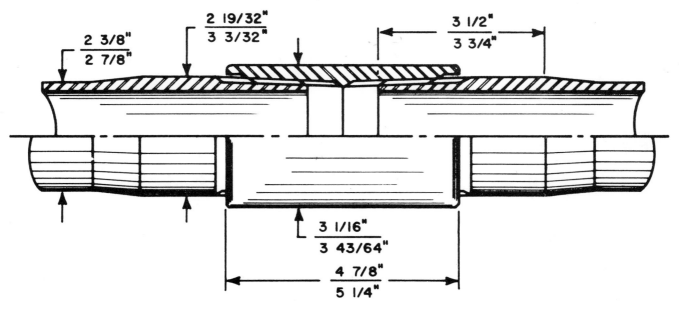

FIG. B

To arrive at the optimum shape, two considerations were of prime importance. First, the physical dimensions of the design should be such that rubber extrusion or flow under pressure will be as small as possible consistent with the design conditions.* Second, the amount of force required to pass the tubing through the rubber will be as small as possible consistent with the other conditions. This second premise is based on the theory that the force required to pass tubing through the rubber is a significant influence on the amount of service wear due to friction between the stripper and the tubing.

The physical properties of the rubber compound will have an important influence on both of the above considerations. Later in this report it will be explained how the compound was varied to meet the desired conditions.

Design procedure

In order to proceed systematically, various portions of the stripper rubber were considered separately and in the following order:

1. Throat seal length and diameter
2. Containing ring
3. Horizontal sections
4. Rubber material specifications

The two sizes under consideration were $2\frac{3}{8}$ in and $2\frac{7}{8}$ in. Preliminary investigation and consideration revealed that the $2\frac{7}{8}$ in size was more critical because, assuming the same throat seal length, the volume of rubber to be displaced in passing a $2\frac{7}{8}$ in coupling is 39% more than the volume of rubber to be displaced in passing a $2\frac{3}{8}$ in coupling; and the load required is obviously a direct function of the volume of displaced rubber. Also, the seal diameter variance is $\frac{7}{64}$ in greater in the $2\frac{7}{8}$ in size.

* Extrusion is illustrated in Fig. C.

The throat seal diameter can be expressed in terms of nominal tubing size interference; that is, $\frac{3}{16}$ in interference on diameter in a $2\frac{7}{8}$ in stripper rubber gives a throat seal diameter equal to $2\frac{11}{16}$ in. Since the frictional resistance force and consequently service life or wear is a direct function of the throat seal interference and the throat seal length, it was suspected that some optimum dimension for these two quantities did exist. On the other hand, it was obvious that these quantities (optimum) are essentially empirical in nature because of the large number of influencing factors. Loudermilk* conducted an investigation of these dimensions and concluded the following:

Optimum throat seal diameter interference – $\frac{3}{16}$ in
Optimum throat seal length   – $1\frac{1}{2}$ in

His results indicated that more interference or greater seal length did not appreciably increase wear life due to an increasing rate of wear.

The containing ring at the top of the stripper rubber is primarily a foundation piece and guide bushing. It serves as a base to which the rubber form can be moulded. The OD configuration of this unit is determined by the configuration of the bowl which carries the stripper rubber in service. The bore of the ring and the shape where it is bonded to the rubber had to be determined. The necessary bore is a function of the maximum diameter the stripper rubber must be able to pass. It is important to have the minimum possible bore in order to minimize the seal gap which the rubber must bridge when sealing on the minimum diameter. The other interior shape of the ring was chosen to provide adequate bond area. The circumferential vertical groove provides for maximum rubber bond area. The initial field tests indicated a need for the horizontal communicating holes to increase the mechanical bond strength between the rubber and the ring.

The interior and exterior shape of the rubber, or the horizontal sections, were arrived at as a result of the following considerations:

1. Sufficient rubber section in the area of the containing ring to provide adequate tensile strength and minimize rubber extrusion under pressure.
2. A rubber section in the area of the containing ring that will allow maximum flexure of the lower section of the stripper rubber.
   (*Note:* A condition of compromise was necessary to satisfy the conflict between 1 and 2.)
3. Entrance tapers to the throat seal area in order to provide reasonably smooth passage of the irregular diameter tubing string.
4. A structural insert at the lower end to reduce rubber extrusion at high pressures. The value of this was determined by tests that are detailed in the Appendix.

The requirements of the rubber compound are a series of contradicting physical properties as follows:

1. High modulus in order to minimize extrusion.
2. Low modulus to provide ease of passing the irregular diameters.
3. High ultimate tensile strength.
4. Good abrasive and tear resistance.
5. Low compression set properties.
6. High resistance to hydrocarbon attack.

Details of the rubber compound were developed in conjunction with the molded-rubber vendor in the following manner. Initially a compound was chosen which represented some compromise of each of the physical properties required. This was tested to determine the weakest property. The compound was then altered to strengthen the weak point, retested, altered again, retested, etc. As an example, tests revealed one sample to have poor abrasive resistant property. The rubber compound

* L. E. Loudermilk, *Engineering Report*, The National Supply Company, Houston, Texas, July 15, 1959.

was changed to use a much finer carbon black ingredient. The result was much improved absasive resistance, with little sacrifice of the other desired properties.

The nature of the tests outlined above was twofold. The rubber molder conducted various tests in accordance with ASTM Standard D735-52. Final refinement of the rubber compound was accomplished by conducting field service tests to evaluate the service wear life. Due to the wide range of service conditions and the inability to secure identical field conditions, the results of these tests were only relative for any set of conditions, but each test by itself served as a general guide in arriving at the final rubber compound.

Conclusion

The drawing, 516370-7A (Exhibit 1) dated 16 June 1960 is the design resulting from the work outlined in this report. Field service reports on the use of this product indicate that remarkable durability and long service life have been achieved.

EXHIBIT 1   Final design

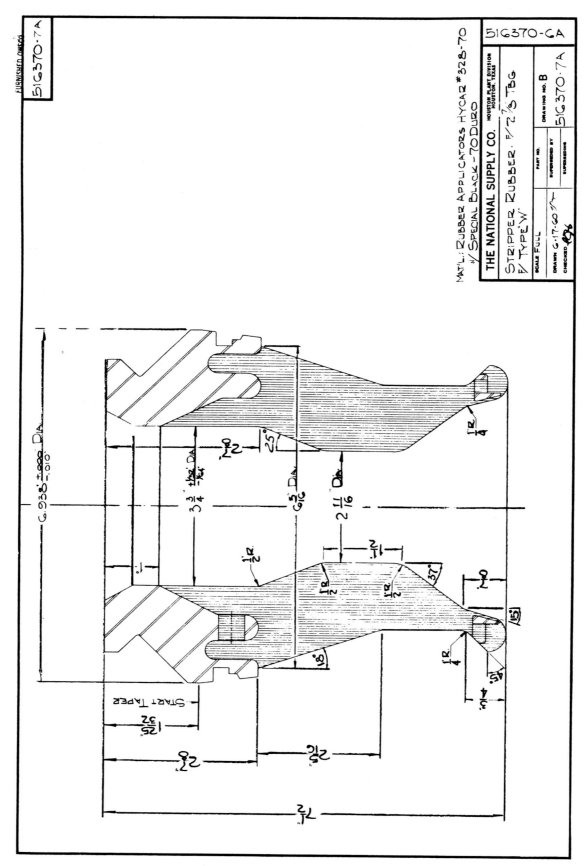

Final Design

Appendix – Test results

Three series of tests were conducted to determine the optimum shape as outlined on page 68.

Test series #1

This series of tests consisted of measuring the vertical extrusion in the seal gap. The test set-up and measurements are shown in Fig. C.

Test series #2

This series of tests consisted of measuring the load required to pass a collar through the stripper rubber. The test set-up shown in Fig. D was arranged in a Baldwin–Southwark Universal Testing Machine where the collar penetration in inches and the applied load in pounds were taken as direct readings. A typical plot of these readings is shown in Fig. E.

Conclusion : In comparing the two increment increases in D, it became apparent that the $D = 6\frac{5}{16}$ in was the more desirable since the lesser reduction of extrusion at $6\frac{1}{2}$ in was at the expense of considerable load increase.

Table 1.   Series #1 and #2

D^* inches	Series #1		Series #2
	P psi	Z† inches	L avg. lb.
$5\frac{7}{8}$	3 000	$1\frac{1}{8}$	1 165
	6 000	$\frac{7}{8}$	
$6\frac{5}{16}$	3 000	$1\frac{3}{8}$	1 300
	6 000	$1\frac{3}{16}$	
$6\frac{1}{2}$	3 000	$1\frac{7}{16}$	1 740
	6 000	$1\frac{5}{16}$	

Table 2.   Series #1 and #2

D, inches	Reduction of extrusion %	Increase load %
$5\frac{7}{8}$–$6\frac{5}{16}$	36	12
$6\frac{5}{16}$–$6\frac{1}{2}$	19	34

* See Fig. D for definition of D.
† See Fig. C for definition of Z.

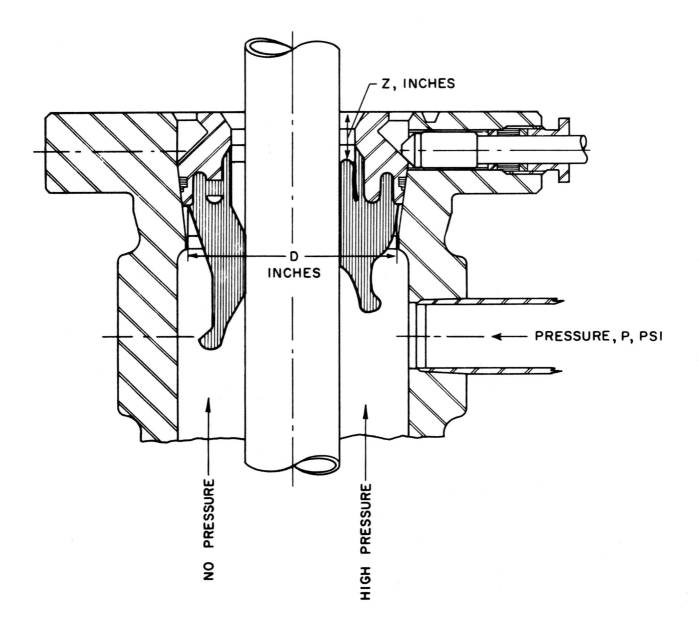

FIG. C

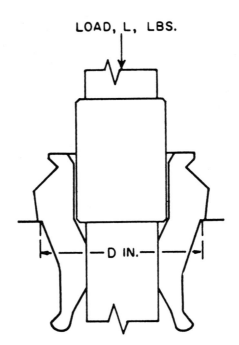

FIG. D

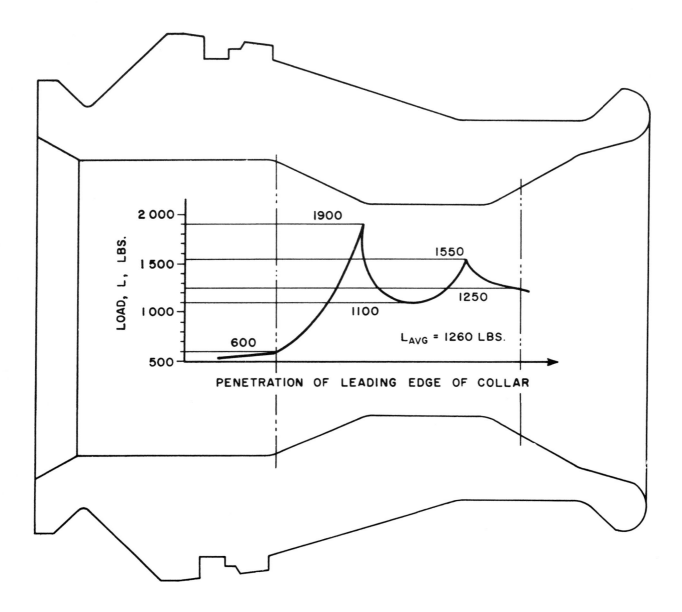

FIG. E

Test series #3

The first prototype stripper rubber that was built did not contain the metal ring in the lower lip. In a test set-up the same as outlined in Test series #1, this stripper rubber was pressured to 8 100 psi at which pressure the bottom lip completely inverted (see Fig. F). Although 8 100 psi is considerably above the maximum required working pressure, the test did point out the need to investigate the effect of a ring in the bottom.

The set-ups were the same as Test series #1 and #2, Fig. C and D.

Results

Stripper rubber	Extrusion			Load	
	P psi	Z inches	Reduction %	L^* ave. lbs.	Increase %
Without lower ring	3 000	$1\frac{3}{8}$			
	6 000	$1\frac{3}{16}$	—	1040	—
	8 100	Failure			
With lower ring	3 000	$2\frac{1}{16}$	50		
	6 000	$1\frac{7}{8}$	58	1440	38
	10 000	1	—		

Conclusion: The addition of the lower ring resulted in much improved extrusion characteristics of the rubber. The load increase was also significant but tolerable.

FIG. F

PART 3

Designing an oil-well stripper rubber

In 1959 I returned to the Houston plant, National Supply Division, Armco Steel Corporation after having a number of field engineering assignments. At the time of my return I was given the job of design engineer and in this job was responsible for various projects assigned by my immediate superior, Mr C. R. Neilon, Chief Engineer of the Houston plant.

I was reassigned to this job in Houston because the product on which I was working as a field engineer was discontinued as a Houston plant product and transferred to another division of the company. This particular product was a reciprocating hydraulic engine which, when coupled to a conventional reciprocating oil-well pump, provided an assembly for pumping oil wells with the so-called prime mover downhole rather than on the surface.

In addition to myself there was another engineer, L. E. Loudermilk, who was reassigned to the Houston plant engineering staff from the Hydraulic Engine Project.

The number of field complaints regarding the existing Type E Stripper Rubber had at that time reached a point where they could no longer be ignored. They had been growing in frequency and intensity for the past 4 or 5 years in direct relation to the greater service requirements demanded by the changing drilling and completion practices of the oil companies. Instead of well pressure conditions of 10 to 100 psi, the conditions were now as severe as 4 000 psi. Instead of a requirement to pressure seal with only vertical motion of the tubing string, the requirement now was to seal the tubing string as it rotated. In most cases the Type E design could simply not withstand these service conditions; at best it had a very short life span. Loudermilk was therefore assigned, under the very close supervision of Neilon, to investigate the durability and life characteristics of the Type E Stripper Rubber with the purpose of improving it. This particular product, the Type E Stripper Rubber, was designed by Neilon in about 1950 before he was appointed Chief Engineer. There were no important changes or design modifications of this stripper rubber in the intervening eight or nine years. Although I was not directly involved in the project at that time, I remember that it was Neilon's feeling that the stripper rubber could be improved and made acceptable by some relatively small changes in dimensions or proportions. Loudermilk and Neilon designed a testing device which incorporated the hydraulic engine with which we were all familiar. This device consisted of a hydraulic engine mounted vertically above a conventional Type E tubing head. Attached to the bottom end of the hydraulic engine piston rod was an adapter containing a female $2\frac{3}{8}$ in OD tubing thread.

Using the above described testing device they established the following procedure in an attempt to evaluate the stripper rubber. A 2-foot long piece of tubing was screwed into the engine piston rod adapter. To the bottom of this piece of tubing was screwed a tubing coupling. Below the tubing coupling was screwed another 2-foot long piece of tubing. The proportions of the test fixture were such that when this tubing assembly reciprocated with the hydraulic engine the tubing coupling would pass through a stripper rubber held in the Type E tubing head. A means for providing a steady stream of cooling water on top of the stripper rubber was incorporated in the test fixture. In addition, it was possible to introduce air pressure into the tubing head beneath the stripper rubber. The reciprocating hydraulic engine was controlled to operate at 25 cycles per minute; that is, 50 passes of the tubing coupling through the stripper rubber each minute, 25 in each direction. Thus the stripper rubbers were tested by measuring the amount of throat seal diameter wear or dimensional increase as a function of time or the number of passes of a tubing collar through the stripper rubber, and also by visually observing air pressure leakage (the degree of bubbling) at the stripper rubber, again as a function of tubing collar passes. For approximately four months Loudermilk tested standard Type E Stripper Rubbers and various shape or configuration changes as well as competitive stripper rubber designs. In all of the tests and design changes that were evaluated, the underlying object was to improve the stripper rubber performance without changing any basic overall dimensions. This was deemed important at that time because any changes would necessitate a complete redesign or new design of the Type E assembly, since the stripper rubber was only one of a number of components, as illustrated on the attached parts sheet (Exhibit 2). Essentially, the design variables that were evaluated were:

1. Throat seal diameter.
2. Throat seal length.
3. Thickness of horizontal sections at various points along the entire length of a stripper rubber.

Test models for these various configurations were secured by either machining a standard stripper rubber which was manufactured by compression molding or, when this could not give the desired shape, by producing a part to rough dimensions by mandrel wrapping and then doing final machining.

In a period of approximately four months Loudermilk conducted approximately 50 tests. Of these, approximately ten were standard parts, approximately five were competitive designs, and approximately thirty-five were modified Type E designs. At this point in the programme Loudermilk was transferred from the Houston plant and I was given the responsibility for continuing the effort to improve the stripper rubber design. Before Loudermilk left he wrote an engineering report which summarized the approximately 50 tests he had conducted. One of my first efforts was an attempt to analyze the results of his testing and find some direction in which to proceed. The test result data which he reported and inspection of the models actually tested presented a rather confusing picture. In the final analysis, most of the evaluation of shape modifications which he did had to be discarded because the rubber in many of these models contained small sub-surface voids and folds caused by the rubber moulding technique. Subsequent machining of these models would expose these imperfections to the surface where they had a critical effect. The introduction of this variable resulted in radically inconsistent test results. Unfortunately, the significance or importance of these discrepancies was not apparent until near the conclusion of the test program.

In addition, I was not completely satisfied with the reliability of the testing fixture; that is, the reciprocating hydraulic engine device. Primarily, I was concerned with the inability to maintain alignment between the engine and the tubing head which contained the stripper rubber being tested.

For the above reasons it seemed apparent that a completely new approach should be considered. For one thing, outside the considerations of reliability, the testing fixture was quite an awkward mechanism.

All things considered, it would take approximately two days to test one part. I started to wonder if the value of a particular shape could not be determined by the amount of load or pounds of force required to pass a tubing collar through the stripper rubber; the amount of wear should be a function of the amount of force required to pass the collar through the stripper rubber as a result of the seal interference. We had in the plant laboratory a Baldwin tensile testing machine which provided a simple and quick means of measuring the force by direct reading in pounds. At this point I started testing the standard Type E and modifications of the standard Type E design by plotting the load characteristics as shown in a typical plot, Fig. E, of my report

EXHIBIT 2   H-3E tubing hanger

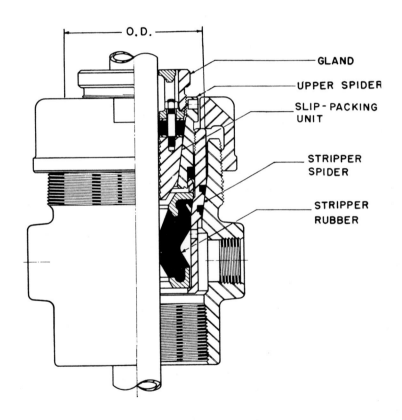

TUBING HANGERS FOR NATIONAL TYPE "E" WELLHEAD EQUIPMENT

O.D.

GLAND

UPPER SPIDER

SLIP-PACKING UNIT

STRIPPER SPIDER

STRIPPER RUBBER

EXHIBIT 3

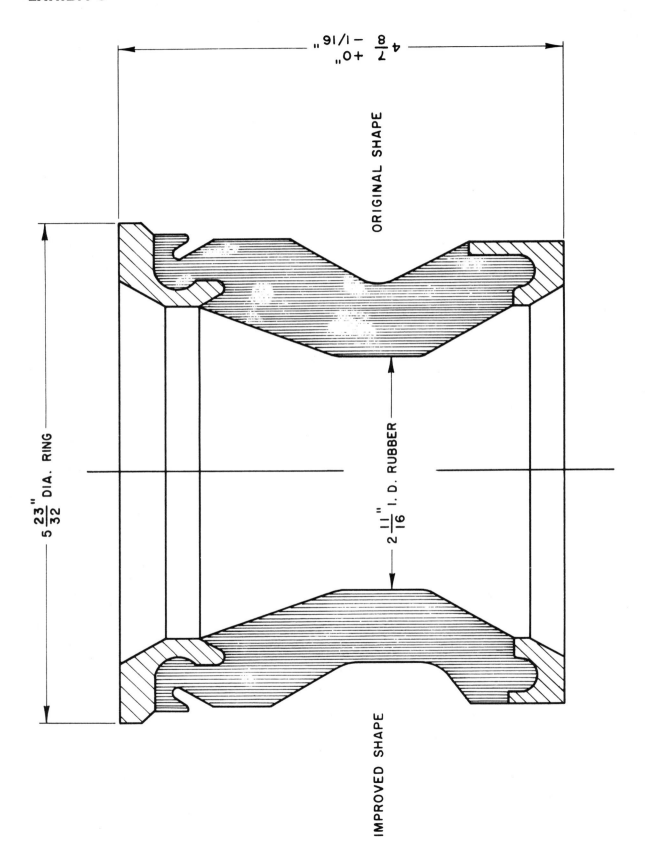

ORIGINAL SHAPE

IMPROVED SHAPE

$4\frac{7}{8}''$ +0" $-1/16''$

$5\frac{23}{32}''$ DIA. RING

$2\frac{11}{16}''$ I.D. RUBBER

(page 74). The result of this work was shown on a Drawing 516381 (Exhibit 3), the section to the left of the center line shows the improved shape, and the section to the right of the centre line shows the original shape. I do not have the load characteristics of these two shapes available to me at this time, but as well as I can recall, the load was reduced by approximately 30%. I remember that the configuration of the lower containing ring as shown to the right of the center line had a very significant detrimental effect on the load characteristics. I recall that on one test model we removed the lower ring entirely and again reduced the load characteristics significantly. However, for some reason that at this time I do not recall, we could not get by without that lower ring. It had something to do with the mating components of the tubing head assembly. At this point, the revised design was tested on the hydraulic engine device and it was concluded that it represented some measure of improvement. However, extrusion tests similar to the test series #1 described in my report on pages 71–2 and subsequent field testing of this revised design indicated that it still could not meet the service requirements demanded of it. I felt, however, that it was the best that could be accomplished within the limitations of the basic overall dimensions. Therefore, at this point I wanted to redesign the stripper rubber utilizing the whole tubing head bore ID which meant that the OD of the stripper rubber could be increased approximately $1\frac{1}{8}$ in. Mr Neilon did not encourage this approach. At the time I thought the major reason for his objection was pride in his previous work and an unwillingness to accept that it had been outdated. However, Neilon's assistant, Mr W. W. Word, agreed with me that we should attempt to improve the product by major design. Therefore, he took the responsibility of encouraging me to work on a redesign without the specific approval of the Chief Engineer. This, in essence, was the start of the design project documented in the engineering report. All of the early work consisted of making various layouts which incorporated various ideas and experiences accumulated during the program to improve the standard Type E stripper Rubber design. This work was done for Mr Word without the knowledge of Mr Neilon. About the time we were ready to make some parts of this new design Mr Neilon took a combination business and vacation trip which kept him away from the office for about four weeks. As soon as he left, Mr Word authorized the expenditure of money to machine molds and manufacture the prototypes of this new stripper rubber design.

By launching a crash program we were able to secure parts in about three weeks. The first load and extrusion tests of the prototype parts were completed before Mr Neilon returned. The results of these were encouraging enough to inform him of what we had done when he returned. At first he was not enthusiastic but when he considered all of the circumstances at that point he did not object to the project. An illustration of just one of these circumstances is as follows: The highest static pressure carrying capacity of the redesigned standard Type E was something less than 4 000 psi at which point the stripper rubber ruptured; the first prototype of the redesign withstood static pressures in excess of 6 000 psi. So, having gotten approval from Mr Neilon, it remained only to see the design project to its conclusion.

The bulk of the ensuing work involved modifying the interior and exterior shapes of the stripper rubber as dictated by the load and extrusion tests. In addition, it was clear that various rubber compounds and rubber durometers had a significant effect on the load and extrusion characteristics. These parameters were altered to give the best combination of characteristics while at the same time attempting to maintain good abrasion characteristics of the rubber compound. We relied entirely on the experience of the rubber compounder to produce good abrasion resistance since this could finally be evaluated only in service.

It was at this point in the design project that the lower metal ring was added to the design as described in the report. After completing the above described modifications which were arrived at essentially empirically by means of load and extrusion tests, it was decided to manufacture a small number of pieces and field test them.

The company's sales department in southern Alabama had complained of losing a considerable number of well-head sales to competition as a result of the poor performance of the Type E Stripper Rubber. We therefore decided to field test the redesigned stripper rubber in that area. So, with unwarranted enthusiasm, I made a trip to Citronelle, Alabama.

In spite of the fact that I was supposedly bringing a new and much improved product, the sales department was hostile because of the previous problems they had experienced with the old Type E design, and because the Houston plant had been unsympathetic to their problem. The drilling company people were hostile partly by nature and partly because they were of the old school who had drilled millions of feet of oil well without the aid or advice of any young mechanical engineer, whom they suspected did not know the engines from the slush pumps. The natives were hostile because I was obviously a Yankee Carpetbagger uninvited to their rather decrepit and backward community. In this atmosphere I proceeded to supervise the installation and evaluation of the first stripper rubber, of which I was quite proud.

The first one was installed about 2 a.m. and performed satisfactorily for about four or five hours in a rather uneventful test. At this point I went back to my motel for some rest, extremely proud and confident. I returned to the drilling site about noon only to be completely shattered by the sight of my prize stripper rubber lying on the ground almost totally disintegrated. Needless to say, the critical eyes and wise I-told-you-so's were not easy to endure. In fact, careful analysis of the ruined part and descriptions of the circumstances that caused the failure when I was not there caused me to believe that they had deliberately abused the part to insure the failure that was experienced. So, I prevailed on them to try a second part. They agreed to this second test primarily because they were confident it would be no more successful than the first. In fact, the second part failed more rapidly than the first, while I was at the drilling site.

I took the two failed parts and retreated to my motel along with my punctured pride. After regaining my composure and analysing the failures it became apparent that the main problem was poor bond between the rubber and the top containing ring. I called Houston from Citronelle and discussed this problem with them; we decided to do two things. First, the rubber moulder was advised of the difficulty and he in turn indicated to us that the steel rings should have first been coated with an adhesive to insure a good bond. Second, we decided to increase the total area of bond between the steel ring and the rubber by adding the circumferential vertical groove shown in Fig. C of the report and on the working drawings (Exhibits 4–7) in Part 4. In addition, the horizontal communicating holes were added to the steel ring to improve the design.

I spent the next two or three days in Citronelle waiting for the Houston plant to ship me some more pieces which incorporated the above described improvements. Needless to say, this was a painful wait. The parts did finally arrive and were placed in service. Fortunately, the performance of these parts was excellent, exceeding not only that of the old Type E design but also that of the best competitive stripper rubbers.

I wasted no time leaving Citronelle, Alabama, with some recovered pride but mainly a good deal wiser. After I returned to Houston another larger lot of prototype parts was produced and distributed to other parts of the country for field evaluation. Again, the field reports of these parts were very successful. The design project was successfully concluded when steel castings of the two insert rings were produced and rubber molds designed to cooperate with these cast rings put into production.

PART 4   Working drawings

EXHIBIT 4   Working drawing

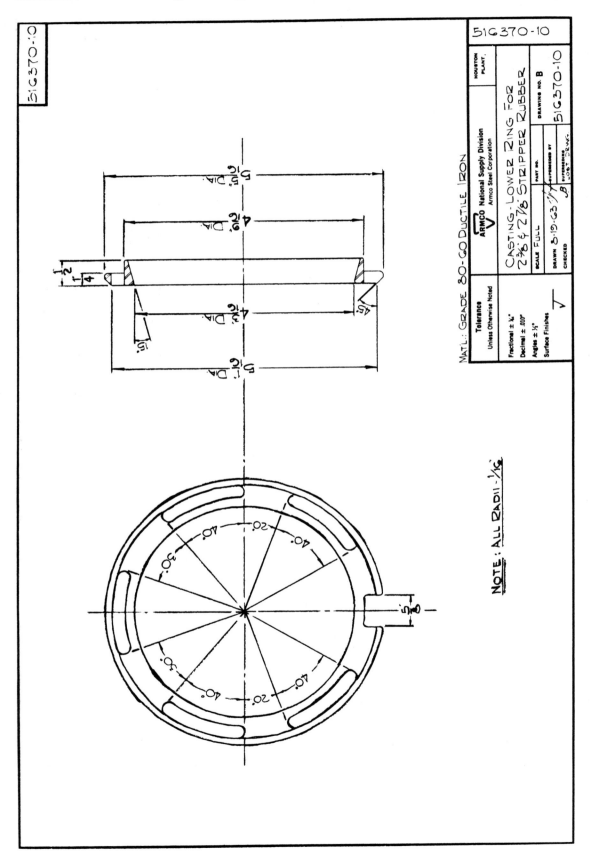

Lower Ring

EXHIBIT 5   Working drawing

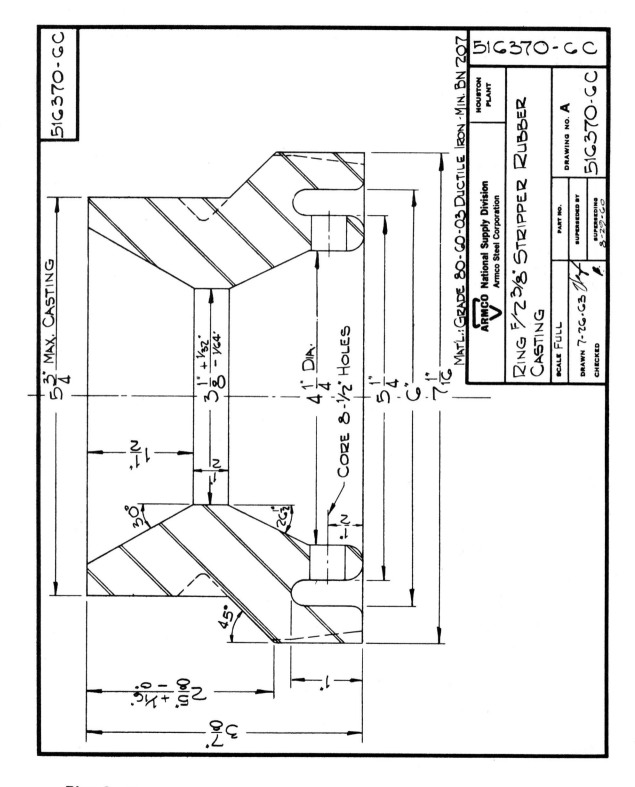

Ring Casting

EXHIBIT 6   Working drawing

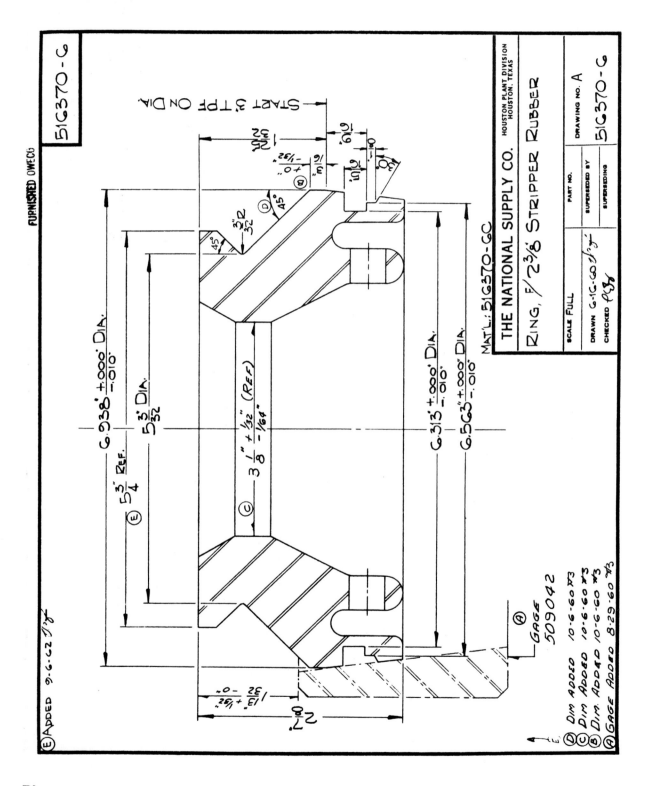

Ring

EXHIBIT 7   Working drawing

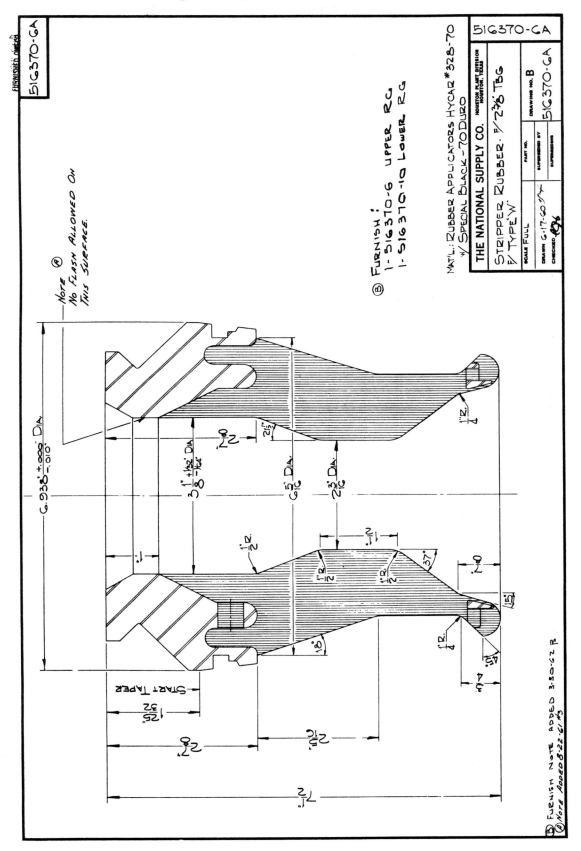

Assembly

SUGGESTED STUDY

The following questions are suggested as examples of problems which may be raised by this case:

1. At the Freshman level:
    (a) What was the problem which Mr Loudermilk tried to solve? State it in terms of given inputs, desired outputs, and constraints.
    (b) Did Mr Bickel change the problem statement? Identify and count the iterations in this design. Evaluate Mr Bickel's formal report. Is it perfectly clear? Perfectly concise? Is anything missing or excessive?

2. At the Senior level:
    (a) What are the criteria for evaluating stripper rubbers? Can you devise a quantitative method of rating competitive stripper rubbers (i.e., a criterion function)? Invent a stripper rubber which will expand as much as the design shown in Fig. A(i), but will extrude much less.
    (b) Which of Mr Bickel's decisions was the most important one? Which one was next in importance? What were the rational bases for these decisions?

3. At the Graduate level:
    (a) What variables did Mr Bickel consider in his testing program? Were they the ones you would use? In your opinion, were they sufficient?
    (b) How do you rank the relative cost and difficulty of field testing, analytical studies and laboratory testing?
    (c) Are there material properties which Mr Bickel did not consider which could have helped him in the design of a better oil-well stripper rubber?

Case 3

Design of a concrete reservoir for liquid oxygen

by H. C. Kornemann

Prepared with support from the National Science Foundation for the Stanford Engineering Case Program. Assistance from the Linde Company, a Division of Union Carbide Corporation, is gratefully acknowledged.

This case is available in pamphlet form as ECL 164 from the Engineering Case Library, Room 500, Stanford University, Stanford, California 94305.

PART 1   Liquid oxygen reservoir

In 1954 the Linde Company, a Division of Union Carbide Corporation, placed in service a 660 000 gallon reservoir for liquid oxygen. Its oxygen gas capacity was 75 000 000 ft³ (N.T.P.).* This was the successful culmination of over four years of pioneering effort which involved the interaction and cooperation of many persons in various departments of several companies. The reservoir continues in service as of 1970. The knowledge gained from this work demonstrated the suitability of concrete for use as a container for liquefied gases and as a structural material at a temperature of $-297°F$ ($-147°C$), the temperature of liquid oxygen.

This work was initiated because business conditions required larger capacity reservoirs which would have a lower unit cost than obtainable from designs available in 1950. Another incentive for developing the concrete reservoir was that tonnage quantities of strategic materials such as copper, everdur, stainless steel, and steel would not be required.

This was a 'captive' project of the Linde Company from its conception through the development, design, construction and operational stages. Linde made use of the available experience of the Preload Corporation and contracted with them to build the concrete reservoir.

In addition to the Preload Corporation, three departments of Linde were involved; namely, (1) Tonawanda Engineering Laboratory (see Exhibit 1), (2) Construction and Design Department, and (3) Production Department. Geographically, the project involved personnel in New York City; Tonawanda, New York; Chicago; East Chicago; Indiana; and Pittsburgh.

Everybody involved in the project was very much aware of the reservoir's safety requirements and of their own safety responsibilities. The quantities of liquid oxygen being stored were so large that it was necessary to assure and even guarantee that the reservoir could be operated safely and that there would be no disaster which might be caused by liquid oxygen escaping from a damaged or ruptured reservoir. Linde and the Preload Corporation were very much aware of the disaster in Cleveland, Ohio, which resulted from the failure of a large storage tank for liquefied natural gas during the 1930s. That disaster had set back the LNG industry by over 25 years.

Beginning of program

During the summer of 1950, Mr Pete Riede and Mr John Beckman made an engineering study to determine the economics and feasibility of building liquid oxygen reservoirs of 2 200 000 gallon capacity, ten times larger than the reservoirs installed at that time. The study included many possible design configurations but had one common characteristic – all the inner containers were made of a metal suitable for liquid oxygen temperatures. In August 1950, Mr George Boshkoff and Mr Pete Riede from the Linde Engineering Laboratory at Tonawanda, New York, had preliminary discussions with Linde New York Management† in the 42nd Street Union Carbide Building on the engineering study. It appeared that the estimated costs of the very large metal reservoirs would not be as low as desired.

After the office discussions were concluded, the meeting adjourned to a nearby bar. The idea of using concrete as a reservoir material for liquid oxygen was initiated at that time. Several possible designs were sketched on cocktail napkins.

Mr Pete Riede, Mr John Beckman, Mr Ed Kent, and Mr Charles Fails spent the next few months in obtaining available information on the properties of concrete and considering tanks of various configurations. They quickly learned that no information on

---

* Cubic feet at normal temperature and pressure –70°F (–21°C) and 760 mm Hg absolute pressure.
† S. B. Kirk, President; L. A. Bliss, Vice President; P. B. Pew, Vice President.

the physical properties of concrete at $-297°F$ was available. A 5-ft diameter × 4-ft high concrete pipe section was grouted on to a concrete slab. It was then filled with and held liquid nitrogen ($-320°F$ ($-160°C$)) without disastrous results. This simple 'quick and dirty' test provided the necessary encouragement to continue the concrete reservoir program.

The initial configurations of concrete reservoirs were based on the liquid container being made of either a single or multiple *reinforced* concrete vertical cylinders of large diameter installed in an outer reinforced concrete shell. The space between the liquid containers and outer shell was to be filled with powder insulation. It soon became apparent that reinforced concrete was not the answer. Estimated costs were not competitive with metal tanks. The estimated amount of reinforcing steel required was almost equal to that of an all metal tank. There was also grave concern as to whether the concrete or reinforcing steel might crack or rupture due to temperature strains.

The above problems led to the idea of using *prestressed* concrete for the reservoir. Concrete is prestressed by imbedded wires which are maintained in tension. This takes advantage of the relatively high compressive strength of concrete by imposing compressive loads on those parts of the concrete which would normally be subjected to tensile loads. Prestressed concrete is, therefore, not stressed in (net) tension whereas ordinary reinforced concrete is.

In the fall of 1950, the Linde Tonawanda Engineering Laboratory contacted the Preload Corporation, who were leaders in the design and construction of structures and reservoirs made of prestressed concrete. During the previous 17 years it had designed and built more than 550 prestressed concrete tanks and reservoirs for water, sanitary installations, petroleum products and chemicals.

The Preload Company is a company that engineers and constructs. In fact, 100% of the work that it builds is built to its own design. Preload also operates with a number of selected licensees in foreign countries. During the 1951–54 period, the annual sales in the United States by the Preload Company, in terms of its own construction activities, was about $2–2·5 million. The tanks were generally used for water or for waste treatment. Although tanks have been used for the storage of petroleum products and for other process applications on a limited scale, the main market is for water storage and waste treatment and water treatment. During the 1951–54 period, the average size of prestressed concrete tanks was probably in the range of 1·0 to 1·5 MG. Tanks of 10·0 MG had been constructed prior to that time. Over the years the average has increased to where the average sized tank is in the range of about 2·5 MG. The reason for this is the broader acceptance of prestressed concrete and its use on larger tanks.

The Preload Corporation's experience was with reservoirs operating at atmospheric temperature. It had no experience with operating temperatures of $-297°F$. Discussions between Preload and Mr Pete Riede of Linde indicated that it might be possible to build a large prestressed concrete reservoir which would be safe to operate and have economic advantages. The Preload Corporation, especially their Chief Engineer Mr Fornerod, was intrigued with the $-297°F$ temperature application. It would be an engineering challenge and would make available to Preload a large new marketing opportunity.

Messrs Riede, Beckman, Kent, Fails, and others in the Linde Engineering Laboratory therefore implemented a one-year program to determine seven essential physical properties of five types of concrete made from thirteen mixes. The effects of moisture, thermal cycling, and aging were also determined. Most of the tests were conducted with liquid nitrogen at a temperature of $-320°F$ rather than with liquid oxygen because of its greater safety and because the concrete reservoir design might also be used for liquid nitrogen. In most cases the physical properties were improved at $-320°F$ provided that the concrete was dry when cooled to that temperature. For example, even the small amount of frost and moisture which condensed on concrete when warmed in the air from −320°F would cause surface cracks and spalling of the concrete when it was cooled again to −320°F during the thermal cycling tests. The essential physical properties of the concrete finally used are listed in Exhibit 2.

Messrs Beckman, Kent, and Fails of Linde also investigated the type of prestressing wire to be used at liquid nitrogen temperatures. Normally stainless steel is used for such temperatures, as carbon steel becomes embrittled. They were very pleased to discover that the cold drawn carbon steel wire used by Preload at atmospheric temperature would also be suitable for use at −320°F. The Preload wire had comparable fair elongation and impact values at both room and liquid nitrogen temperatures. The Preload wire had much higher tensile strength, lower impact values and was more notch sensitive than stainless steel wire. The coefficient of expansion of the Preload wire was almost identical with that of concrete whereas stainless steel was about 40% greater. This was important because of the large temperature changes involved. See Exhibit 3 for physical properties of prestressing wire.

EXHIBIT 1   Linde Tonawanda Engineering Laboratory

LINDE TONAWANDA ENGINEERING LABORATORY

In 1951 the Linde Tonawanda Engineering Laboratory had about 160 personnel. About 100 of these were engineers and scientists. The following table lists the Linde Engineering Laboratory personnel who were active in the concrete reservoir project and the relationship of the Laboratory to Linde New York Office Management.

New York Office   - President - S. B. Kirk

                           Vice President - L. A. Bliss - M.E. - 28 years Linde Service
                           Vice President - P. B. Pew   - M.E. - 25 years Linde Service

Tonawanda Engineering   - Manager - G. J. Boshkoff - M.E. - 25 years Linde Service
     Laboratory
                           Assistant Manager - M. A. Dubs - Ch.E. - 8 years Linde Service

                           Division Engineer - P. M. Riede - M.E. - 13 years Linde Service

                           Engineer - J. H. Beckman - M.E. - 2 years Linde Service
                           Engineer - E. C. Kent   - M.E. - 8 years Linde Service
                           Engineer - C. F. Fails   - M.E. - 1 year Linde Service
                                    plus

Other Supervisors and Engineers with M.E., C.E., E.E., and Ch.E. degrees, technicians, chemists, and draftsmen participated as required.

Mr. Pete Riede's contacts with the Preload Corporation were mainly with:

        Mr. J. J. Closner, Vice President of Marketing
        Mr. Fornerod, Chief Engineer

EXHIBIT 2   Physical properties

Age Tested Days	No. Thermal Cycles 70°F to -320°F	Compressive Strength - psi		Modulus of Rupture - psi	
		70°F	-320°F	70°F	-320°F
30	0	6560	14 500	536	833
60	0	7130	13 300	694	1020
60	10	6250	13 300	212	470
100	25	6530	11 300		

Maximum Allowable Compressive Stress specified for design = 1/2 6000 = 3000 psi

Maximum Allowable Flexural Stress specified for design = 400 psi

Compressive Modulus of Elasticity = 4×10^6 psi at 70°F and 4.5×10^6 psi at -320°F

Coefficient of Thermal Expansion = 4.7×10^{-6} per °F for temperature range 70° to -320°F

Coefficient of Thermal Conductivity = 8 $Btu/ft^2/hr$,°F/in for temperature range of 70°F to -320°F

Coefficient of sliding friction between concrete and concrete lubricated with graphite = 0.12. Value specified for design = 0.5.

Density of Concrete = 147 lb/cu ft

Permeability = Negligible under a 31 ft. head of liquid oxygen

Nominal 6 000 psi, type 1 – 28-day concrete

EXHIBIT 3   Physical properties

WIRE

	Ultimate tensile strength - psi		Impact values ft lb	
	70°F	-320°	70°F	-320°F
Preload Cold Drawn Carbon Steel	230 000	301 000	23	24
303 Stainless Steel (for comparison)	116 000	160 000	36	41

PRELOAD PRESTRESSING WIRE

	Horizontal 0.162" diam. reduced to 0.141" diam.	Vertical 0.196" dia.
Maximum Allowable Stress, psi	152 000	173 000
Initial Applied Stress	140 000	170 000
Losses	20 000	20 000
Design Stresses	100 000	140 000

Coefficient of Thermal Expansion - 10^{-6}/°F - for Room to -320°F temp. range

    Preload Wire   = 5.0   Everdur   =   7.6
    Low Carbon Steel = 4.9   18-8   =   7.0

Conductivity Values - Btu/hr./ft^2, °F/in - for Room to -320°F Temp. Range

    Perlite Powder Insulation   = 0.24
    Foamglas   = 0.38
    Perlite Concrete   = 0.6

Ultimate Compressive Strength of Perlite Concrete = 250 psi

Ultimate Compressive Strength of Foamglas   = 150 psi

Maximum Internal Pressure developed by powder insulation = 40 lb/ft^2.

Heat of Vaporization of liquid oxygen = 92 Btu/lb.

Allowable soil loading = 5 000 lb/ft^2.

Densities - lb/cu ft.

    Perlite Powder Insulation = 6.5
    Foamglas   = 10
    Liquid Oxygen   = 71

Other materials in reservoir

PART 2   266 000 gallon prototype

To confirm the validity of their design and also make known possible unexpected deficiencies which should be corrected before larger capacity reservoirs were constructed, the Linde New York Office and Tonawanda Engineering Laboratory Management decided to design and build a prototype concrete liquid oxygen reservoir of 266 000 gallon capacity. This decision was made in 1951.

Linde contracted with Preload to design and build the liquid oxygen reservoir. The design was a cooperative effort in which Preload supplied their experience with prestressed concrete tanks and Linde their low temperature information. Preload quoted a fixed price to Linde. This was subsequently increased somewhat as the scope of the work increased.

Preload's arrangement with Linde during the development, design, and construction of the first LOX tank LR-30 evolved out of the research and development that Preload had done under contract with Linde which preceded the design and construction of the LR-30 tank. This contract provided for Preload to undertake certain design work and investigations and for Linde to provide certain tests and investigation of material properties to be acquired in order to implement the design. The agreement also provides that any developments or patents that evolve would be owned by Linde. The patent did issue with respect to the prestressed concrete tank, and in the 1960s Preload took license from Linde for this patent. For the LR-30 tank specifically following the development phase Preload undertook the work on a lump-sum design and construction contract.

In the engineering area, Mr M. F. Fornerod was Chief Engineer of Preload at that time as he had been for a number of years. Mr Fornerod left the company about 1953 however. Design engineers in connection with the project included Mr Felix Dushnick, who continues with Preload at the present, and Mr Herbert Weiner and Mr Nicholas Rouzsky. Mr Weiner is no longer connected with the Preload organization, although Mr Rouzsky is with the licensee in Spain.

J. J. Closner was assistant general manager at the time, responsible for both engineering and construction in sales activities.

Appended drawing, Exhibit 4, is a cross-section of the reservoir and its connections. The reservoir consisted of an outer prestressed concrete vessel 46 ft 4 in OD and 49 ft 4 in in height, with a domed roof and flat floor. The inner prestressed concrete vessel was 38 ft ID and 35 ft 4 in in height. It had a domed roof and thin stainless steel floor which rested on a concrete-Foamglas insulated foundation. The sides and top were insulated with Perlite powdered insulation. Piping and safety devices for operation were provided. A metal liner was not required or provided to seal the concrete walls of the inner vessel. However, a carbon steel bar ring was cast into the wall of the inner vessel near the top so that, if it was found to be necessary in the future, a stainless steel liner could be easily installed. The gas phase working pressure was 8 in water.

The inner vessel for liquid oxygen resembled an inverted concrete tumbler. The cylindrical wall rested on and was free to move or slide on its footing ring. This prevented destructive stresses from occurring between the wall and footing during cool-down or warm-up of the reservoir. The bottom of the inner vessel was sealed by a 20 gauge stainless steel membrane which rested on insulation. The inner vessel was to be suitable for a maximum temperature gradient of 20°F ($-6°C$) at the bottom of its 8 in thick wall and a 10°F ($-12°C$) gradient at the top of the wall. These conditions might exist during cool-down and warm-up. During normal operation the gradient would be about 10% of these values. Six thousand psi compressive strength concrete was selected for the inner vessel because it had high compressive strength, the best resistance to temperature cycling, greatest durability, lowest shrinkage upon setting, negligible permeability, and the advantage of the corresponding high tensile strength.

Horizontal prestressing

The cylindrical wall was prestressed by tightly winding it with 0·162 in wire, with tension induced by drawing the wire through a 0·141 in dia. die. The resistance caused by the aperture of the die being smaller than the wire induces the stress in the wire by elongation and by raising the temperature of the wire. The die is fixed on the prestressing machine as you can see in some of the literature and photographs. The machine drives itself around the tank by engaging an endless chain driven by sprockets from an engine on the machine. The initial stress of the wire was 140 000 psi, but to allow for plastic flow and concrete shrinkage the value chosen for design was 100 000 psi. A $\frac{3}{4}$ in thick pneumatic mortar coating was applied over the wire to protect it and maintain most of the wire prestressing in the event that a few wires broke. Exhibits 5 and 6 show the wire-winding machine applying the prestressing wire to the concrete wall. Exhibit 7 is a description of the machine and its operation.

Bottom of the wall of the inner vessel

The first foot of the wall was prestressed by means of 39 wires to apply a compressive prestress in the 8 in thick concrete of 634 psi.

A 20°F (−6°C) temperature gradient across the wall would produce a tensile stress on the inside of the wall and a compressive stress on the outside of the wall when the vessel was being cooled down. The stresses would be reversed during warm-up. The stress due to the 20°F temperature gradient was calculated to be 212 psi.

$$\text{Concrete stress} = \tfrac{1}{2} \times E \times \Delta T \times a$$

$$E = \text{modulus of elasticity}$$
$$\Delta T = \text{temperature gradient}$$
$$a = \text{coefficient of thermal exp.}$$
$$= \tfrac{1}{2} \times 4{\cdot}5 \times 10^6 \times 20 \times 4{\cdot}7 \times 10^{-6}$$
$$= 212 \text{ psi}$$

Prestressing eliminated the 212 psi tensile stress in the concrete.

The 30 ft 11 in head of liquid oxygen would have developed a tensile stress of 550 psi in the concrete. This tensile stress was counteracted by prestressing.

Part of the compressive prestress was also used to counteract the friction forces of about 250 psi which might develop during temperature changes as the concrete wall slid on the concrete footing.

The number of wires and required prestressing were reduced as the height above the footing increased.

Top of the wall of the inner vessel

The top foot of the wall was prestressed by means of 42 wires to apply a compressive prestress in the concrete wall of 682 psi. This was used to counteract the thrust of the concrete dome plus insulation and the 10°F (−12°C) temperature gradient during cool-down and warm-up.

Vertical prestressing

The inner vessel wall was vertically prestressed by casting vertical bundles of 12 hard drawn carbon steel wires 0·196 in diameter, located about 4 ft apart (see Exhibit 8). The wire was threaded through a thin vertical carbon steel tube with loops formed in the wire at the base. A tube for pressure grouting extended from the steel tube, which was around the vertical prestressing wire, to the outside of the wall. Eight days after the wall was poured, the wires were stretched, keyed into position and grout was forced into the vertical carbon steel tubes. The tensile stress developed was 182 000 psi which permitted a design stress of 140 000 psi.

The prestress induced in the concrete was 132 psi to which were added 9 to 41 psi dead loads.

The maximum vertical stresses would occur during cool-down and warm-up. They would be caused by prestressing, temperature gradients, and friction forces on the footing.

Stainless steel floor

This was the third major novel element of the concrete liquid oxygen reservoir. The others were discussed previously; namely, the use of concrete and carbon steel wire at −297°F.

The floor of the inner vessel was 20 gauge (0·0375 in) type 304 stainless steel. It was welded at the periphery to a 1 in $\times \frac{3}{16}$ in stainless steel angle which was welded to a one-foot-high skirt. This, in turn, was welded to a 1 in $\times 2\frac{1}{2}$ in carbon steel ring which was cast into the inner concrete wall to complete the bottom seal. Carbon steel was chosen for the ring because its coefficient of expansion was about the same as concrete.

The stainless steel floor accomplished the following:

1. It simplified the design of the concrete wall of the inner vessel. The wall was free to slide on its footing. The inner vessel and footing were designed to be cooled through a temperature range of 410°F (90°F to −320 °F) or 228°C (32°C to −196°C). Its 38 ft diameter would change 0·8 in. This would be the maximum relative movement between the wall and footing and would depend upon their relative speed of cool-down. The friction coefficient was minimized by troweling in graphite on the top of the concrete footing before it hardened.
2. It provided a tight metal floor. This could not be assured by a concrete floor tied into the cylindrical wall. There would have been a very good possibility of the floor and/or the tied-in concrete joints cracking from excessive and uneven temperature stresses.
3. It provided positive welded metal joints to all pipe connections to the bottom of the inner vessel. During cool-down the Everdur piping would shrink about 1 in between the point where it was welded to the stainless steel floor and metal sleeves imbedded in the wall of the outer vessel. Changes in the pipe length were compensated for by means of bends in the piping.
4. It provided a collecting pan so that liquid oxygen, which was initially added through the floor, would be uniformly distributed and evaporated over the floor area. The stainless steel floor would cool down first. Its diameter would be reduced by $1\frac{1}{4}$ in and cause bending stresses to develop in the stainless steel skirt as the bottom of the concrete wall and its 1 in $\times 2\frac{1}{2}$ in imbedded carbon steel ring would cool down much more slowly.
5. The uniform distribution and evaporation of liquid oxygen on the stainless steel floor facilitated the uniform and gradual cool-down of the concrete inner vessel, insulation and foundations. Stresses in the concrete were thus minimized.

The stainless steel floor and the liquid oxygen were supported by a 7 in slab of Perlite concrete which, in turn, rested on a 40 in thick pad made of Foamglas blocks. These also supported a concrete footing ring which carried the weight of the concrete inner vessel. The Perlite concrete slab had some insulating value. It also protected the fragile Foamglas from thermal shock and mechanical damage from the stainless steel floor and piping. The Foamglas blocks were supported on the 4 in concrete floor of the casing. The concrete floor was free to move vertically with respect to the casing wall. Finally, a 1 ft layer of compacted sand was provided between the casing floor and ground.

The heat leak through the insulated foundation was calculated to be 3·4 Btu/hr ft², 4100 Btu/hr – about 24% of the heat leak for the entire tank. The heat loss into the ground might eventually freeze the ground and damage the foundations, insulation and the reservoir. In order to avoid freezing the ground, a grid of electric heating cables was installed on the ground below the compacted sand. Five cables with a total heating capacity of 4 000 watts (13 600 Btu/hr) were installed with a thermo-switch to maintain the ground temperature at 35°F (1°C). This arrangement was considerably less expensive than providing a space under the insulation and outer vessel for circulating atmospheric air under the casing.

Other items of interest in design

The outer vessel was also made of prestressed concrete and provided a space of 36 in along the sides and 46 in at the top for Perlite insulation powder. The prestressed section of the outer vessel was prestressed by means of 4 500 lb of hard drawn carbon steel wire (1·2% of the weight of the concrete) whereas the wall of the inner vessel was prestressed by means of 7 200 lb of hard drawn carbon steel wire (1·9% of the weight of concrete in the wall).

In order to provide additional safety, the amount of steel used for prestressing the outer vessel was increased somewhat above the amount normally provided for atmospheric tanks. Also, all the manholes located near the bottom were sealed. These features would allow the outer vessel to act as a safety device to hold liquid oxygen in the event a major leak developed in the inner vessel.

Pipe connections were provided to add and remove liquid, for venting of gas and liquid, and to measure the contents of the reservoir. Thermocouples were installed at strategic points – at foundations, under the insulation, at the dome, and on the inside and outside of the wall of the inner vessel, to measure temperatures during cool-down operation and warm-up.

The heat leak into the reservoir was calculated to be equivalent to the daily evaporation of 0·2% of the reservoir's capacity. This was so low that if the barometric pressure increased rapidly the vapour pressure of the liquid oxygen would be less than atmospheric pressure, evaporation would stop, and atmospheric air containing moisture would enter the reservoir through vents. Atmospheric air entering the reservoir would be objectionable and dangerous as the moisture in the air would freeze in the vent piping and plug it. Also, the purity of the oxygen in the reservoir would be reduced by the nitrogen in the air. In order to prevent this, vents were provided which would relieve oxygen to the atmosphere but would close and prevent atmospheric air from entering the reservoir. In order to avoid a vacuum, means were provided to add oxygen gas to the inner vessel when required to assure a slight positive pressure there.

As the reservoir was to be installed on sloping ground – over an abandoned coal mine – a drainage ditch was provided around the outer casing and connected to a 10 000 ft³ collecting basin. Such a basin was installed as a safety measure to collect liquid oxygen in the event both the inner and outer vessels failed.

Construction

The reservoir was installed in a Linde liquid oxygen producing plant near Pittsburgh, Pennsylvania. The Preload Corporation, built the concrete portions, including the concrete forms, and prestressed the concrete. Linde, through its Construction and Design Department, provided and installed the stainless steel floor, internal piping, and the external piping to connect the reservoir to the Linde Plant piping system.

As the reservoir was to be used in liquid oxygen service, it was necessary that all the aggregate used for the concrete be free of carbonaceous materials. Neither could the usual 'form' oil be applied to the inside of the concrete forms to facilitate their removal. A mixture of water-glass and graphite (which was satisfactory for oxygen service) was used to coat the forms. Most of the graphite was removed by brushing prior to adding liquid oxygen to the reservoir.

The entire 91 yd³ of 6 000 psi concrete, required for the wall of the inner vessel, was poured continuously during a single day to eliminate possible cracks and leakage. This was considered quite a feat because the high strength concrete, which had only a small amount of water, flowed with difficulty.

Construction started during June 1951 and was completed in March 1952.

Operation

The reservoir was placed in operation and filled with liquid oxygen by the Linde Production Department with the aid and consultation of the Linde Engineering Laboratory, who specified and prepared the operating instructions. The inside of the reservoir and oxygen piping were thoroughly cleaned for oxygen service and dried before oxygen was added.

Liquid oxygen was added to the stainless steel floor at a rate such that the maximum temperature gradients through the inner vessel wall would be no more than 20°F at the bottom and 10°F at the top of the wall. During the first seven days of cool-down, liquid oxygen was added slowly and the level was kept below the top of the 12 in stainless steel skirt of the floor. See Exhibit 9 for temperatures during cool-down.

The maximum thermal gradient across the wall six inches above the footing was 17°F. It occurred 12 hours after liquid oxygen was first added to the floor. Twenty-two hours after filling started, the maximum temperature gradient of 14°F (-10°C) was measured at the 5-ft level. At the 28-ft level, the maximum temperature gradient of 10°F occurred eight days after the start of filling when the liquid level was about two feet above the footing.

The maximum vertical thermal gradient of 31°F (0°C) per ft occurred between the 6-in and 5-ft levels on the seventh day of filling. Between the 5- and 28-ft levels, the maximum gradient of 6°F (-14°C) per ft did not occur until 21 days after start of filling.

Filling started in March 1952 and the reservoir was finally filled in May, 62 days later. Of the 37 675 000 ft³ (N.T.P.) of liquid oxygen added to the reservoir, 3 870 000 ft³ were required to cool down the concrete, insulation, etc., and 3 417 000 ft³ were required for normal heat leak.

Later in 1952 the reservoir was emptied and warmed up over a 50-day period. Inspection showed the inside concrete wall to be in satisfactory condition. Several minor leaks were repaired in the welds of the stainless steel floor and the lower 7 ft of the concrete wall were painted with water-glass.

The prototype reservoir was then turned over to the Linde Production Department for normal operation in November 1952. Operation was satisfactory.

660 000 gallon reservoir

Early in 1953, Linde authorized the Preload Corporation to build a 660 000 gallon reservoir for liquid oxygen at its East Chicago, Indiana, production plant. This reservoir was 67 ft OD and 53 ft height. The inner vessel was 56 ft 3 in ID and 36 ft high. The design was essentially the same as for the prototype reservoir. However, the thickness of the stainless steel floor of the inner vessel was increased to 16 gauge and an 'improved' joint was provided between the concrete wall and stainless steel skirt.

The reservoir was placed in service in 1954. In 1958 the joint between the concrete wall and stainless steel skirt, which had developed an objectionable leak, was permanently repaired by welding to the skirt ring a stainless steel curb, 6 in wide by 18 in high, and filling it with concrete. The reservoir continues in satisfactory condition as of the present date.

US Patent No. 2,777,295, covering the design of the concrete reservoir for liquefied gases, was granted to Messrs L. A. Bliss, P. M. Riede, and J. H. Beckman. The Preload Corporation was licensed by the Linde Company to use this patent and has built additional prestressed concrete reservoirs for liquefied gases.

In recalling the project Mr J. J. Closner, now Vice President of Preload, said, 'Very little came out of the work we did with Linde until the early 1960s. We found no markets for the prestressed concrete LOX tanks since their initial cost was somewhat more than stainless steel tanks at the time, and the size of the tanks were small. However, in the early 1960s the American Gas Association became interested in the possibilities of using prestressed concrete for the storage of liquefied natural gas for peak-shaving plants. Preload undertook a further development contract with the American Gas Association and the Institute of Gas Technology to accomplish this investigation and to add and to extend the information which was made available from Linde at that time. This resulted in the use of prestressed concrete tanks for the storage of liquefied natural gas. Of most interest to us is that these tanks are in the range of 600 000 to 800 000 barrels. Our studies are now taking up to LNG tanks beyond 1 000 000 barrels of capacity.'

H. C. KORNEMANN

Dated: 21 August 1970

EXHIBIT 4

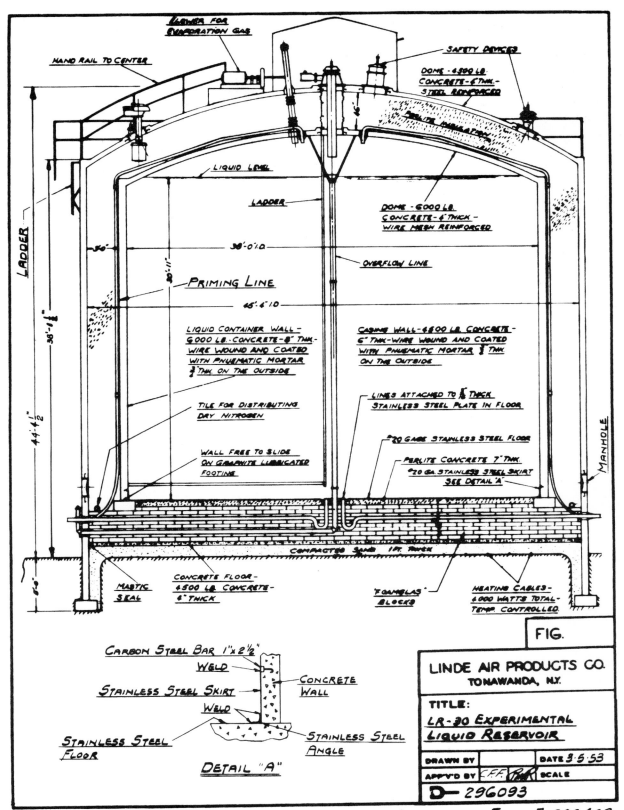

DETAIL "A"

LINDE AIR PRODUCTS CO.
TONAWANDA, N.Y.

TITLE:
LR-30 EXPERIMENTAL
LIQUID RESERVOIR

DRAWN BY
APPV'D BY
DATE 3-5-53
SCALE

D- 296093

3-5-53 REPRODUCED FROM B-286499

EXHIBIT 5   Horizontally prestressing

The cases being prestressed with 0·162 in dia. high-carbon steel wire, 28 January 1952. The lower carriage is pulled from left to right by a sprocket which engages the chain about the tank

EXHIBIT 6   Wrapping the casing

The lower portion of the casing being wrapped by drawing the wire through a 0·142 in dia. die. The wire-winding machine is powered by air.

EXHIBIT 7   Description of wire winding machine

The self-propelled wire winding machine (see Exhibits 5 & 6) was suspended by cables from a traveling carriage on the top of the wall. The carriage was pivoted from a center post which was bolted to the dome and held in proper alignment by cables. The operator spaced the wires by raising or lowering the wire winding machine with an air motor located on the upper carriage.

The wire winding machine (see Exhibit 6) consisted of a vertical pipe, the end of which contained the 0.142 in diameter wire reducing die. The vertical pipe was fastened to a frame and an air motor driven sprocket gear which meshed with a heavy roller chain that encircled the wall. A 400 lb roll of 0.162 in diameter wire, the end of which had been previously reduced to 0.142 in diameter, was placed on the machine, the elongated end of the wire was threaded through the die and fastened to an expansion shield bolt anchored in the concrete wall.

The 140 000 psi initial stress in the wire was obtained by simply pulling the wire through the die as the air motor driven wire winding machine wrapped the wire around the wall.

Exhibit 6 shows the wire winding machine, with the die below the frame, winding the wire upward near the end of winding operation.

Wire winding was started near the top of the wall and progressed downward. The die was above the machine because the chain drive was below the machine. This arrangement made it less likely to nick or damage the wire. In order to wrap the bottom of the wall, it was necessary to reverse the position of the die, see Exhibit 6, and wind upwards.

The ends of the wire from the various rolls were joined together (after pasing through the die) by means of carbon steel collets. Clamps were used to bind several turns of wire together and were placed on the windings whenever the machine stopped and whenever a new roll was added. The clamps were left in place so that if a wire should break the number of turns which would loosen and have to be replaced would be minimized.

EXHIBIT 8   Prestressing wires

Vertical bundle of 0·196 in dia. galvanized prestressing wires. Note the location of the grouting tube. The bar ring and skirt is in position against the inner form. Note the holes drilled through the bar ring

EXHIBIT 9   Initial test of LR-30 reservoir

INITIAL TEST OF LR-30 RESERVOIR

SOUTHEAST WALL TEMPERATURE GRADIENTS

Amount Of Liquid Oxygen ~ Million Cu Ft

Amount Liquid In Reservoir

Total Liquid Transferred

Earth Under Center Of Floor

Wall, 28' Above Floor

Gas At Dome Level

Center Of Dome

Wall, 5' Above Floor

Wall, 6" Above Floor

Outside Temperature

Inside Temperature

Days From Start Of Fill

Temperature ~ °F

SUGGESTED STUDIES

This case study gives the instructor an opportunity to introduce the student to the design of equipment involving thermal stresses, in particular, the storage of cryogenic liquids. The first part of the case introduces the student to the background of the design and places before him the fundamental information on which the design is to be based. The second part of the case gives the student the final design.

Based upon reading the first part of the case class discussion can be centred around:

1. The preferred storage tank configuration.
2. The merits of concrete for low temperature application.
3. Where and how prestressing could be applied.
4. The significance of the carbon steel reinforcing wire having the same coefficient of thermal expansion as concrete. Is this important in non-cryogenic use ?
5. Criteria for establishing the prestressing required and working stresses in the reservoir.

The first part of the case can also be used as a basis for design projects or design analysis problems. By giving the students some of the information available in part 2 of the case such as over-all dimensions, the student can be asked to determine the following:

1. Maximum temperature gradient in reservoir walls.
2. Steady state stress gradient in reservoir walls.
3. Thermal stress in reservoir walls.
4. Prestressing in reservoir walls.
5. Tank filling procedure.

Part 2 of the case can be assigned for reading so that the students can compare their analysis and design with that of the case.

Class discussion on Part 2 of the case can be centred on:

1. Method of applying prestressing and control of prestressing.
2. Why prestressing is different at different parts of the wall?
3. Why vertical prestressing?
4. Problem of joining stainless steel to carbon steel.
5. The merits and significance of the design details discussed in the case.

Case 4

A four barrel step-and-repeat camera

by J. McMills, G. R. Mehta, and F. A. Wile

Prepared under the supervision of Professor R. F. Steidel, Jr., in partial fulfilment of their requirements for the Master of Engineering in Mechanical Engineering at the University of California, Berkeley. It has been edited and shortened by Geza Kardos.

This case study has been prepared and written with the aid of a grant from the Ford Foundation for the demonstration and development of a new Master of Engineering program in design.

This case is available in pamphlet form as ECL 181 from the Engineering Case Library, Room 500, Stanford University, Stanford, California 94305.

Paul Piper was hired by the Friden Research Center to design, develop and produce a microphotographic camera for the production of final stage integrated circuit masks. Piper's problems are described from the management decision to make instead of buying the camera through the design to the testing and modifications of the unit.

In the mid-sixties the Friden Division of the Singer Corporation designed, built, and marketed their first all electronic calculator. The electronic unit was such a success that in the first quarter of 1968 they discontinued development of all new electromechanical calculators. Keen competition from Japanese and American electronics firms led Friden to decide that it would be necessary to develop their own integrated circuit capability. The use of integrated circuits for electronics also offered obvious technical and cost advantages over other methods.

The electronics industry had started by assembling circuits using vacuum tubes, wires, etc. By the 1950s, this evolved into utilizing solid state components transistors, diodes, and printed circuits, etc. in the assembly. Micro-miniature electronics appeared in the early 1960s and began to hit its stride in the mid-sixties. The principal advantage of this third generation of electronics was in the manufacture of integrated circuits. Up to this point, electronic components had been packaged as individual units and connected together to form the necessary circuits. With integrated circuits, the individual components are fabricated simultaneously with their inter-connections. Integrated circuits will contain 100 to 1 000 complete circuits in a unit the size of an existing transistor. The benefits are so great that, as Paul Piper of the Friden Research Center points out, 'If the electronic firms in this country don't get on the stick, they are going to be out in the cold. We are going to see, in the next few years, big companies going bankrupt if they don't convert, and that time is almost too late.'

It is inherent in the manufacture of integrated circuits that the circuit designers and manufacturers be familiar with and produce the transistors and other devices that make up the circuits. The state-of-the-art of circuit manufacturing is given in the following abstract by A. Loto from the *Engineering Journal*, December 1966.

Microphotographic Methods in Modern Electronic Manufacture

Figures A and B give some idea of the scale of modern silicon devices and microcircuits. The high frequency transistor chips shown in Fig. A make relatively inefficient use of the material since only about 0.1% of the crystal volume plays an electronically active role, it being impractical to cut, handle, and make electrical connection to much smaller chips. The microcircuit shown in Fig. B represents more efficient material usage since it is only about six times the area of the transistor chip but contains 4 transistors, 4 resistors, and 22 junction diodes together with their interconnections.

In order to fabricate a device such as a transistor from a crystal of silicon it is necessary to introduce certain elemental impurities into the crystal in selected areas to generate the pn junctions. It is also necessary to make electrical connection to the separate zones of the crystal so produced. By far the most widely used technology is that known as planar diffusion. In this method the starting point is a highly polished slice of single crystal silicon a few thousands of an inch (mils) thick and about 1 inch in diameter. A thin skin of the natural silicon dioxide is grown on the polished surface by high temperature oxidation and an array of minute holes are etched through this skin. When the silicon is subsequently exposed at elevated temperature to an atmosphere containing a desired impurity species, the oxide skin acts like a stencil and allows the impurity to diffuse into the silicon crystal lattice only through the etched holes. The holes may be closed by oxide regrowth and the process repeated, through a new array of holes in spatial relationship to the first array, using a second impurity species. The silicon dioxide remains on the surface of completed devices for protection and insulation, electrical connection being made through further holes etched in it. The contacting metal is applied by vacuum plating over the entire surface and is subsequently formed to the desired pattern by etching away unwanted areas.

In order to make manufacture economically feasible a large number of devices are processed simultaneously on each slice of silicon, which is separated into discrete dies by diamond scribing only after all processing is completed. For separate devices such as diodes or transistors as many as 3 500 may be fabricated in one square inch of crystal while 600 microcircuits could be accommodated in the same area, representing the equivalent of maybe 20 000 components per square inch.

Since all the etching processes for silicon dioxide and the contact metal are controlled photographically it will be evident that the fabrication of a device requires the use of a series of photographic negatives corresponding in number to the number of etching processes involved, each one having a multiplicity of micro-images precisely located to permit them to be printed sequentially onto the silicon crystal. Each pattern must, therefore, be positioned with respect to others in the series to within tolerances dictated by the particular design. Line widths and clearances commonly used on microcircuits are 0·5 mils while higher frequency transistors may go down to 0·1 mils and laboratory prototypes have been made with some dimensions of less than 0·05 mils. Displacement of any one pattern of a series would result in devices with degraded electrical characteristics or complete failure due to shorting of the rectifying pn junctions. A set of masks for a transistor normally includes four plates whereas a silicon microcircuit requires six or more for its manufacture.

A common starting point in the various methods of preparing masks is the preparation of a set of greatly enlarged patterns of a single device or circuit. Such a set consists of one pattern corresponding to each etching operation in the manufacturing process drawn between 100 and 500 times the final size of the device in order to achieve the necessary accuracy. Such patterns are normally prepared on a high precision drafting machine capable of holding dimensions to within 1–2 mils over an area of 30–40 in square. In order to obtain the highest image contrast, these patterns are normally prepared as transparencies to be illuminated by transmitted light during photographic reduction. For this purpose a commercial film, consisting of Mylar coated with a thin strippable film of photographically opaque ruby colored plastic, is widely used. The pattern is cut through the plastic coating which is stripped off where not required.

In the majority of processes the next phase is the preparation of a reduced image of the large patterns, usually to about 3 to 10× final device size by one or more stages of photographic reduction. For this purpose a special type of camera is used which is typified by rigid construction and the provision of means for the precise adjustment of copy to plate distance and lens focusing (Fig. C). The originals are illuminated by large area mercury vapour discharge lamps having a filtered output in the middle green of the visible spectrum. Such equipment is preferably operated in a dust-free air conditioned location, free from vibration and with temperature constant to ±1°F.

The next step in the process generally involves the multiple printing of each negative onto a single photographic plate. Reduction down to final size may also be carried out at this stage, resulting in the final mask, or postponed to a final stage reduction of the larger multiple image plate. The type of equipment used will be quite different according to which approach is chosen.

The final size step and repeat method has the advantage that microscope type optics can be used to give high image resolution, since only one device area need be covered at one time. However, very precise mechanical construction is required since mechanical errors are reproduced on a one-to-one basis **and the lens** focus must be held to within a few microns over the entire printing area **for** optimum image quality. Basically such a machine consists of a reducing optical projector mounted to print onto a plate held on the table of a high precision *xy* positioner together with suitable equipment to control the table position for each exposure.

The alternative approach of multiple printing at about 10× final size usually involves the use of a step and repeat contact printer. An advantage is that any positional errors will be scaled down during the final reduction but the large image area to be covered during final reduction (typically at least 1 in diameter circle) and the continually higher resolution demanded by the device designer severely tax the capabilities of the finest lenses and it seems inevitable that this system will, in time, be entirely superseded by the final size step and repeat methods in which the mechanical positioning problem is more readily solved.

A useful technique which permits precision of alignment independent of the precision of repeat spacing is known as simultaneous stepping. In many devices the active regions occupy only a small fraction of the total area of the silicon crystal (Fig. A). In such cases all the patterns required to fabricate the device can be accommodated within one repeat area on the crystal. If the original patterns are cut as a single group they may be reduced and stepped to produce a single mask plate which can be used for all etching operations. At each process step only one of the patterns within the group is actually used, the others producing redundant images outside the active device area. This principle is illustrated in Fig. D. Since the dimensions of each group are identical, alignment of one device automatically gives alignment of all others even if the step and repeat spacing is entirely random, provided that there has been no relative rotation between the negative and the plate during the preparation of the mask.

FIG. A   A common housefly gives scale to some silicon U.H.F. transistor chips. Inset: Plan view of a single chip

The advantages of simultaneous stepping can be extended to devices which occupy the whole available area of the crystal (e.g. microcircuits) if the separate patterns of the group are spaced apart a distance greater than the total length of the mask. In this case each pattern forms its own separate array which may be recorded on separate plates on a common carrier. The machine shown in Fig. E achieves this end by having four separate reducing projectors each containing one pattern negative mounted over a common *xy* positioner carrying four photographic plates.

In transferring the images from the masks to the semiconductor surface, use is made of photoresists. Their function is to provide an adherent stencil through which the semiconductor, its oxide or an overlying metal film can be etched. They consist of organic materials which can be applied in solution by spin coating, dipping, or spraying, and which react to light in one of two ways. One type is photo-polymerized to render it insoluble in the exposed areas while unexposed materials can be washed away in a solvent developer. This type is most widely used to etch holes through oxide leaving most of the oxide in situ. The other type as coated is insoluble in the developer but is rendered soluble by light exposure. This type is frequently used when most of the etched material is to be removed leaving only islands or narrow lines as in the case of metal contacts and interconnections. The use of both types as described permits the masks to be all of the same contrast, i.e. all having opaque image areas on a clear background. This simplifies visual alignment and eliminates an additional photographic process which would be necessary to produce masks having clear patterns on an opaque background.

Photoresists have their maximum sensitivity in the near UV and violet end of the spectrum and are generally exposed using mercury or xenon discharge lamps. Exposure of the photoresist is practically always carried out by contact printing which entails precise alignment of each mask with the preceding set of images while the mask is very close to, but not touching, the photoresist, followed by making positive contact for exposure in order to obtain the best quality image transfer. In the most critical cases an immersion liquid may be used to fill the gap between mask and photoresist in order to minimize loss of resolution by refraction (variations in silver content cause slight surface contours on the mask emulsion) and printing of spurious images by diffraction at surface defects and foreign particles. In any case, wear on the delicate gelatin emulsion is inevitable and masks must be discarded frequently to minimize disastrous loss of yield in the multi-stage semiconductor process. In order to overcome this problem many manufacturers now use masks consisting of etched metal films on glass substrates made by photolithography from the camera exposed emulsion masters. A ready made metallized glass plate, precoated with photoresist for the fabrication of such masks, will probably be the next commercially available photographic material created specifically for the electronics industry.

It will be obvious that the economic manufacture of semiconductor devices and microcircuits is critically dependent on the utmost attention to detail at every stage of a long and complicated process. The extreme sensitivity of semiconductors to impurities on the submicrogram scale places godliness a poor second to cleanliness in semiconductor manufacture. The aligned etching of six or more arrays each of hundreds of flawless micro-images, an extremely difficult feat even in the laboratory, is demanded daily of the microcircuit manufacturer on a mass production scale.

Friden Research Division

Friden, as a subsidiary of Singer Corporation, has many operating divisions. Each division was responsible for its own research, engineering, production, and for

Fig. B Part of a slice of diffused silicon microcircuits before cutting. The diffused areas are various shades of grey. The light pattern consists of metallic interconnections

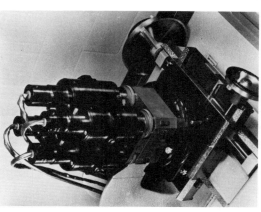

FIG. C Precision reduction camera designed for electronic work (*Photo: courtesy H.L.C. Engineering*)

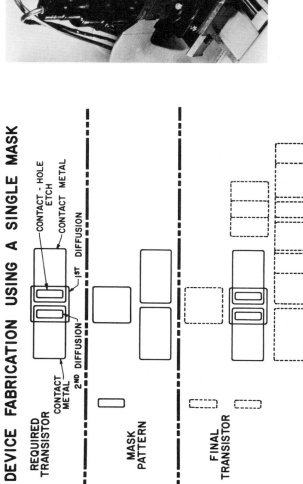

FIG. E A photorepeater incorporating four projector barrels which permit simultaneous stepping of four patterns. (*Photo: courtesy David W. Mann Company*)

DEVICE FABRICATION USING A SINGLE MASK

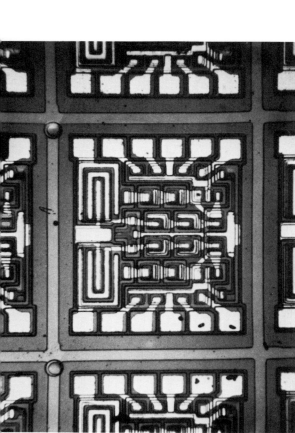

FIG. D To produce the transistor (top) the same group of patterns (centre) is printed in a new position at each process step (bottom). Identical displacements occur at every similar group on the mask

showing a profit on its total operation. Early in 1966, Mr L. P. Robinson, the Friden Vice-President who had been the driving force behind the electronic calculator, saw the need for a research center completely divorced from the operating divisions, devoted to state-of-the-art research. At first Friden management rejected the idea, since such a center could not show a profit and would thus be an uncontrolled expense burden on the other divisions. Finally after some revision of his original plans Mr Robinson secured Friden and Singer Corporate approval.

The Friden Research Division was formed in June 1960 for the purpose of becoming expert in the field of integrated circuit design and manufacture but not necessarily to develop immediately useful processes. It had been agreed by management that only by thoroughly understanding the field of integrated circuits could a fully automated manufacturing facility be eventually built, and the cost of this research would have to be recovered by having a more economic production facility.

Mr Robinson remained a Friden Vice-President but also became Director of Research. There were six major areas of interest, each with a director (Fig. F). Mr Barney D. Hunt joined the division on 31 October 1966, to become Director of Microcircuits. His responsibilities included microphotography, circuit design and all integrated circuit processes.

The operation of the Research Division was based on Friden's decision to produce integrated circuits internally. While companies were available which did produce integrated circuits to specifications, it was apparent that integrated circuits in Friden products would be an important item and a large part of the cost. The cost of purchasing custom circuits from outside firms as well as length of delivery time would make that option prohibitive. The decision as to whether production equipment would be made or puchased was a compromise. Equipment which was not available or which was critical to the production process would be developed by Friden. Although available equipment could probably be purchased cheaper than developing it, Friden's experience showed that the time required to fully understand critical components of purchased equipment was usually equal to the time to build it. Designing and manufacturing the equipment carried the added advantage of producing a better understanding of the variables involved. The non-critical production equipment would be purchased.

The Camera Project

Since the photo reduction process was essential to microcircuit manufacture

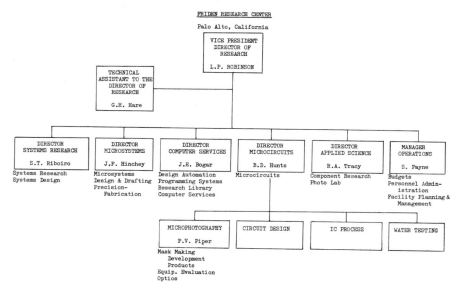

FIG F   Organization chart

EXHIBIT 1   Process flow diagram for the manufacture of integrated circuits

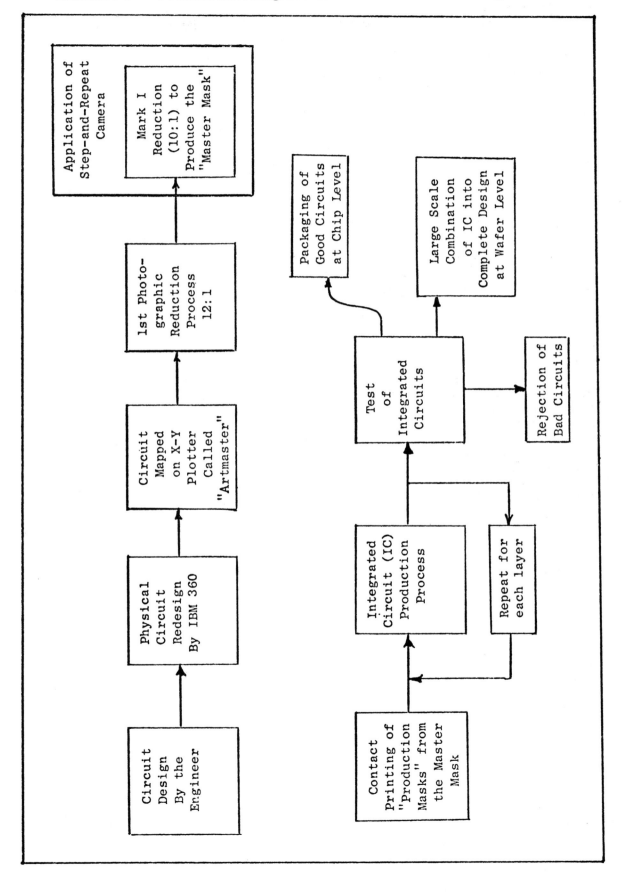

EXHIBIT 2

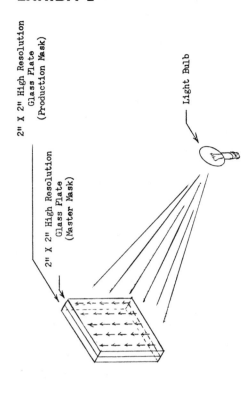

CONTACT PRINTING OF PRODUCTION MASK FROM MASTER MASK

2" X 2" High Resolution Glass Plate (Production Mask)

2" X 2" High Resolution Glass Plate (Master Mask)

Light Bulb

CONTACT PRINTING OF SILICON WAFER FROM PRODUCTION MASK

2" X 2" High Resolution Glass Plate (Production Mask)

Light Bulb

Silicon Wafer

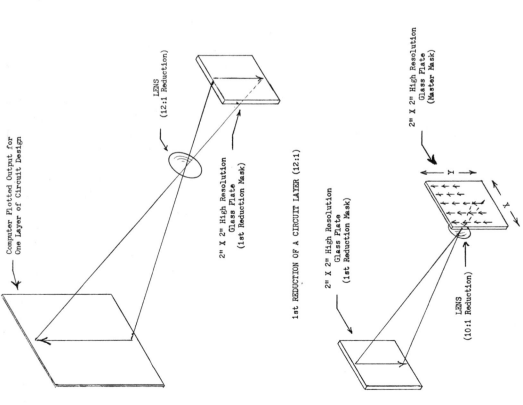

Computer Plotted Output for One Layer of Circuit Design

LENS (12:1 Reduction)

2" X 2" High Resolution Glass Plate (1st Reduction Mask)

1st REDUCTION OF A CIRCUIT LAYER (12:1)

2" X 2" High Resolution Glass Plate (1st Reduction Mask)

LENS (10:1 Reduction)

2" X 2" High Resolution Glass Plate (Master Mask)

2nd REDUCTION OF A CIRCUIT LAYER PLUS STEPPING (STEP AND REPEAT CAMERA--10:1)

(Exhibits 1 and 2) it was decided to build an extremely accurate step-and-repeat camera for the making of the final production masks. The camera was to replicate the circuit masks from the first reduction. The use of four lenses would provide identical replication of four different masks. Because of the knowledge to be gained in the optics of integrated circuit technology, this camera was one to be built without hope of profit.

By 14 November 1966, a first formal presentation of a masking **system was** made to management based on a feasibility study conducted by G. Hare and A. Rowland. Some of the conclusions drawn were eventually proved to be in error because of insufficient knowledge of optics. As a result, Paul Piper, a mechanical engineer, was hired on 5 December to take charge of microphotography development. He was to implement the design and production of the step-and-repeat camera.

On 15 December Paul Piper attempted his first definition of the problem. He decided that a successful four barrel step-and-repeat camera is dependent on the designer's ability to overcome the following problems:

1. Registration, which is the alignment between corresponding images in each mask. The distance between any two consecutive images or patterns in any array must be identical in both the x and y directions (Exhibit 2). The camera should be able to repeat the position at every operation to better than ± 50 micro-inches. The movement of the x–y stage must be controlled within this tolerance and the motion must be orthogonal.
2. The environment in which the camera is to operate must be controlled. It must first be dust free. The presence of even a one micron dust particle could cause faulty image generation. Temperature and humidity changes could affect the dimensional stability of the elements and consequently precise control of temperature and humidity are necessary.
3. Because of the precision required the camera must be isolated from vibration. The isolation must eliminate relative motion between object and image plane.
4. The reduction ratio must be extremely accurate and identical in all four barrels. With a 10:1 reduction the optical system must be able to resolve a minimum line width of 0·2 mils with high contrast and good edge definition over the entire useful image area, approximately 0·360 in diameter.
5. The image distortion in each of the barrels must be within the pattern tolerances. The alignment between the optical and mechanical axis must be as accurate as possible in order to match the resulting mask with other members of the set.

Piper realized that the heart of the system would be the lens. He immediately started an intensive literature search to bolster his knowledge of optics. This search continued throughout the project. He decided that before any further design could be carried out, an optical system must be selected. The optical system would consist of: the illumination system composed of the light source, the condenser, the light filter, the shutter, the filter holder, the light housing, and the system support; and the optical column composed of the lens, the lens holder, the object mask holder, the lens object separation column, and the barrel support housing.

Early in January 1967 Piper had the purchasing department start a search for a suitable lens. He had set the requirements as a 10:1 reduction with a resolution of 0·2 mils minimum line width with good edge definition; the image field was to be approximately 0·36 in (object field 3·6 in). They sent the request for quotation (Exhibit 3) to approximately 30 manufacturers.

At this time, Roy Nakai, a mechanical engineer, was assigned to work under Piper as a junior engineer. As a result of their studies Piper and Hunt recommended to management that Friden purchase a microphoto-repeater camera from an outside source. This would provide an immediate reliable source of masks for research. This proposal was rejected and the original plan to build was upheld.

Preliminary specifications were set by 25 January (Exhibit 4). These were the sole

design constraints placed on the camera at the outset of the project. Preliminary sketches were made. To estimate the cost of the project Piper divided the design and construction into six major areas: optical barrel; positioning mechanism; support system; image cassette; instrument packaging; and overall evaluation. Each of these areas was further divided into materials cost (Exhibit 5) and manpower (Exhibit 6). The materials cost included raw materials plus purchased finished parts. Manpower estimates included time for engineers, draftsmen, technicians, and machinists (Exhibit 7).

The design and construction of a four barrel step-and-repeat camera, Project 355, was approved 26 January. This marked the official start of the project. Because the research division needed masks immediately, Project 362, a one-barrel first reduction camera was initiated on 31 January. Because of the higher priority of Project 362 it consumed almost all of Piper's and Nakai's attention, completely halting any work on Project 355 for about three months.

All lens quotations were received by early February. The most favorable reply was from the Elgeet Company, Rochester, New York. They proposed to produce four lenses to specification for a total of $6 138. Since performance was guaranteed this seemed an ideal source. A purchase requisition for the lens was submitted to management for approval, but before it could be approved, the Elgeet Company went bankrupt. The search for a lens supplier started again.

The construction of the first reduction camera, Project 362, was completed 6 April and it was producing experimental microcircuits. This freed Piper's team to focus their attention of the four barrel step-and-repeat camera. Piper revised his cost estimates 10 April (Exhibit 7).

A Nikon lens was obtained on loan from the supplier for evaluation. It turned out to have excellent resolution and distortion qualities. The evaluation prompted Piper to devise a proper test program for all future lenses. Four Nikon lenses would cost $5 200. When Piper submitted a purchase requisition for these lenses, his request was turned down by management. The reason given was that the lenses were needed for a specific purpose; the Nikon lenses were considered to be 'more lens than needed'. This rejection caused Piper to reflect that even if Elgeet had not gone bankrupt the purchase of the Elgeet lenses would not have been approved by his management. Following the rejection of the Nikon lens, a series of other lenses were tested with little success.

A Schneider Xenon lens was purchased in the latter part of April for $61. It proved to have exceptional performance. Four additional lenses were purchased immediately contingent upon Friden's testing and accepting the lenses. Three of the lenses were acceptable, the fourth was rejected and returned. Another lens was furnished in an attempt to complete the order but it too was rejected. Following this rejection, it was noted that the satisfactory lenses were of the same serial number series, while the rejected lenses were not. The Schneider representative was asked to ship all his remaining lenses in the accepted serial number series for evaluation.

On 1 June Piper called his team together in an attempt to pin down the necessary physical dimensions and to start detail design. The team now included Harold B. Wells who had been hired two weeks earlier as a designer.

Starting from its most basic form, the microphoto-repeater was to have four lens systems. The image plane was to move in two orthogonal directions, x and y, perpendicular to the optical axis. The object plane was to be fixed. The assumption was made that Schneider Xenon lenses would be used and that all would perform as well as the lenses in hand (Exhibit 8).

The first decision made was to fix the relative location of the optical components. Three configurations were available: 'horizontal orientation' with the optical axis parallel to the horizontal plane; and two vertical orientations (Exhibit 9); 'image plane down' with optical axis vertical and the image plane below the object plane; and 'image plane up' with optical axis vertical and the image plane above the object

EXHIBIT 3   Copy of a letter requesting a lens quote

Friden Inc.

RESEARCH DIVISION
OAKLAND · CALIFORNIA
A SUBSIDIARY OF THE SINGER COMPANY

4 January 1967

James Archer
Space Optics General
1855 W Street 169
Gardena, California 90247

Dear Mr. Archer,

This letter is a follow up to our telephone conversation of 3 January. As we discussed, Friden has a requirement for lenses for use in a step and repeat process to produce master masks for use in the fabrication of integrated circuits.

For production efficiency and mask contrast sharpness reasons, it is highly desirable to use commercially available lenses. In that most of these resists are not sensitive within the range of the photosensitive resists, it is necessary to have a lens system designed which is sensitive somewhere within the wavelength bandwidth of 310 - 460 millimicrons.

We at Friden are currently involved with the design and ultimate construction of a multi-barrel (four to six barrels) step and repeat camera system and are at the point where we must decide upon a lens system before continuing with the current design. Therefore, it is necessary for us to obtain performance, delivery and cost data for such a lens system.

Since Friden is not in the lens design business, I shall only give the essential requirements for our needs and leave the design decisions and "trade offs" to the experts. Our basic requirements are:

1. Magnification - 1/10 (Reduction of 10 to 1)

2. Resolution - Lens system must be capable of resolving a minimum line width of 0.2 mils with high contrast and good definition results anywhere within the useful image field.

3. Image field - The diameter of the useful image field (D_F) at a reduction of 10 to 1 for each lens system must be somewhere within the range of $0.15 < D_F < 0.36$ inches. It is desirable to have D_F closer to 0.36 inches than 0.15 inches.

4. Wavelength - Lens systems must be designed to function somewhere within the wavelength bandwidth of 310 - 460 millimicrons. The design frequency should correspond to that of a commercially available illumination source. It is desirable to have the lens system corrected for a reasonably large bandwidth about the design point.

(cont...)

5. Matching - It is far more necessary to match each lens system (four or six systems) to each other than it is to eliminate all distortions.

6. Cost - Naturally, the costs are to be minimized. However, should design decisions or "trade offs" occur that significantly increase the cost while improving the lens performance from a marginal to an acceptable situation, they are to be identified. Nonrecurring (engineering) costs are to be separated from that required for the manufacturing process. Provide quotes for the four each and six each lens systems.

7. Delivery - Show delivery schedule as referred to receipt of order.

8. Additional information -

In addition to cost and delivery data, please supply as complete as possible design goals which correspond to the following specification items.

a. Focal length
b. Maximum relative aperture.
c. Aperture scale.
d. Standard magnification 1/10, (10 to 1 reduction)
e. Object size
   1) maximum
   2) useful
f. Image field
   1) maximum
   2) useful
g. Working distance from object to image plane.
h. Wavelength
   1) design point
   2) bandwidth
i. Aerial resolving power (value used to be referenced to same contrast (intensity) ratio.
j. Aperture efficiency at image corner.
k. Distortion of image corner.
l. Lens system mounting.
m. Construction
n. Dimensions
   1) maximum diameter
   2) maximum length
o. Weight
p. Ambient temperature range.

I would appreciate hearing from you as soon as possible. Should it be necessary, you can contact me at (415) 557-6800 Ext. 578.

Thank you.

Sincerely,

FRIDEN, INC. -RESEARCH

Paul V. Piper

PVP: s

EXHIBIT 4   Basic design requirements

	Specification	Jan 1967 Conceptual Design First	June 1967 Second	June 1968 Final Design
1.	Reduction ratio	10:1	10:1	10:1
2.	Resolution/contrast/ edge definition.	This system must have the capability of resolving a minimum line width of 0.2 mils with high contrast and good edge definition, at all points within the useful image field.		Same
3.	Image field.			
	a. Maximum useful image square	0.250 x 0.250 in	0.170 x0.170 in	Same
	b. Maximum useful image diameter.	0.354 in	0.240 in	Same
4.	Object mask size.			
	a. Basic glass plate size.	4 x5 in	2 x2 in	Same
	b. Lip of plate	1/8 in	15/128 in	
	c. Object mask size	2.5 x2.5 in	1.7 x1.7 in	
5.	Image mask size			
	a. Basic size of glass plate	3 x3 in and 2 x2 in	2 x2 in	Same
	b. Lip of plate.	1/8 in	15/128 in	Same
6.	Type of photo-sensitive plates	Photo Resist on ehrome coated glass plates and high resolution glass plates.	High resolution glass plates.	Same
7.	Image positioning capability			
	a. Max. stepping error	$\pm$ 50μ in	$\pm$ 500μ in	Same
	b. Repeatability	$\pm$ 10μ in	$\pm$ 50μ in	$\pm$ 20 in
	c. Min. stepping increment.	0.001 in	Min chip size	0.010 in
	d. Max. depth of focus variation.	$\pm$ 25μ in	$\pm$ 50μ in	Same
	e. Range of increments	0.01 -0.33 in	Fixed scale	Increments of minimum chip size
	f. Type of stepping operation	Automatic	Manual	With semi-automatic control
8.	Camera environment.	Self contained dust-free filter system using dry N$_2$ atmosphere with automatic temperature control.	Dust free with constant temperature ($\pm$1°F) and constant humidity.	

EXHIBIT 5   Materials cost estimate (optical barrel only)

Task: Optical Barrels (4 each)

Description	Quantity	Unit cost $	Total cost $	Non. cap. cost	Capital cost	Cost est. method
1. Illumination Column						
A. Light source						
1. Bulbs (xenon)	4	40.00	160.00			C
2. Bulbs (Press. Hg)	4	30.00	120.00			C
3. Sockets (and Matches both bulbs)	4	6.50	26.00			C&BG
4. Back reflectors, cover and support	4	25.00	100.00			BG
B. Condenser	4	80.00	320.00			C&BG
C. Filter	4	40.00	160.00			C&BG
D. Filter holder	4	10.00	40.00			BG
E. Shutter						
1. Mechanical parts	4 (set)	10.00	40.00			BG
2. Actuater	1	32.00	32.00			C
2. Optical Column						
A. Lens (Catadioptical reflective system) Elgeet Optical Co. Inc.						
1. Non-recurring optico mech. eng.			3 578.00			TQ&LQ
2. Lens system	4	640.00	2 560.00			BG
B. Lens holder	4	30.00	120.00			BG
C. Object mask holder	4	25.00	100.00			BG
D. Obj. mask and lens separation Column	4	12.00	48.00			BG
E. Object shims	4 (sets)	10.00	40.00			BG
TOTAL COSTS			7 444.00			

C – Catalogue price   LQ – Letter quote
BG – Best guess   TQ – Telephone quote

EXHIBIT 6   Manpower estimate (optical barrels only)

Task: Optical barrels - (4 each)

Subtask description	Engineers (Hrs)	Drafting (Hrs) Designer	Detailer	Technician (E&M)	Machinist
1. Illumination column					
1. Design	48	20	40	–	–
2. Procurement	10	–	–	–	–
3. Fabrication	–	–	–	16	48
4. Assembly and checkout	4	–	–	32	–
2. Optical column					
2.1 Object Mark Holder (Includes shims)					
1. Design	40	24	40	–	–
2. Procurement	4	–	–	–	–
3. Fabrication and Assembly	–	–	–	4	40
2.2 Lens and lens holder	8	8	16	–	–
2. Procurement	16	–	–	–	–
3. Fabrication and assembly	–	–	–	6	48
4. Checkout-testing (adjustment)	–	–	–	4	–
2.3 Object mask and lens separation column					
1. Design	3	2	4	–	–
2. Procurement	2	–	–	–	–
3. Fabrication and Assembly	–	–	–	1	8
TOTALS	125	54	100	63	144

EXHIBIT 7   Cost summary–manpower consumption

Original — Date: 1/25/67 — Total $60 859

	Optical barrel	Positioning mechanism	Support system	Cassette (2)	Instrument packaging	Over-all evaluation	Total man-hours	Total cost ($)
Engineering hours (1 hr = $18)	135	172	27	20	373	240	1 367	24 606
Drafting hours (1 hr = $10)	154	114	42	48	308	–	666	6 660
Technicians' hours (1 hr = $8)	63	144	7	8	422	80	724	5 792
Machinists' hours (1 hr = $15)	144	60	40	96	124	–	464	6 960
Material cost ($)	7 444	3 040	650	100	2 800	–	–	16 841

Revised — Date: 4/10/67 — Total $57 297

	Optical barrel	Positioning mechanism	Support system	Cassette (2)	Instrument packaging	Over-all evaluation	Total man-hours	Total cost ($)
Engineering hours (1 hr = $18)	199	268	48	20	43	240	1 188	20 384
Drafting hours (1 hr = $10)	154	114	42	48	308	–	666	6 660
Technicians' hours (1 hr = $8)	63	144	7	8	422	80	724	5 792
Machinists' hours (1 hr = $15)	154	84	60	96	124	–	508	7 620
Material cost ($)	7 560	3 040	650	100	2 800	–	–	16 841

EXHIBIT 8   Schneider Xenon lens

1. Focal length = 28 mm (nominal), (E.F. = 29.4 mm $\pm$ 0.2 mm)

2. Maximum aperture = F 2

3. Aperture scale: 2, 2.8, 4, 5.6, 8, 11, 16

4. Magnification = 1/10

5. Object size = 287 mm (92 mm with resolution better than 300 lines/mm

6. Image size = 28.7 mm (9.2 mm with resolution better than 300 lines/mm)

7. Resolution = 400-500 lines/mm

8. Overall working distance (obj. - im. dist) = 355.7 mm (14.01 in)

9. Wavelength = 366-687 mμ

10. Mounting = thread

11. Dimensions:

    Max. diameter = 37.5 mm

    Max length = 23.8 mm

12. Weight = 67 grams

13. Cost = $61

EXHIBIT 9   Orientations

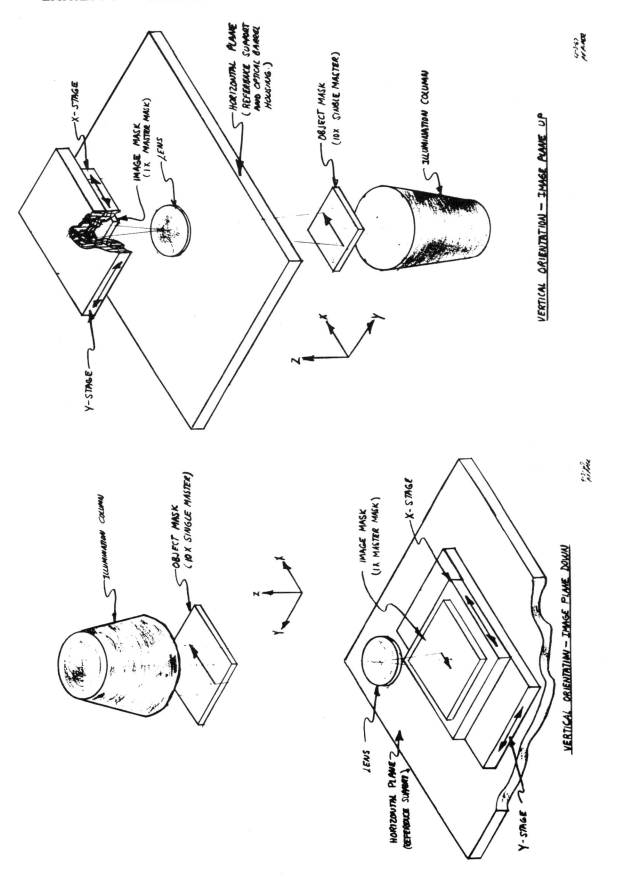

plane. The image plane down configuration was standard for most commercially available microphoto-repeaters.

The image plane up configuration was chosen as the best. It promised the most compact construction of the three, particularly for the short focal length lens. The geometry was simple. This configuration requires the emulsion on the image plate to be facing downward, therefore, it is potentially dust free. The image cassette being on the top provides a convenient working height. Light shielding could be provided with the image plane up configuration without the use of bellows.

On the basis of the above a preliminary schematic was made (Exhibit 10). The combined optical barrel and object holder was to be attached to a granite reference block which is isolated from the main frame. The image cassette would be attached to the x–y stage. This stage is to give controlled motion in two exactly orthogonal directions parallel to the horizontal plane. The first reduction masks on 2 in × 2 in glass plates act as objects and were to be located securely in a drawer at the bottom of the optical barrel.

The lens was focused by moving the lens in the vertical direction. Gimbal mounting about two axes would permit alignment with respect to the x–y plane. The light source was to be two rows of mercury vapor lights. A diffuser plate was to provide a uniform light intensity with variation of less than 5% over the entire object plane, eliminating the need for a condenser system.

Consideration was given to controlling the exposure time by simply switching the light source off and on. It would have been effective and economical, but this would have limited all barrels to the same exposure. Anticipating that different exposures may be desired at each barrel because of different object film density, a spring-loaded solenoid operated leaf shutter was included for each barrel. To detect faulty operation a photocell was to be included on each shutter triggering an optical and audible alarm.

The function of the step-and-repeat camera is to produce a master mask which consists of many small images of a 'first reduction mask', arranged in a regular pattern. The function of the four barrel step-and-repeat camera is to produce four master masks, all with the same size reduction and the same pattern. Since the images are to be superimposed in production it is important that the relative location of the repeated image be identical from mask to mask. Two types of errors are possible in the stepping process. Reproducible error which is constant, is dependent on the geometry and construction of the stepping mechanism. The reduction of the magnitude of this error results in increased cost. Reproducible error will be the same from mask to mask and does not have an adverse effect on manufacture of masks but may affect later processing of circuits, such as cutting and testing. Random error varies with each step. Its presence can be attributed to operator skill, system backlash, etc. The use of four barrels eliminates the effect of random error between a set of four masks which are made concurrently. The x–y stage was to be driven by servo motors manually controlled. Two drive rates would be required: gross motions from step to step and fine motions for exact location. A positioning grid was to be attached to the fixed stage and a reticule microscope to the moving stage.

The four glass plates which shall become master masks would be housed in a cassette which has metal shields to protect the masks from light during transport.

To operate the system the cassette is loaded on top of the x–y stage in the square opening. The cassette shield is retracted. Initially the x–y stage is positioned over the reference point using the reticule microscope. Depending on the direction of motion required, the corresponding step motor is operated. The coarse motion is accomplished automatically by the electronic control. The exact location is then set by viewing the positioning grid through the microscope and manually controlling the servo motor at the fine translation rate. The ability of the operator to match the reticule to the positioning grid is the main source of random error. After exact location the x–y stage is locked to the reference block. The shutters are then operated

EXHIBIT 10   Schematic of the final design

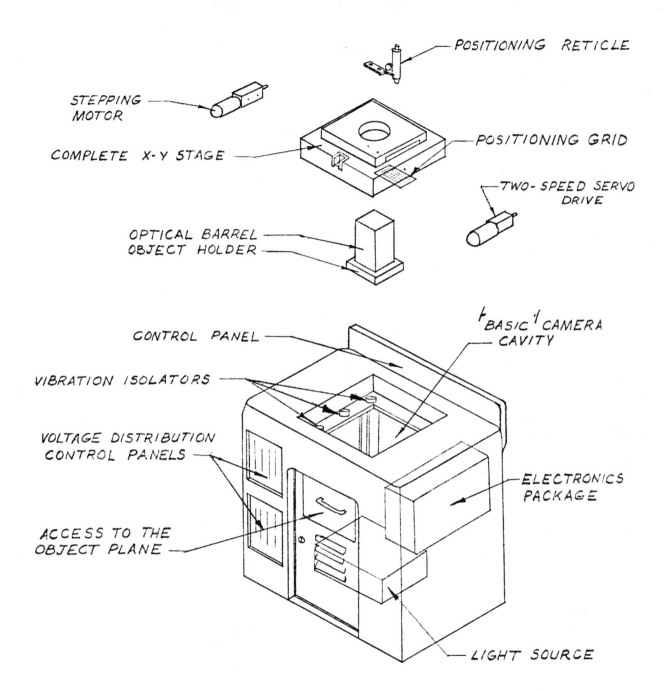

POSITIONING RETICLE

STEPPING MOTOR

COMPLETE X-Y STAGE

POSITIONING GRID

TWO-SPEED SERVO DRIVE

OPTICAL BARREL OBJECT HOLDER

BASIC CAMERA CAVITY

CONTROL PANEL

VIBRATION ISOLATORS

VOLTAGE DISTRIBUTION CONTROL PANELS

ELECTRONICS PACKAGE

ACCESS TO THE OBJECT PLANE

LIGHT SOURCE

exposing images from the first reduction masks on to the four production masks. The stepping and exposing operation is repeated as many as 100 to 1 000 times to complete one set of production masks.

The basic camera had to be a rigid unit isolated from the building and other external vibration. Three sets of isolators were to be used, each designed to eliminate a frequency range. The set nearest the camera would eliminate the high frequencies and the set farthest from the camera would eliminate the low frequencies.

Temperature effects were to be eliminated by maintaining a constant temperature of 68° in the critical regions. The heat generated by the electronic equipment and lights was removed by separate forced-air systems, thermostatically controlled. The region near the cassette must be maintained at $68°F \pm 1°F$. To facilitate temperature control it was decided that the unit would be housed in a temperature and humidity controlled room.

The camera was to be built into an integral frame housing the camera, its necessary electronic package, and the power supplies. The electronics package was to operate the camera, the position and exposure controls, as well as the interlocking safety and alarm systems. Piper felt that 'A device that is inherently difficult to operate is nearly useless regardless of the care which has been taken in designing the major components.' The operating controls and their location would therefore be selected with human factors in mind so that 'natural operation' is achieved. Although not incorporated in the present design, consideration will be given to future incorporation of an automatic positioning system. The present design would incorporate a fully automatic 'origin return'.

Several useful indicating systems would have to be incorporated. An audible and visual alarm for camera malfunction was desirable. An x–y position indicator relative to the origin, an exposure meter, and lights indicating sequence operation were desired. Malfunction interlocks would prevent double or false imaging.

The layouts were finished and outline drawings completed 16 June. Many of the working drawings had also been prepared. The design was then submitted to management for approval. It was rejected because the unit was too large. The exact reasons for this rejection were never explicitly 'pinned down' but Piper felt that the Friden management expected a more compact camera because of their familiarity with commercially available units. Such units were advertised as being 24 in high by 18 in square. However, these units did not contain any of the electronics, controls, or power supplies which are purchased as separate assemblies. Piper felt that there was prejudice against his design because it was self-contained.

As a consequence of the rejection the specification was revised on 19 June (Exhibit 4). The new specification reflected the desire for a smaller unit by reducing the required mask size by one-half. The deadline for the completion of the new design was September. Roy Nakai was given the task of designing the x–y guide mechanism. The accuracy of the x–y stage was considered critical to the success of the unit.

At that time the Research Division hired its first model-shop man. This indicated their commitment to manufacture their own hardware.

Six Schneider Xenon lenses were received on 18 July. They were all accepted. Piper now had ten satisfactory lenses. The discovery of these 'better than average' lenses could be called a lucky chance, since other lenses made by Schneider Xenon did not meet the specifications. The evaluation and resolution matching of the Schneider Xenon lenses was completed in August. The four lenses with resolution and distortion patterns as nearly equal as possible were chosen for use in the camera.

To document this important selection and testing procedure Piper and Rowland who had helped perform the tests prepared a formal report. This turned out to be the first formal report documenting results of tests in Friden Research Division. The report was completed and submitted 9 October.

Meanwhile the preliminary redesign for smaller physical size was completed 25

EXHIBIT 11   Conventional *x–y* **positioning stage**

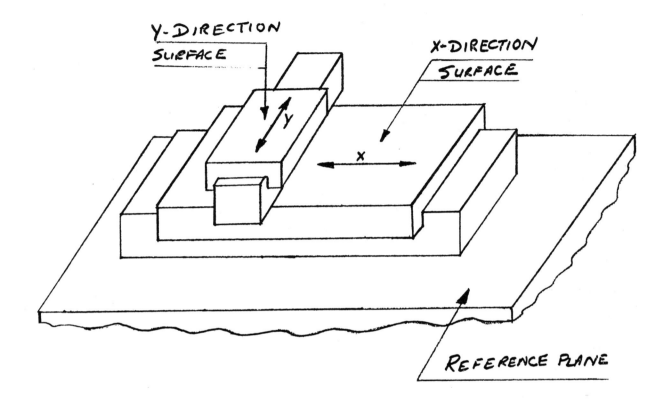

July. Although the maximum mask size had been reduced by one half, the overall physical dimensions had changed only slightly. This was due to the fact that the electrical and electronic components had not changed. In spite of this the design was approved by management.

In early August the fabrication of parts was begun and by September the design had proceeded sufficiently that a model-shop man was assigned to work exclusively on Project 355 for Piper. By 1 October all working drawings were finished and production was in full swing.

Nakai's design for the x–y stage was a conventional one (Exhibit 11) and Piper did not feel it would fulfil the accuracy requirements. This led to a conflict within the design group.

Nakai was finally removed from Project 355 on 11 October and Piper took over responsibility for the design himself.

Piper was faced with two alternatives to achieve the desired straightness and orthogonality. He could modify the existing and accepted design by using extreme tolerances or devise an entirely different, inherently more accurate approach. The conventional mechanism requires that the motions be perpendicular to one another and, at the same time, that the planes of their motion be exactly parallel. This imposes critical precision and alignment problems during manufacture and assembly. The conventional design could only be made to work at great expense. Therefore, Piper elected to devise another approach.

By taking a fresh approach Piper devised a kinematically different mechanism which separated the elements controlling the parallel plane motion from the elements controlling orthogonality. The use of air bearings gave smooth free movement that locked simply by turning off the air. By the first week in November the design of the x–y stage was completed and Do-All Science Center Inc. was given the contract to produce it.

Piper now turned his attention to the x–y stage drive. This mechanism must be capable of imparting incremental linear motion with precision commensurate with the positioning tolerance of the image plane and the travel time for gross steps must be reasonable. He concluded that the mechanism must incorporate both a coarse rapid travel and a fine positioning motion. One conventional system using stepping motors (Exhibit 12) produces rapid travel by speeding up the stepping rate for rapid travel. This would require additional costly and complicated electronic controls. An alternate system produces rapid travel by over-driving with a separate motor for high speed and uses the stepping drive for fine positioning. The high speed required in these mechanical components leads to excessive wear and loss of precision.

After examining many alternatives Harold Wells came up with an ingenious design early in December. He used a stepping motor and an electromagnetic clutch-brake and by applying a differential drive he was able to achieve the following performance:

Steps per revolution of motor	160
Linear output per step, coarse mode	0·0005 in
Linear output per step, fine mode	0·0000028 in
Repeatability	± 0·000020 in

The mechanism was directly coupled to the drive motor and the entire unit packaged into a space 4 in by $2\frac{1}{2}$ in diameter.

Piper accepted the completed x–y stage at Do-All on 19 January, and personally carried it back to the Friden Research Center (Exhibit 13). The components were made of granite with a surface flatness better than 20 micro-inches for the air bearings. All the plates had a central square cut-out to accommodate the optical barrels.

In January 1968 Piper had to revise his cost estimates upward a second time. He now estimated that material cost would be $19 700 and labour a total of 130 man weeks. Piper was informed at this time that the firm intended to design and build a

EXHIBIT 12   Conventional linear servo-drive system

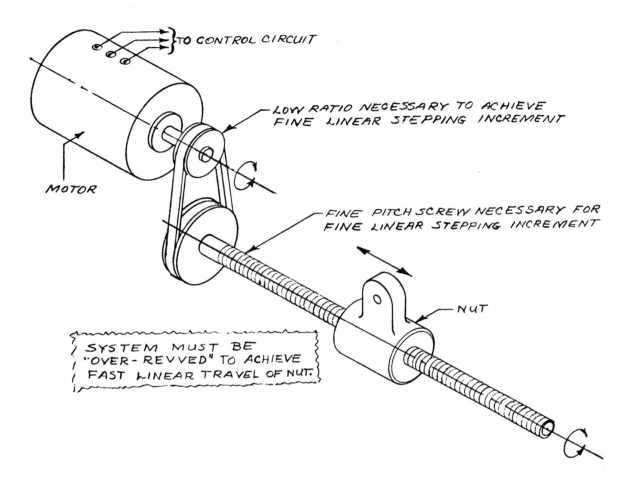

TO CONTROL CIRCUIT

LOW RATIO NECESSARY TO ACHIEVE FINE LINEAR STEPPING INCREMENT

MOTOR

FINE PITCH SCREW NECESSARY FOR FINE LINEAR STEPPING INCREMENT

NUT

SYSTEM MUST BE "OVER-REVVED" TO ACHIEVE FAST LINEAR TRAVEL OF NUT.

second generation of the step-and-repeat camera to capitalize on their experience with the first. The new camera was to incorporate automatic positioning.

Epilogue

Piper found that the transition from working drawings to finished hardware presented many problems he could not have foreseen. They not only affected the completion date but left their mark on the success and efficiency of the project.

On 2 March 1968 Friden Research moved from Oakland to a new building in Palo Alto. The dust associated with the new building delayed the assembly and alignment of the camera. The change in locality meant an almost complete change of machine shop personnel. This break in continuity essentially halted production of machined parts.

The movers dropped the assembled x–y stage onto a concrete floor during the move to Palo Alto. It was not damaged but it had to be checked again. During the installation of a mounting bolt into the main base of the x–y stage, a piece was chipped off the bearing surface. It had to be carefully epoxied back into place and the surface hand-honed to its original flatness. Problems in the positioning control logic were encountered. Although not entirely expected, Piper felt that it was not unusual with new electronic circuitry. Due to these and other problems the total manpower consumption jumped to 151 man weeks by 17 May 1968, 20 man weeks greater than the last estimate.

By the end of June the camera was sufficiently aligned to allow preliminary tests. The tests showed that parts of the basic camera design and construction were inadequate. The object holder repeatability was unsatisfactory and an error had been made in fabricating the lens barrels. The image distance was 2 in longer than specified. These required a major rework. After considering all the alternatives it was decided to convert to individual pulse Xenon light sources with a condenser and exposure control for each barrel. The object carrier was eliminated in favour of a simple slide holder for each barrel.

With the modifications and corrections underway, Piper looked forward to the completion of the project and contemplated the design of the second generation camera. He felt that he was in a position to do a much better job on the new camera. The intelligence necessary to design the automatic stepping feature would be a direct result of experience gained from designing rather than purchasing this first camera. He felt this vindicated management's decision to build rather than purchase the first camera. A great deal of knowledge and experience had been gained in the field of optics and the requirements and limitations of the reducing process. Such valuable information could not have been gained if a camera had been simply purchased. Piper felt that when Friden went into production on microcircuits the necessary microphoto-repeater would be purchased but the experience gained with their own design would be invaluable in selecting the most suitable unit to give the best automated facility.

EXHIBIT 13   **The** *x–y* **stage**

EXHIBIT 14   **The cassette**

EXHIBIT 15   The lens-object holder

EXHIBIT 16   Shutter mechanism

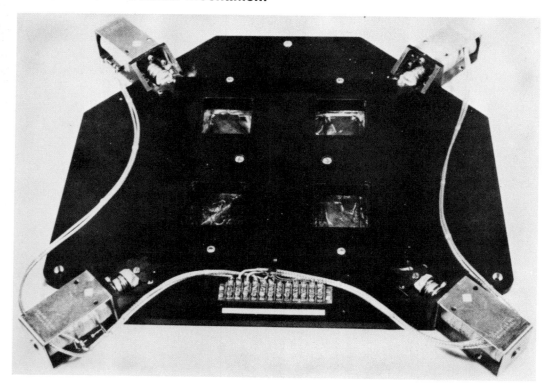

EXHIBIT 17   Finished camera

SUGGESTED STUDY

This case records the events in the design and production of a camera for making microcircuit masks. It can be used to show the student how management decisions (make or buy), costs, luck (Schneider Xenon lens), interpersonal relations (Nakai), and extraneous events (moving) can affect a project. In reading the case it can be seen that while technical excellence was mandatory, many of the critical decisions were governed by other considerations.

Class discussions can be centred around:

1. The merit of the make–buy decision.
2. Management's rejection of the expensive lens.
3. The need for formal reports in a company.
4. Could the unexpected problems have been anticipated and avoided? How?
5. What are the sources of errors in the photo-reproduction process?
6. Who should be charged with the increase in cost?
7. Was the building of the camera as useful as indicated?
8. Schedule for the next generation camera.
9. On the basis of his performance as a project engineer, how would you rate Mr Paul Piper?

The detail solutions in the design have been deliberately omitted from the case. The design of a suitable x–y stage, the design of a drive mechanism or the location and selection of vibration isolators can be assigned as student design problems or projects.

Case 5   The Hydro-Constant Pump

by Albert Klain, Roy Ikegami, and T. Sandukas

Prepared in partial fulfilment of their requirements for the Master of Engineering in Mechanical Engineering at the University of California, Berkeley. It has been edited and shortened by R. F. Steidel, Jr. We are indebted to Mr Arthur McGee and Mr Philip Bunnelle of the Central Engineering Laboratory of the FMC Corporation.

This case study has been prepared and written with the aid of a grant from the Ford Foundation for the demonstration and development of a new Master of Engineering program in design.

The problem

Successful marketing of a product can many times generate its own problems. The 'Hydro-Constant' pump is a perfect example of this, and in the summer of 1960, Mr Gene Sickert, Manager of Indianapolis plant of the Peerless Pump Company had the problems to prove it. The Hydro-Constant pump, a system designed to provide constant pressure flow for variable demand by varying pump impeller speed, had been successfully marketed for $2\frac{1}{2}$ years in the Chicago area. Since the market was beginning to extend nationally, Mr Sickert was faced with redesigning the control mechanism to eliminate certain features which were tolerated while sales were local, but would impede national sales.

The Chicago Housing Authority, in 1956, built a number of low rent high-rise apartment buildings and needed a constant pressure water supply which would boost the local water system. The Housing Authority was dissatisfied with the several systems used in high-rise buildings to maintain sufficient water pressure on the upper floors. The principal system in use at the time was hydro-pneumatic in nature, consisting of a tank filled with air and water, into which water was pumped. The air was compressed in the process, providing the pressure to maintain sufficient water service on upper floors. Intermittently, a water pump supplied water to repressurize the tank. This system had several drawbacks. It required a large tank, which in a commercial building is costly in terms of space. Since the air was under pressure, it tended to dissolve in the water. Corrosion and mineral deposits necessitated frequent maintenance.

As the Housing Authority investigated other means of boosting pressure, one system appeared promising. It used a variable slip coupling incorporated in a centrifugal pump drive to vary the pump speed. In 1957, the Housing Authority contacted the Peerless Pump Company with a proposal to build this system.

Peerless Pump Company and the Hydro-Constant pump

The Peerless Pump Company is part of the FMC Corporation, an international organization with annual sales approaching nearly one billion dollars in the areas of machinery, chemicals, defense, fibers, and film (rayon, polypropylene, etc.). As a leading pump manufacturer, producing vertical turbine pumps, process pumps, horizontal split-case pumps as well as other special application pumps, Peerless Pump has annual sales in the order of 30 million dollars. While it is true that individual companies compete quite successfully in particular areas, the Peerless Pump Company probably controls more of the total market than any other company.

Peerless engineers designed a system using standard water pumps, electric drive motors, and an American Standard 'Gyrol' variable fill fluid coupling. The American Standard unit consisted of a fluid coupling, scoop tube and a constant-speed charging pump mounted in a housing. The charging pump continuously put oil into the coupling, while a scoop tube which entered through the end of the coupling, skimmed off oil in the coupling to a constant level. By raising and lowering the scoop tube, the level of fill in the coupling was changed, controlling the output shaft speed.

Besides mating these three elements, the engineers designed a relatively simple means for automatic operation. A small electric gear motor was mounted on top of the American Standard housing and was connected to the scoop tube through a bell crank and drag link. The motor was operated by a pressure switch at the pump discharge. If the discharge pressure was outside the operating zone, the pressure switch would energize the motor, moving the scoop tube to adjust the amount of slip. As soon as the feedback loop sensed that discharge pressure had returned to the operating zone, the pressure switch stopped the motor. This operation was slow, since the motor took nearly a minute to complete a ninety degree revolution.

The Hydro-Constant pump, as it was called, consisted of an electric motor and coupling made by American Standard, a control center designed and furnished by the Lexington Co., and the pump and base for the unit, the latter two being the only parts actually manufactured by Peerless Pump. Barber-Coleman supplied the pressure control switch, damper motor, and linkage to the Lexington Co. for the control center. This resulted in a ratio of $1 worth of Peerless manufactured parts for every $4-5 of items purchased from others, which posed a serious problem in trying to improve the profit margin for higher sales volume.

The configuration of the scoop tube linkage presented another big problem area. It was necessary to ship the linkage from the factory disassembled and due to the many adjustments which had to be made to meet the particular operating conditions, it was necessary for a Peerless engineer to assemble the pump at the installation site. Three points of adjustment were necessary to set the range of tube stroke, plus the setting of limit switches to control the extremes of damper motor rotation. None of these could be made at the factory without losing the set adjustments when the pump was dismantled for shipment.

Difficulty in making adjustments for changing operation conditions created another problem. When the overall building demand changed, or a different pressure setting was desired, a Peerless engineer or service man would have to make the necessary set point changes. As long as all the installations were in the Chicago area, near the headquarters of the Peerless Pump Company in Indianapolis, sending an engineer out to each site was not a serious problem. However, this could not be tolerated if sales were nation-wide. Another obstacle to increased sales was the general makeshift appearance of the unit, which ran counter to the company's image of producing only quality products.

The Central Engineering Laboratories and the decision to repackage

The Central Engineering Laboratories is within the corporate office of FMC and provides research and development services on a contract basis to other divisions of FMC as well as outside customers. While each division has its own engineering group, the Central Engineering Laboratories normally undertakes jobs which are beyond division capability, such as problems in control, electronics, industrial design and systems studies. Since the divisions are under no obligation to have the Laboratories do the work, it is up to the Laboratories to solicit work from them, in many cases.

In the summer of 1960, Mr Frank Sako was visiting FMC's Machinery Division, of which Peerless Pump Company is a part. His job was that of liaison for Central Engineering to search out jobs from the divisions which the Laboratories might handle. While visiting Peerless, he learned of the Hydro-Constant pump problem.

As Mr Sickert studied the problem, he began to feel that outside help might be needed. The design and construction of pumps has its own highly specialized technology, and his engineering staff was geared to meet this need. Mr Sako suggested that Central Engineering was well qualified to handle the job because of their broad range of capacities and would be glad to undertake the redesign, especially if it meant a continuing development program on the part of the Laboratories.

The first direct communication with Central Engineering was by way of a letter from the Peerless Pump Company, on 28 June 1960, written to Mr Oldenkamp, Manager of Central Engineering. The heart of the letter was a request that '... CEL look into the possibility of designing a unitized control arrangement for the Hydro-Constant'.

Following another visit to the Peerless plant, Mr Sako brought back to Central

EXHIBIT 1   Gyrol fluid drive

Components and Operation

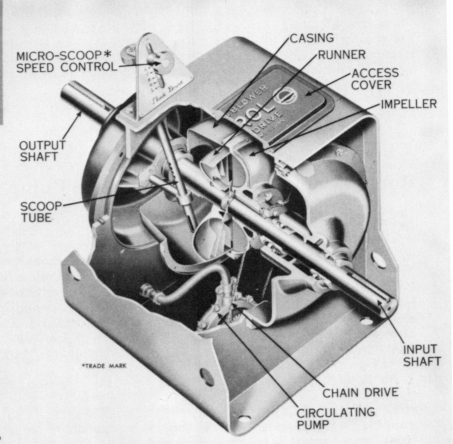

MICRO-SCOOP* SPEED CONTROL

OUTPUT SHAFT

SCOOP TUBE

CASING
RUNNER
ACCESS COVER
IMPELLER

INPUT SHAFT

CHAIN DRIVE
CIRCULATING PUMP

*TRADE MARK

MICRO-SCOOP* SPEED CONTROL
For Manual Operation

LEVER SPEED CONTROL
For Automatic Controller Operation

Components

The Type VS Class 2 Gýrol Fluid Drive, described in this bulletin, is available in four sizes—1 thru 25 HP—speeds to 3600 RPM. It is an enclosed, self contained unit.

Housing—A fabricated steel housing acts as an oil reservoir as well as an enclosure for the rotors, bearings, oil circulating pump, and dual tip scoop tube.

Rotors—The impeller and runner are of precision die cast aluminum construction with the rotating casing of formed steel construction.

Shafts and Bearings—The input and output shafts are each carried by two anti-friction ball bearings mounted in an end bell casting. Both assemblies are factory aligned as a package requiring no field adjustment. The bearings are grease lubricated through two alemite fittings, one located on each end bell. This Gýrol Fluid Drive is designed to take care of its internal thrust only.

Pump—A circulating turbine-type oil pump, driven from the input shaft, supplies the necessary volume of oil for the working circuit regardless of the output speed and direction of rotation.

Scoop Tube—A dual tip scoop tube, located between the runner and casing, provides stepless speed control for either direction of rotation.

Control—A Micro-Scoop or a Lever Speed Control is furnished, thus making the Class 2 unit easily adaptable to accurate manual or automatic operation.

Coolers—Coolers may or may not be required on the 1 thru 25 HP units, depending on the size and type of load. When a cooler is required either an air-oil cooler or a water-oil cooler is furnished.

Operation

There is no mechanical connection between the input and the output members in a Gýrol Fluid Drive. Instead power is transmitted smoothly and shocklessly from the impeller to the runner by a vortex of oil. By regulating the amount of oil in this vortex, the speed can be adjusted.

The flow of oil in the Gyrol Fluid Drive is begun by the circulating pump, which is driven at constant speed by the input shaft. It pumps the oil from the reservoir at the bottom of the housing to either first an external oil cooler, if used, and then to the rotating casing or else directly to the rotating casing.

Oil entering the rotating casing is acted upon by centrifugal force caused by the casing rotating at the input speed. This centrifugal force throws the oil outward against the side of the casing and into the impeller and runner, or working circuit, where it takes the form of an annular ring.

The amount of oil in the working circuit is regulated by the scoop tube. The scoop tube removes the oil from the casing and empties it into the oil reservoir at the bottom of the housing where it is ready to begin the cycle once more.

By using either the Micro-Scoop Speed Control or the Lever Speed Control the scoop tube is raised or lowered in the casing. This in turn, sets the level of the oil in the working circuit since the oil tends to seek the same level in the entire assembly. The scoop tube is designed to give fast response for both increase or decrease of output speed as required. This method of speed control also means that a smooth, stepless speed regulation over a wide range is possible for either a constant or variable torque type of load.

Engineering an outline of the Peerless proposal of work to be undertaken by the Laboratories. The objective was to be a 'repackaging' of the system with consideration given to production, installation, service, and user operation. Mr Sickert enlarged upon these points in a letter to Central Engineering on 2 September 1960. Specifically, it was desired that the control system be redesigned to overcome its several limitations – field assembly of linkage, difficulty in changing set point, costly field service and make-shift appearance. It was desirable for Central Engineering to design a way to vary the linkage by adjusting a single dial, which was patentable. In addition, he requested that the study and pilot design be finished and available for production by March 1961.

Repackaging the Hydro-Constant Pump   PART 2

The proposal to repackage

When it became apparent that the Peerless Pump Company would give the job of 'repackaging' the Hydro-Constant Pumps to Central Engineering Laboratories, Mr Philip Bunnelle was assigned as project engineer. He was to be solely responsible for any design work done on the part of Central Engineering. However, at this early stage of the program, Mr Bunnelle found himself working closely with his immediate supervisor, Stewart Thompson, Manager of the Laboratories Mechanical Design group, and William Adams, Assistant Manager of the Laboratories, because of their great interest in the project.

Mr Bunnelle received his education from the University of California, having received his Bachelor of Science degree in Mechanical Engineering at the Berkeley campus in 1949, followed by a Master of Science degree in agricultural engineering from the Davis campus the following year. After doing research and teaching at Davis for three years, he worked for the John Deere Co. at their tractor plant in Waterloo, Iowa. In 1957, he joined the Central Engineering Laboratories. Before being named project manager, he had done work on various food processing equipment.

Now that the scope of the project had been established, Mr Bunnelle had to develop the proposal and request for appropriation to be sent to Peerless Pump. First, he expanded the problem statement into a series of definite operations, and from an estimate of man hours associated with each operation, a cost figure was determined. Being realistic, Mr Bunnelle knew he could not anticipate every problem area, nor could he know exactly how many hours would be needed for the steps he had listed. To cover these unknowns, he added a sum to the preliminary cost figure. The final figure of $10 000 ($1 000 for engineering analysis, $2 000 for development exploration, $7 000 for development model design) was then approved by Mr Thompson. Peerless Pump was to be billed for the work at standard rates, plus all material costs, on the same basis as if it were an outside company. By 8 September, the proposal of work to be done, along with the appropriation request, had been submitted to Peerless for their approval.

The proposal for engineering work contained a program with four points in which Central Engineering would:

1. Engineer a standardized control package which is easy to adjust and finished looking. It would be adaptable to both electric and hydraulic control.
2. Design any other components so that the system could be assembled and pre-adjusted at the factory.

3. Supply curves and instructions for setting it up and adjusting it.*
4. Do related work as the need arose including liaison with World Motor concerning their Dynamotor drive.

Prototype drawings were to be finished by the first of May 1961, while pilot models were to be in the field by spring 1961.

Reevaluation

In a memo of 23 September, Mr Adams made the announcement that Peerless Pump had not accepted the $10 000 request. When asked why Peerless had turned it down, an engineer at Central Engineering said, 'The way Peerless R and D work is financed, a certain amount is appropriated each year for all new product development, and there are always more projects than money. They were taking money out of their own development funds for CEL to do the work. So they were trying to whittle the cost down to save money for other projects. We came in first with a request which we thought was needed for a complete job.'

The decision was made by Mr Sickert. His feeling at this time was that an all-electric drive, such as that proposed by the World Motor Co., using their Dynamotor motor, would give better performance and be less costly. Also, the top management of Peerless Pump did not feel that the Hydro-Constant sales potential was sufficient to warrant a $10 000 development program. However, Mr Sickert did agree that Central Engineering could do the job of 'repackaging' the Hydro-Constant if the cost could be held to $3 000 to $4 000. In this manner, if both systems proved to be successful, Peerless Pump Company would have two systems to offer in the variable speed pump market.

Peerless Pump had become quite interested in World Motor's Dynamotor as a means of varying pump shaft speed. This drive was a high slip induction motor powered through magnetic amplifiers. The current to the control unit could be varied by a slidewise potentiometer in a pressure transducer. The entire control system was simple and consisted entirely of solid state elements.

In late August 1960, Mr Sickert offered his company's help to World Motors in developing a constant pressure, variable demand pump system using the Dynamotor drive, in exchange for an agreement giving Peerless Pump exclusive distribution of it for two years. The answer from World Motors was negative. However, they were willing to give Peerless Pump the right to put a Peerless nameplate on all the equipment in the system for a period of time.

During the beginning of October, Mr Bunnelle had an occasion to discuss the Dynamotor drive with a World Motor's representative. His first impression from this meeting was that it appeared well suited for this type of application.

The idea of using this drive was later dropped. The maximum speed of the Dynamotor was low compared with that of a standard electric motor. As a result, Dynamotor would require a larger pump for high pressure-flow installations compared to that required by the higher speed Hydro-Constant. The controls for the Dynamotor would have to be bought from World Motors while those for the Hydro-Constant could be made by Peerless Pump.

For other reasons, the decision not to use the Dynamotor proved to be correct. World Motors, in trying to convince Peerless of the merits of their drive, had whittled down their price until it was barely competitive with the drive using American Standard's coupling. However, when Peerless finally started to build their own drive, they were able to do so for considerably less. Even more important, World Motors was later forced to withdraw the Dynamotor from the market because of problems which had developed with the drive.

There was a feeling on the part of the management of Central Engineering that it was not worthwhile to undertake the redesign project if only $4 000 were involved. It takes too much time and money just to get started – learning new technology and

* Disguised name.

determining the requirements of the design – and if Central Engineering's interest in the field was to end with redesign, it simply was not worth the involvement. However, Mr Bunnelle, Mr Adams and several of the other Central Engineering people looked upon this redesign as a chance for the Laboratories to enter into the variable pumping field. Their problem was to justify doing $4 000 worth of 'repackaging' by showing that this would lead to continued work in the field.

During the early years of the Hydro-Constant, there had been one or two installations in which electrical controls were not tolerated. These had been for small town water systems, where the water supply was furnished by perhaps one well. A power failure, such as might be caused by a fire, would disrupt the flow, so a gasoline engine was used to drive the coupling. To eliminate the scoop tube actuating motor, an all hydraulic control system was developed. Basically, it consisted of a diaphragm actuator connected to the scoop tube. The actuator was connected to the pump discharge, and was actuated directly by the discharge water pressure. A pressure regulator and bleed orifice were used to regulate the range of tube travel. This system worked reasonably well, except the orifice tended to become fouled with algae and mineral deposits.

From this crude system, the concept of an all hydraulic control system was born. Mr Bunnelle did not know how the design might ultimately develop, but he felt sure that it would greatly improve the present electromechanical controls. Just as important, it would justify proceeding with the present redesign by offering the prospect of Central Engineering being asked to develop an all hydraulic system.

Peerless Pump was receptive to the idea of supporting a continuing development programme on the part of Central Engineering towards an all hydraulic system. Management of Central Engineering was satisfied that they would have a continuing interest in the field, and gave the go-ahead for resubmitting a proposal for 'repackaging'.

It was now up to Mr Bunnelle to prepare an appropriation request for less than half the amount of the original request. His approach to the problem was summed up when he said, 'If you can't get the kind of money to do the job you want, you settle for what you think you can get and hope you can do it for that amount. Or, you hope you can ask for more funds later, or get funds from somewhere else to finish the job.' Mr Sako commented, 'You say it takes $10 000 to do the job and they say all they have is $4 000. So you take it and do as much as you can. You hope by the time you spend $2 000, it looks so good that they will come up with the balance. It's the getting started that is hard.'

By 6 October 1960, Mr Bunnelle had prepared another appropriation request for $4 000, with $2 000 for development exploration and $2 000 for development model design. The stated objective was to repackage the drive to:

1. Place all speed control elements in the package.
2. Design the controls for partial or complete pre-setting at the factory with simple adjustments.
3. Have the controls perform their existing functions in conjunction with a wall-mounted control panel.

In the accompanying cover letter, Mr Oldenkamp reiterated the fact that Central Engineering expected to continue with the design of all hydraulic controls and planned to submit a proposal later. Within a week, approval had been received from Peerless Pump. However, Mr Bunnelle was requested to submit a more detailed outline of work (see Exhibit 2).

As he assessed the problem, there would be three areas of work involved in repackaging the Hydro-Constant. The first was synthesizing a suitable linkage to position the scoop tube. Second, a simple means of adjusting the controls was needed. Finally, a prototype would have to be assembled to assure proper performance. Before any of these steps could be undertaken, it was necessary to investigate the relationship between the system variables.

EXHIBIT 2   Statement of work on Hydro-Constant project

FOOD MACHINERY AND CHEMICAL CORPORATION

CENTRAL ENGINEERING LABORATORIES

1105 COLEMAN AVENUE - SAN JOSE - CALIFORNIA

DATE   October 31, 1960

FROM   P. R. Bunnelle

TO   Ray Volpatti, PP-I

cc: G. D. Sickert
       R. E. Kummer
       W. J. Adams, Jr.
       C. A. Romano
       W. S. Thompson

SUBJECT STATEMENT OF WORK ON HYDRO-CONSTANT PROJECT
         CE-9663

Dear Ray:

The purpose of this letter is to enlarge upon the statement of work given in our Project Appropriation Request dated October 6, 1960 which was recently submitted to Mr. Sickert.

In this phase of the work we will re-engineer the electro-mechanical control for Gyrol sizes 107 and smaller for simpler installation, fewer adjustments, better appearance, etc. We will place all speed controls in a package mounted directly on the Gyrol drive. We will design the control to permit partial or complete pre-setting at the factory with field adjustments, if any, being simple and direct. The control will be capable of performing all present functions of motor starting, stopping, etc. The only essential control component external to the package will be the 24 volt power supply, which can be mounted in the motor control enclosure. A prototype unit should be completed within two and a half months.

After this unit has been tested and approved by Peerless, we will adapt it for use on the larger drives, at least through the 100 HP size.

Finally, we will prepare installation and operating instructions for these units and submit a report summarizing the work done, along with the production drawings of the control.

P. R. Bunnelle

PRB:jc

System design

PART 3

System analysis

The head, H, which a centrifugal pump can develop is determined by its size, the flow rate, Q, and impeller speed.

Each pump manufacturer publishes curves relating these variables. A typical set of pump characteristic curves is shown in Exhibit 3. At two specified pump shaft speeds, the head-flowrate relationship is plotted for several impeller sizes. Lines of constant efficiency, e, indicate the ratio of water horse-power to pump shaft power input, P:

$$e = \frac{Q\gamma H}{\omega} \tag{1}$$

Given a specific impeller size, the relationship between head and flow can be plotted for any number of impeller speeds, using the similarity relationships:

$$Q \propto rpm_{impeller} \tag{2}$$

$$H \propto rpm^2_{impeller} \tag{3}$$

Thus, points on the originally given iso-speed curves can be transformed to points on curves for other speeds. The result is a family of iso-speed curves instead of the original two, and might appear as shown in Exhibit 4. Iso-efficiency lines can be drawn with the aid of points of known efficiency on the two original curves.

Returning to equation (1), and using $V = Q/A$, where A is the pipe area:

$$H = \left(\frac{P_2 - P_1}{\gamma}\right) + \frac{Q^2}{2g}\left(\frac{1}{A_2{}^2} - \frac{1}{A_1{}^2}\right) \tag{4}$$

The head, H, can be expressed as a function of the inlet and outlet pressure P_1 and P_2, the inlet and outlet area A_1 and A_2 would be constant and known. A plot of H versus flow rate Q can be made with $(P_2 - P_1)/\gamma$ a parameter. This plot is superimposed over the characteristic curve for the pump in Exhibit 4. This clearly shows why there is a need for a variable speed pumping system when constant pressure at variable flow is the requirement. Neglecting upstream pressure fluctuations, the requirement of constant discharge pressure means that the system should operate along a line such as A–B. However, to do this, it is obvious that pump speed must be variable.

The fluid drive has a one-to-one torque ratio. Since power is proportional to the product of torque and rpm, any loss of power in the drive can manifest itself only as a loss in rpm. Neglecting fixed losses such as bearing friction, the percentage of power lost is equal to the per cent slip, where per cent slip is defined as:

$$rmp_{in} - \frac{rpm_{out}}{rpm_{in}} \times 100 \tag{5}$$

The torque-speed characteristics of a drive are presented in terms of a family of K-per cent slip curves at constant levels of coupling fill (given in terms of the per cent of scoop tube travel). These curves, like pump characteristic curves, are experimentally determined from laboratory tests. K is defined as:

$$K = \frac{\omega}{(rpm/100)^3} \times D^5 \tag{6}$$

$$K = \frac{190 \cdot 5 \times torque}{rpm^2 \times D^5} \tag{7}$$

where hp and rpm are input to the drive, torque is expressed in lb-ft, and D is the profile diameter of the coupling in meters. American Standard uses the profile diameter to specify the coupling size; that is, the Gyrol size 107 has a profile diameter of 10·7 in. A typical set of K curves for American Standard's class 2VS units (including the 107) is shown in Exhibit 5.

Scoop tube position and flow rate

To have a better understanding of the design problem, Mr Bunnelle wanted to see what the steady state relationship was between the scoop tube position and the flow rate. He reasoned that some simplifying assumptions were justified, since the results were just to give him a rough picture of the problem. The last term in equation (4) was to be neglected (i.e., A_1 was equal to A_2, or nearly equal), which, meant that for constant pressure, he would want a constant pump head. In Exhibit 4 operation along a line C–D, determined by the pressure boost.

For these calculations, he used a Peerless 4AB-10 pump with a 10 in diameter impeller, and a Gyrol 107 coupling drive by a 1 800 rpm motor. The procedure was as follows:

1. The pump characteristic curve was plotted, showing several iso-speed lines. Starting with the two given lines at 1 450 rpm and 1 750 rpm, the similarity relationships were used to plot the others.
2. A desired head was chosen. This fixed the operating line on the pump characteristic curve. At points on the line, flow rate, rpm and efficiency were read off the graph. At point 1, for example, the torque to the pump shaft was calculated.
3. Proceeding to the K curve, K was calculated, knowing the coupling input rpm, D, and the torque as calculated above. The per cent slip was also known, given the pump rpm at point 1. Therefore the per cent of scoop tube travel could be estimated from the graph.
4. The flow rate found in step 1 was converted to per cent of discharge at full couple, and plotted versus the per cent tube travel found in step 2.
5. This was repeated (steps 1–3) for several points along the operating line.
6. Steps 1–4 were carried out for pump heads of 40 ft and 80 ft.

From the resulting tube position-flow curve shown in Exhibit 6, one fact was obvious. With the scoop tube initially in the full speed position (zero per cent travel), it could be moved 45% of its travel with less than a 10% change in pump discharge. In the next 25–35% of its travel, the discharge fell sharply to zero. In its steady state performance, the flow rate would be highly insensitive to tube position at high flow rates, and highly sensitive at low rates of flow.

Completing the design

Mr Bunnelle had decided to retain the constant-speed operator motor to position the tube, as had been done previously. When the pressure sensing device detected a change in discharge pressure, a signal was sent to the motor, which then began rotating at a constant rate in the proper sense to change the tube position. Without going into a detailed control analysis, it seemed apparent to him that for best system performance, the change in discharge should be linearly proportional to the rotation of the operator motor shaft, over the entire range. A complete analysis, including transient response and system feedback was not made, due to the limited time.

The problem then became one of finding a linkage arrangement that would transmit motion from the motor shaft to the tube in such a way as to compensate for the discharge-tube travel nonlinearity. This was done on a trial and error basis, using graphical methods to see how close the mechanism generated the desired function. His first thought was to use a drag link mechanism as was used on the original Hydro-Constant. Later, successive trials with a double bell crank mechanism produced the functional relationship between input and output which was reasonably close to that desired.

Simultaneously, Mr Bunnelle discussed the matter with a Peerless engineer who was familiar with the operation of the Hydro-Constant pump. He was particularly interested in knowing if Peerless Pump had encountered any control stability problems due to nonlinearity. He learned that they had only minor problems. Further, this was the first he had heard about the nonlinearity, and Peerless Pump had not tried to correct it in the original design. With this in mind, Mr Bunnelle took another long look at the problem. Perhaps there was no need going to extremes to linearize the shaft rotation–discharge relationship. It was not done during the original design and it apparently had no detrimental effect. His own assumptions were based only on a steady state open feedback loop analysis. Most important, by linearizing the control, he would be forced to use a more complicated linkage.

Relaxing his requirements, Mr Bunnelle found that a simple three-member linkage would perform well. It is shown in Exhibit 8. The range of tube travel could be set by simply adjusting the two nuts holding the pivot block. With the tube in the full-speed position (Exhibit 8), it would move a relatively long distance for each degree of arm rotation. As the arm rotated to the low-speed position, Exhibit 8, it became nearly aligned with the tube. Thus at the low-speed position, there was little tube movement per degree of arm rotation. Mr Bunnelle had thus obtained the compensation he needed to linearize the shaft rotation–discharge relationship (Exhibit 7).

In many installations, the pump was not needed during part of the time. For example, apartment buildings had such a low water demand at 2 a.m., that the water main pressure was sufficient to maintain pressure on the top floors. As discharge pressure rose, the tube would be moved downward to slow the pump and drop the pressure. However, if water main pressure rose sufficiently, or demand was small enough, the tube would be extended as far as it would go, resulting in maximum clutch slip (minimum pump speed). The pump would still be turning slowly even with maximum slip, but would be producing negligible pressure boost. So, it was desirable to shut the pump drive off when pressure could be maintained without using the pump, and also to stop the motor operator when the extremes of tube travel had been reached. The original system provided for this with limit switches located in the motor operator, which were quite difficult to adjust.

Mr Bunnelle noticed that changing the length of the operator arm would affect the tube position in the low-speed (declutched) region, but would have no effect at full speed (fully clutched). He used this fact to provide a simple means of setting the minimum pump speed at pump shutoff. Limit switches were built into the operator motor to shut it off at either ends of a 90-degree revolution. The range of tube travel was set by positioning the pivot block and operator arm. Any adjustment of the minimum pump speed was made just by turning one screw on the adjusting block. This changed the tube position at the low end pump shutoff, but had little effect on its position at high speed, a feature which later proved to be patentable. The adjustment was much simpler to make this way than by changing limit switch settings.

To control the motor operator, he used the same pressure switch as before. This was a three-position, diaphragm actuated switch, which sent no signal if the pressure was within a certain margin. If the margin was exceeded, it would energize the motor operator in the proper direction. The set point (the pressure at which no signal was sent) was adjustable with a knob.

With most of the design now accomplished, Mr Bunnelle sent drawings to the Central Engineering machine shop. There, the controls were fabricated, assembled and mounted on a drive unit supplied by Peerless Pump. He wanted to test the unit before turning it over to Peerless Pump to satisfy himself that it worked properly, but lacking a pump, he could only make sure that the scoop tube moved in and out without the linkage binding.

Officially, the project ended in the middle of December 1960, with his completion of technical report covering the results of the repackaging. The whole project had taken eight weeks, and of the original $4 000 appropriation, only $2 800 had been spent. This was an extremely small amount to spend for a project, even considering that only one engineer plus a draftsman were directly involved. The fact was that eight weeks did not represent the actual time spent on repackaging. Mr Bunnelle was spending part of his time on other projects. Also, as he explained it, 'While I was working on the Hydro-Constant, I was studying pumps, instrumentation and fluid couplings on the side, all in my own time. I had been working on food processing equipment for about two years and I wanted to get into a new area. I was quite interested in it and wanted to do the job, so I was willing to make a little extra effort.'

Peerless Pump had originally complained about the number of items which were bought from other companies, and yet this project did nothing to improve this situation. It had been decided by the Central Engineering people that this was not the immediate problem. Since the market for the Hydro-Constant had not yet been established nationally, the immediate problem was to design an improved product which could be used to test the market. The repackaging of the Hydro-Constant gave Peerless Pump an improved item with which to develop the market, and gave Central Engineering a basic understanding of the problem which could be applied to the design of an all hydraulic system.

EXHIBIT 3   Single stage double suction pumps

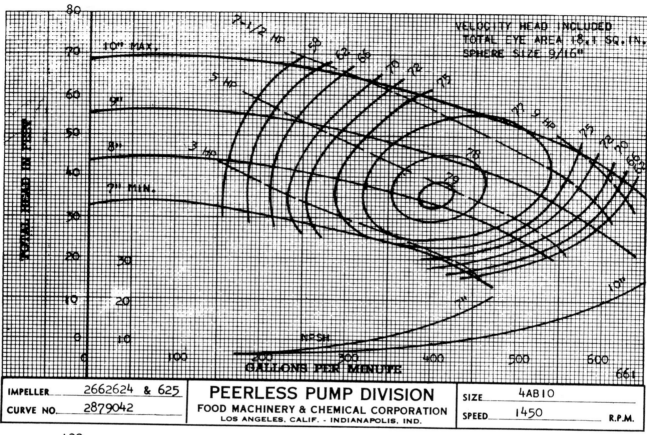

IMPELLER	2662624 & 625	**PEERLESS PUMP DIVISION**	SIZE	4AB10
CURVE NO.	2879042	FOOD MACHINERY & CHEMICAL CORPORATION LOS ANGELES, CALIF. - INDIANAPOLIS, IND.	SPEED	1450 R.P.M.

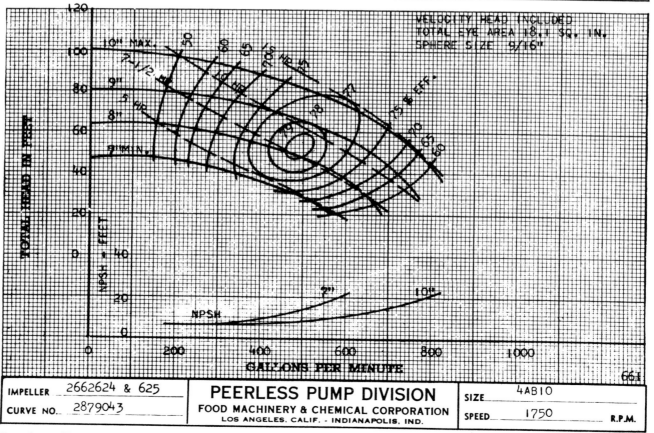

IMPELLER	2662624 & 625	**PEERLESS PUMP DIVISION**	SIZE	4AB10
CURVE NO.	2879043	FOOD MACHINERY & CHEMICAL CORPORATION LOS ANGELES, CALIF. - INDIANAPOLIS, IND.	SPEED	1750 R.P.M.

Peerless #4 AB-10 pump characteristic curves

EXHIBIT 4 **Pump characteristic curves with system head requirements superimposed**

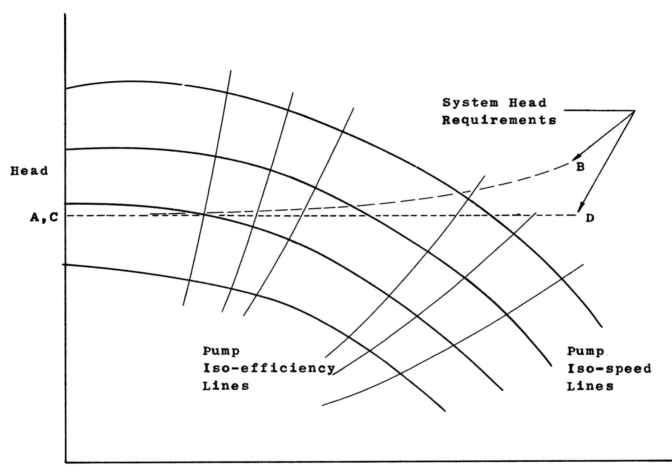

EXHIBIT 5   Coupling K curves

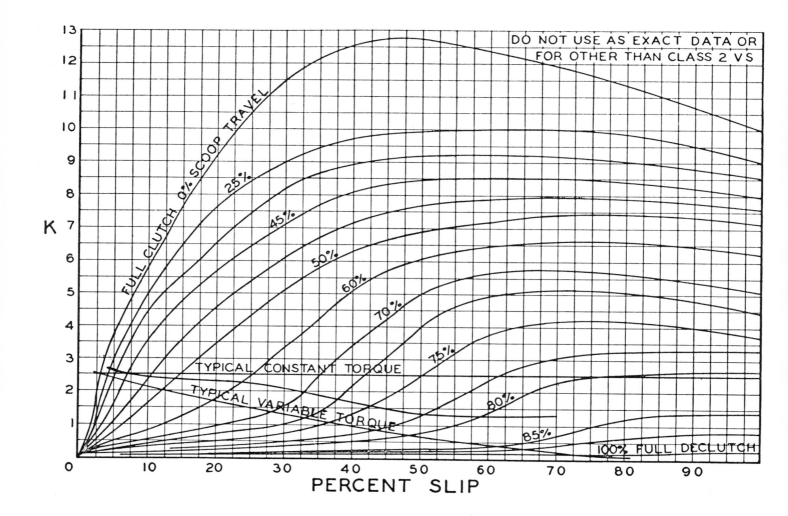

EXHIBIT 6 **Discharge rate versus scoop tube travel**

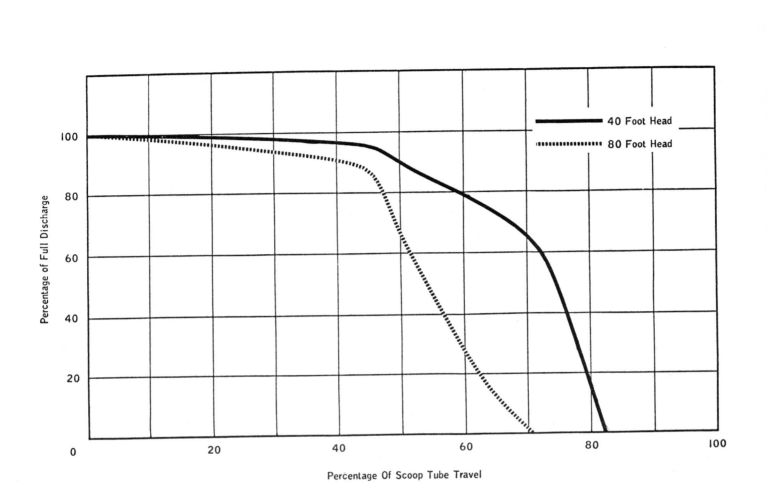

Calculated curves for #4 AB-10 pump and 107 gyrol drive

EXHIBIT 7   **Scoop tube travel versus operator arm travel**

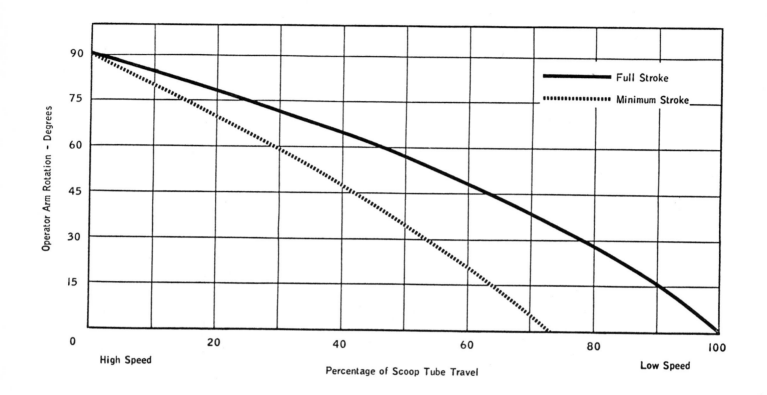

For control unit #operating #107 gyrol drive

EXHIBIT 8   Tube positioning mechanism, high-speed position

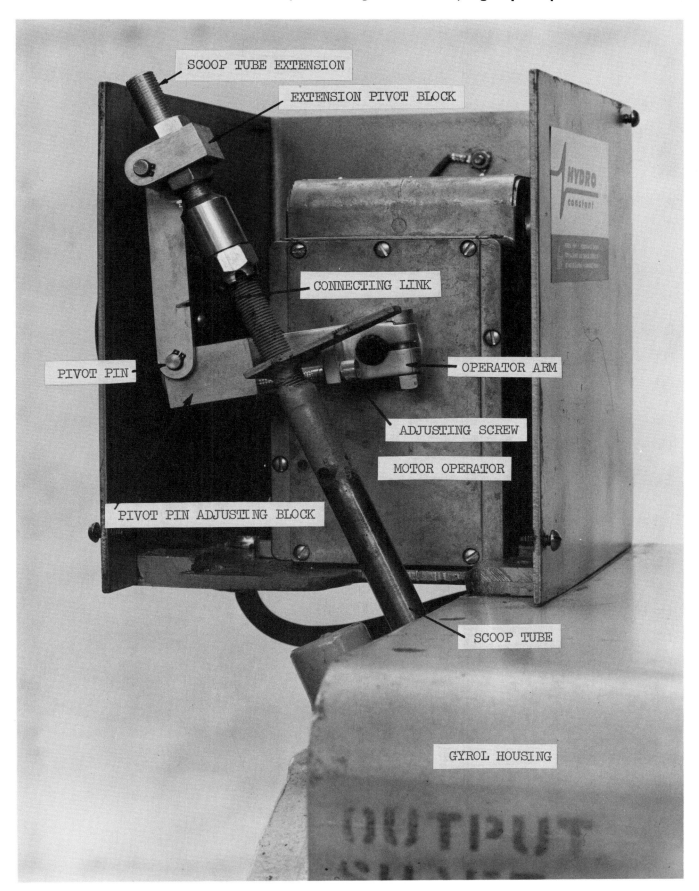

EXHIBIT 9   Tube positioning mechanism, low-speed position

EXHIBIT 10   **Hydro-Constant after repacking**

EXHIBIT 11   Redesign of Hydro-Constant control system

December 28, 1960

W. J. Adams, Jr.

G. D. Sickert

cc: W. S. Thompson
C. A. Romano
P. R. Rumulle

REDESIGN OF HYDRO-CONSTANT CONTROL SYSTEM
CE-9663

Dear Gene:

Enclosed is the summary report on the redesign of the electro-
mechanical controls for the Hydro-Constant System. The prototype unit has been shipped to Indianapolis, along with a set
of revised production drawings. The installation instructions
are included in the appendix of this report. This completes
the work specified in the Appropriation Request signed by
G. F. Twist on October 21.

This work was done at a cost of approximately $2,800. Rather
than close the project, it is recommended that the remaining
$1,200 left in this appropriation be applied to a feasibility
study of a hydraulically controlled drive for the Hydro-Constant
System. We will await your specific instructions, however,
before doing anything further.

Best wishes for the New Year.

W. J. Adams, Jr.

WJA:je

Encl. - Report No. 1471

Letter—Adams to Sickert, Dec. 28, 1960

EXHIBIT 12   **Hydro-Constant control system – GE9663 (i)**

January 9, 1961

W. J. Adams, Jr.

G. D. Sickert
Peerless, Indianapolis

cc: G. F. Twist, Los Angeles
H. A. Oldenkamp
W. S. Thompson
J. A. Abbott
F. Sako
P. Bunnelle

HYDRO-CONSTANT CONTROL
SYSTEM -- GE 9663

Dear Gene:

Following up on the second paragraph of my memorandum of December 28, with regard to continuing work on a feasibility study of a hydraulically-controlled drive for hydro-constant system, we would like to offer the following thoughts for your consideration:

1. We can envision a large potential market for hydro-constant systems to regulate pressure and also flow for many industries. Peerless is to be commended for their pioneering efforts in applying the hydro-constant principle to the building field. The problem of pressure and flow control is common to many industries, including power plants, process and chemical industries, waste disposal, etc.

2. We visualize an opportunity for Peerless to make a substantial reduction in cost of hydro-constant systems by engineering and manufacturing a "packaged" line of units. Instead of repackaging motors, hydraulic couplings, pumps, and control devices, we visualize Peerless manufacturing, in addition to the pumps, the variable speed unit. Attached is a sketch which illustrates the approach we have in mind.

The engineering principle involved in this sketch is to utilize the shaft and bearing system of a motor to support one of the rotating elements of a fluid coupling and the bearings and shaft of the pump to support the other part of the coupling. This would eliminate two shafts, four bearings, and two shaft couplings in addition to eliminating completely the precision alignment problem.

In addition, we visualize the fluid pressure or flow sensing device being built into the pump, which would control the movement of the hydraulic stream that would govern the amount of oil being put back into the fluid coupling to maintain the speed of the pump as a function of pressure or flow.

3. This philosophy or approach is generally what we had in mind in recommending the feasibility study. After considering the

G. D. Sickert - Page 2   January 9, 1960

matter, we hope that you will share our enthusiasm and optimism. Assuming that you do, we would like your comments on the following feasibility study plan:

1. Review of the performance requirements of the most popular size pump used in the chemical and process industry. As a result of sitting in Manufacturing Chemists Association pump standardization meetings I am under the impression that about 85% of the chemical pumps are in the 1-1/2" size category.

2. Establish design specifications and make concept layout drawings.

3. Estimate development costs through step 5. This estimate will include the cost of jury rigging a hydraulically controlled fluid coupling by modifying a commercially available coupling unit.

4. Estimate the manufacturing cost of the complete hydro-constant pressure control package. The flow control sensing means is probably a more sophisticated problem and perhaps it would be better to estimate the development and manufacturing costs separately. It would be advisable, however, to have a design approach in mind so that the pump housing could accommodate either the pressure or flow sensing means.

We shall be looking forward to your comments and suggestions.

W. J. Adams, Jr.

WJA:db

Letter—Adams to Sickert, Jan. 9, 1961

The All-Hydraulic Pressure Control

PART 4

Feasibility study

The previous two letters (Exhibits 11 and 12) indicate the position of the Central Engineering Laboratory on the Hydro-Constant development project. The Peerless Pump Company was quite reluctant to undertake any kind of development because of the market situation. Even the anticipated cost reductions which could be accomplished by manufacturing a complete unit, including both motor and pump, were not sufficient to induce the Company to make any further commitments. The Peerless Pump Company also had a proposal from American Standard (Gyrol fluid drive manufacturer) to standardize on its close coupled Gyrol unit.

Early in 1961, Mr Bunnelle was sent to Belgium on an engineering assignment. When the $1 200 appropriation recommended by W. J. Adams was granted, Mr Baird, a Mechanical Engineer of CEL, was assigned the job of making a feasibility study of an hydraulically controlled drive for the Hydro-Constant system. On 9 March 1961, Mr Baird met with Mr W. J. Adams, in his office, where the objectives of the project were established.

His first objective was to examine the Hydro-Constant system already marketed by Peerless and determine its deficiencies, having in mind a wider market including industrial applications. The second objective was to establish the performance requirements and design specifications for an all hydraulic hydro-constant pressure control system. Third, he was also asked to determine the availability of commercial hardware to meet the selected system requirements. A preliminary design and a cost estimation for the entire system, based on production quantities of from 10 to 50 would be the final step in the programme.

Mr Baird studied the several reports relative to the Hydro-Constant development which had been prepared at different times by Peerless Pump and CEL. The most recent report was that submitted by Mr Bunnelle on the completion of the redesign of the electro-mechanical system. A study of the controls for this system revealed that it would be impossible to market this system with more exacting pressure control requirements.

The redesigned system exhibited a response delay of about one minute for sudden major changes in line pressure. For building water supply application this was perfectly acceptable, since the pressure variation is gradual. This is not the case, with industrial processes. Pressure changes may be sudden, and fast reaction of the control system is imperative. Furthermore, the rate of scoop tube movement, which controlled the rate of change of the pump speed and consequently the pressure, was not proportional to the deviation of line pressure. In the Barber–Coleman electro-mechanical control, the scoop tube moved at a constant rate regardless of deviation of the line pressure from the set point. The no-load rotation rate of the motor operator arm was adjustable over the range of 65 to 130 sec for full travel of the scoop tube. However, at a given setting, the rate was not proportional to the deviation of P_2 from the control pressure. Thus it was not a true proportional control but a modified two-position controller with position feedback. The proportional band, which can be described as the range of P_2 which corresponded to 90° rotation of the motor operator shaft, was pre-set at the factory at 3 psi. This setting could be doubled. The dead band of the instrument was approximately 3–4 psi. These characteristics produced a rather sluggish inaccurate control which was inadequate for future installation of the pump.

Besides these shortcomings, the cost of the control components was high. The assembly of motor operator, mounting bracket, crank arm, connecting link pressure switch, power box and relay cost $335, including $175 for wiring diagrams and supervision. Competition was keen.

Finally, the redesigned system required an electric power source to operate the controls which made it less reliable for certain districts. Mr Baird was asked to design a control system complete in itself, utilizing hydraulic power, and satisfying the requirements of exactness and quick response.

Mr Baird was present during the last phase of the redesign of the electromechanical system. During this time Mr Bunnelle suggested a new method of controlling the amount of fluid in the hydraulic coupling by using a jet splitter instead of the conventional scoop tube arrangement. Mr Baird didn't support the concept. When he started his work for the preliminary design he centered his study around the scoop tube arrangement and a hydraulic servo-valve for positioning the scoop tube. He consulted the different manufacturer's bulletins in the control field. The outcome of this work was the determination of the following performance requirements.

1. The basic function of the control system is to sense the fluid pressure P_2 in the line or near the pump discharge, and convert this signal into a linear movement of the scoop tube maintaining a constant pre-set value of P_2 within $\pm 10\%$.
2. The minimum time required for full travel of the Gyrol scoop must be 15 to 30 sec. The time necessary to change from zero to full speed with the Gyrol drive is about 15 sec.
3. The control pressure P_2 must be adjustable over the range of from 40 to 75 psi. Most of the Hydro-Constant systems installed in 1959–60 were designed for this range.
4. The scoop tube should move at a rate which is proportional to the deviation of the line pressure (P_2) from the control pressure.
5. The control system must be capable of continuous operation for long time periods compatible with the other components of the Hydro-Constant system. Replaceable filter element should require changing no more frequently than every 30 to 60 days. Long-term reliability and low maintenance are desirable.
6. Field adjustments should be simple enough for the average building maintenance personnel to perform.

Mr Baird spent the month of April 1961, working on a preliminary design of the control loop. From an analysis of the performance requirements and a review of technical publications on automatic controls he established the design specifications for the system and formulated the conceptual design of the control loop. This system composed of a pump, filter, pressure operated servo valve, relief valve (if a closed servo valve is used), and a double acting power cylinder is shown in Exhibit 13.

About \$620 of the original \$1 200 allotted for this project had been spent by 5 May 1961. The amount available for completing the job was insufficient. In order to complete his work and remain within his allotted budget, Mr Baird prepared a request for an informal quotation which enclosed figures, but gave only a minimum description of the system with the proposed hydraulic control. This form letter (Exhibit 14) was then mailed to a selected group of eight manufacturers of hydraulic control equipment. Only four of the companies responded.

Proposals for the all hydraulic control system

Proposals for control systems or components were submitted by the following four companies:

1. GPE Controls, Inc. (formerly Askania Regulator Co.)

The unit proposed for the Hydro-Constant application is one of the few general purpose, hydraulic integral controllers manufactured for industrial use.

This unit consisted of an integral pump, relief valve and reservoir which supplied pressurized hydraulic oil to an integral jet pipe servo valve. A manual control point indicator and adjuster and a diaphragm input system was provided when used as a proportional speed floating controller (integral control) the floating rate could be varied by changing the jet pipe counter-balancing spring. A filter was not required, since the oil circuit was completely enclosed. The total price of the controller with 2 in diameter cylinder was $849.

2. Waterman Hydraulics Corporation, Evanston, Illinois

The components included in this proposal were: Servo-valve, pump, relief valve, filter and cylinder.

The only additional engineering required by FMC in using this control system for production Hydro-Constant units, would be in mounting of the components and providing a drive for the pump. The pump and relief valve could be mounted inside the Gyrol case. The component, including miscellaneous fittings would cost $360 for quantities of from 1–10 and $339 for production quantities of from 10–50 units.

3. The Rucker Company, Oakland, California

This company proposed to use a pump, relief valve, filter, 1·5 in diameter cylinder and spool-type servo valve. The cost of the servo valve alone was $488. The total cost of the system in quantities from 10–50 would be about $590.

4. Vickers Incorporated

This company proposed the use of automotive type power steering components in the design of the control system. The components included a vane pump, filter, relief valve and integral cylinder and servo valve.

The unit costs were comparable to those of other manufacturers such as Ford Motor Company. The total cost of the components including miscellaneous fittings would be $166.

To make use of these low-cost, mass-produced components, would require that FMC design and manufacture a diaphragm control spring actuator. This actuator unit could be manufactured for an estimated $70.

Evaluation of proposals for the all hydraulic control system

The second part of this phase of the Hydro-Constant development consisted of Mr Baird's evaluation of the answers to his letter. He made a comparative cost estimation of the different proposals besides examining the mechanical features of the different systems and individual components. He also examined them from an industrial and manufacturing point of view. He summarized his results in his final report submitted on September 1961.

1. Only one 'off the shelf' control system (GPE Controls, Inc.) could be found which would meet the design specifications. However, the cost ($849) was much too high for use with the production quantity Hydro-Constant systems. If other controls similar to the GPE unit were available, the cost would probably not be significantly different.
2. The Waterman proposal offered a good opportunity to obtain a low-cost, pre-engineered prototype control system in a short time period (60 days). Production-type units cost about $400 complete.
3. There was a good possibility that the cost of production unit could be lowered to about $236 by using automotive-type power steering components. This cost was about the same or less than that for the present Barber–Coleman Control system.
4. There is a good possibility that patent protection can be obtained on the complete all-hydraulic pressure control system, and on the design of the diaphragm-actuated servo valve-cylinder combination. There are only a few pressure-actuated servo valves manufactured in the US and none of them exactly fitted the requirements of the Hydro-Constant system, although the design of none of the components is entirely unique, their combination of use may be original.

EXHIBIT 13

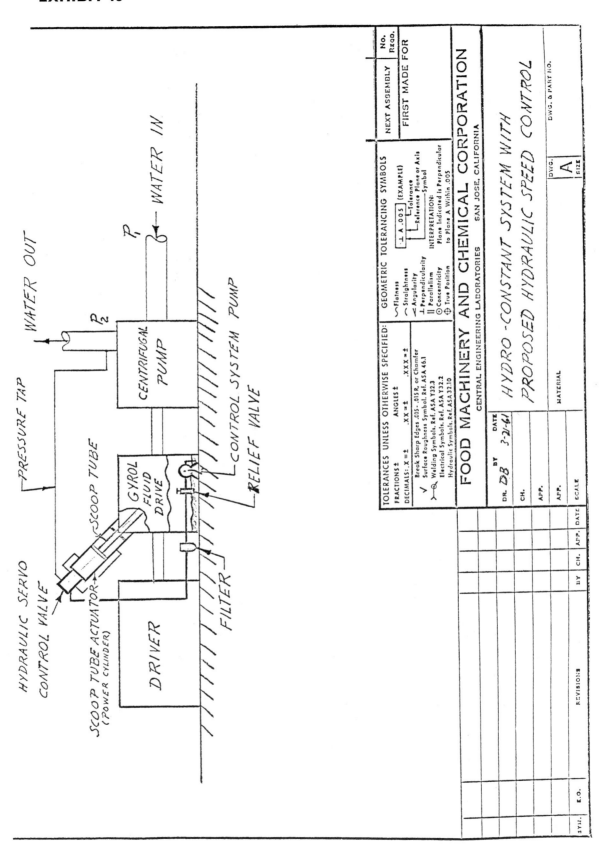

Hydro-Constant system with proposed hydraulic speed control

EXHIBIT 14   Form letter

May 16, 1961

Askania Regulator Company
240 E. Ontario Street
Chicago 11, Illinois

Gentlemen:

Subject: Request for a Hydraulic Control Proposal

This is a proposal request for a hydraulic control system. A schematic sketch of the over-all system is shown in Figure 1, and a sketch of the Integral Control System in Figure 2. The oil reservoir-heat exchanger combination is already available. What is needed is the control system enclosed in the dashed square of Figure 1.

The basic function of the control system is to sense the fluid pressure P_2, and convert this signal into a linear movement of the power piston so as to maintain a constant pre-set value of P_2. The system shown in Figure 1 consists of a pump, filter, hydraulic servo control valve (Figure 2) and power cylinder. The control must operate with the use of hydraulic energy only (no electrical components). A prime mover is available to drive the pump.

The following is a summary of the operating characteristics and performance requirements of the system:

1)   Prime mover speed range .... 1700-2000 rpm

  Prime mover speed regulation ... $\pm$ 10%

2)   Hydraulic Oil ... Mineral base of the type used for steam turbine lubrication, viscosity of about 150 SSU at 100° F.

3)   Power Cylinder

  a)   Maximum axial travel ...... 6 inches

  b)   Maximum axial force ...... 40 lbs.

  c)   Time to complete full travel ... 15 to 30 seconds

Request for a hydraulic control proposal

EXHIBIT 14 *(contd.)*

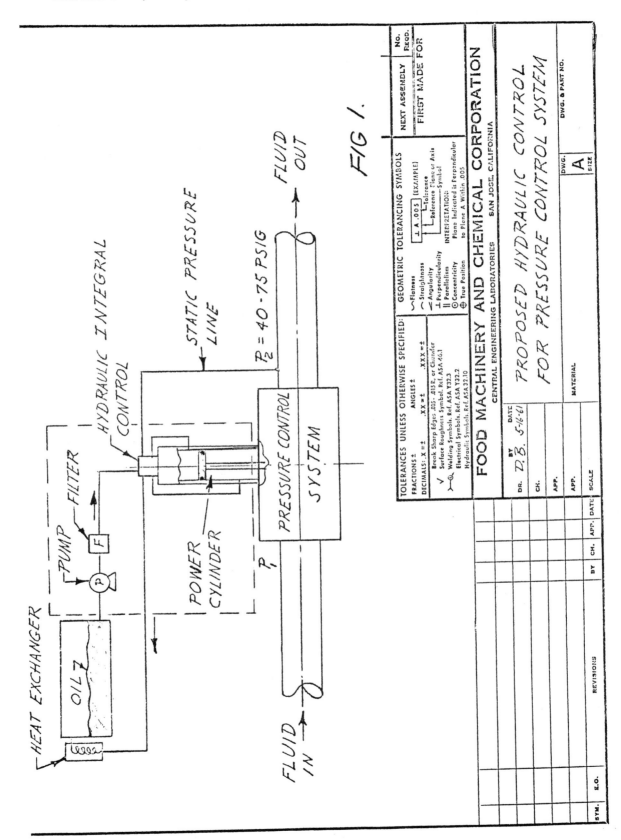

FIG 1.

FOOD MACHINERY AND CHEMICAL CORPORATION
CENTRAL ENGINEERING LABORATORIES   SAN JOSE, CALIFORNIA

PROPOSED HYDRAULIC CONTROL
FOR PRESSURE CONTROL SYSTEM

EXHIBIT 14 *(contd.)*

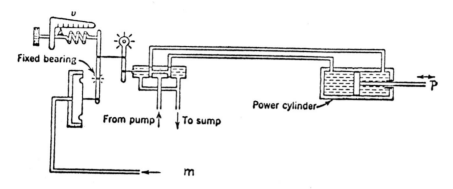

HYDRAULIC INTEGRAL CONTROL

EXHIBIT 15   Letter - Adams to Sickert, Sept. 21, 1961

September 21, 1961

W. J Adams, Jr.

G. D. Sickert
Peerless, Indianapolis   cc: R. Kummer
                                              W. S. Thompson
                                              D. Baird

HYDRO-CONSTANT CONTROL
SYSTEM -- CE 9663

Dear Gene:

Undoubtedly, by now you have received your copy of Doug Baird's
report T-1606 covering the engineering investigation on an "all-
hydraulic" pressure control system to be used in place of the present
"electro-mechanical" control system.

You will note that Doug has recommended a next phase of development involving incorporating the Waterman servo-valve into a
prototype.

Doug is reluctant to recommend the next phase utilizing the automotive type "Vickers" servo-valve and cylinder because he felt
the additional cost to develop the actuator would not be acceptable
to Peerless. On the other hand, I personally feel that it would be
better to not only prove the principles of the "all-hydraulic" control
in the first prototype, but also to prove out the lower-cost components, so as to achieve an engineering model of a hydro-constant
that would demonstrate:

      a.   Improved performance, such as
          quicker response, and

      b.   Lower cost.

If you are interested in having us look into this possibility, we
will be glad to try and estimate the additional costs of development involved.

                                    W. J. Adams, Jr.

WJA:db

EXHIBIT 16   Letter - Sickert to Adams, Oct. 2, 1961

PEERLESS PUMP HYDRODYNAMICS DIVISION

R. F. Foster

E. W. Lundy

G. F. Twist   10-2-61   R. E. Kummer

G. D. Sickert   A. Grant

    Technical Report - T1606   R. B. Brown
    Hydro-Control Unit for Hydro-
    Constant Pumping System   Blind copy:
                                            W. J. Adams Jr.

Dear Pete:

In your letter of 9-22-61 you acknowledged a copy of the subject report
from Central Engineering Laboratories and you asked our opinion of the
market potential, the profit margin, and our suggestions for further development.

I am of the opinion that we have been slow in exploiting this product
and that there are many markets other than the building trades, (such as
sewage, chemical, petroleum, and the industrial fields,) for both pressure
and volume control for which we lack promotional and sales penetration.
Obviously, Sales must have a clear understanding of the capabilities of
the product and then bird-dog new applications, which will require some
Application Engineering expense. The attached tabulation on sales, January 1, 1961 through September 29, 1961, indicates total volume to be
$72,624.07, and on units shipped to date a selling price of $23,643, with
an average margin of 35.6%. Item 21, Order 87692-V, with only a 9.6%
margin was a Westinghouse Rectiflow unit specified by the customer; consequently, this is misleading in the fluid coupling or hydro-constant
analysis.

I am sure you are aware of the promotional advertising appearing in
trade publications on a system of this type being done by Allis-Chalmers
Manufacturing Company and attached is promotional literature being circulated by Flow Engineering. You will also find attached Peerless sales
manual information which was released to field representatives on August 25,
1961.

It was our thinking at Indpls. that conservatively we would sell 100
units per year. On this assumption we worked out a preferential volume
discount with American Standard and they, too, are concerned about our
inability to meet our objective. They have extended our agreement for
an additional six months but unless Sales shows greater enthusiasm for
the market potential we will be on the same basis with American Standard
as our competitors. Our slow start may be the result of lack of pro-
motion and late issuance of field instruction data and these areas, I
believe, require further attention.

EXHIBIT 16 *(contd.)*

Peerless Pump Hydrodynamics Division

G. F. Twist
Re: Technical Report - T1606
     Hydro-Control Unit for Hydro-   P 2
     Constant Pumping System   10-2-61

I am reluctant to curtail development work for fear that some other organizations may accelerate their programs and obtain a patentable device
for control of the mechanism; however, unless our Sales organization shows
some initiative, our development efforts and funds can be wasted. Personally, I prefer to see Central Engineering spend its efforts on the
Vickers Servo-Valve and Cylinder rather than the Waterman Servo-Valve for,
although development would be somewhat more costly, it appears it will
result in a lower unit cost and quicker response.

I'm for development, so leave you with that perplexing problem of which
comes first? Development or sales?

                                        Sincerely yours,

                                        G. D. Sickert

me
Att.

Developing the Hydro-Constant pump

PART 5

Fluid coupling drive

The all-hydraulic pressure control system suggested by Mr Baird was never developed. At the time he submitted his final report, Peerless was not willing to finance any further development. When Peerless decided to continue with the Hydro-Constant development, the job was re-assigned to Mr Phillip Bunnelle who had originally worked on the project and had returned from Belgium. Mr Bunnelle decided to develop his original design and not use that proposed by Mr Baird.

During the time of the original repackaging, Mr Bunnelle had felt that the fluid coupling drive, which Peerless was purchasing from American Standard for the Hydro-Constant, was much larger than necessary. The coupling was designed to drive big blowers running at low speeds and was overdesigned for driving pumps. Peerless was also losing money due to low production, since American Standard sold a limited number of these couplings, more or less made to order.

He also felt that a scoop tube was a very inefficient means of controlling the fluid level in the coupling. In order to effectively control the level of fill in the coupling, the scoop tube had to be moved 4 in, and required a force of approximately 10 lb, applied through a slow servo-motor. The system was thus very slow acting, requiring a long time to respond to changes in discharging pressure. To get a faster responding drive, a faster control of fluid level was needed.

In order to circumvent the problems of the American Standard coupling, Mr Bunnelle thought that Peerless should buy the bare coupling from American Standard and develop some other means of controlling the level of fill (see Exhibit 17a). He first considered using variable-opening bleed orifices on the periphery of the coupling. By using a constant charging rate and moving shutters to vary the bleed orifice sizes, the bleed rate and thus the level of fill in the coupling could be controlled. This approach was discarded because no simple means of positioning the shutters could be developed.

Looking through the Gyrol Fluid Handbook, he came upon the idea of varying the charging rate and maintaining a constant bleed rate from a coupling used for marine application (see Exhibit 17b). As can be seen from the figure, the marine fluid drive is enclosed in a stationary casing which is connected to a sump tank. The motor-driven pump delivers the charging fluid from the sump tank through a hole in the pinion shaft into the working circuit of the drive. The drive emptied through calibrated nozzles provided in the periphery of the secondary rotor housing. By regulating the quantity of oil in the working circuit, the drive could be used to provide adjustable speed. Mr Bunnelle felt that the level of fill in the coupling could be controlled by diverting all or any portion of the charging fluid flow and keeping the size of the bleed orifices constant.

While still in the process of repackaging the original electromechanical unit, he submitted a disclosure on 5 December 1960 to Mr Adams, and Mr Baird. In this disclosure he proposed to vary the charging rate by moving a nozzle which discharged the fluid against a fixed deflector plate, but he also suggested the possibility of having a fixed nozzle and movable deflector (see Exhibit 18).

On 9 January 1961, Mr Adams sent a letter with the disclosure to G. D. Sickert of Peerless, Indianapolis. He suggested that for Peerless to seriously compete in the Hydro-Constant field, they should engineer and manufacture a packaged line of units. Instead of repackaging motors, hydraulic couplings, pumps, and control devices, Peerless should manufacture, in addition to the pumps, the variable speed drive utilizing as few purchased components as possible. He proposed that Peerless allow CEL to do a feasibility study of an all-hydraulically controlled drive for the Hydro-

Constant system following P. R. Bunnelle's approach. The feasibility study would estimate development costs up to and including the cost of building and testing a hydraulically controlled fluid coupling by modifying a commercially available unit. Depending upon the results of the tests, it was planned that this would be the first phase of a continuing development programme.

At first Mr Bunnelle thought that this would be a crude but effective means of controlling pressure for domestic water supply systems where wide pressure margins and slow rates of change could be tolerated. Intermittently for the next six months, he spent some time calculating the time constant, the linearity of the response as the charging rate was varied, the heat rejection characteristics, a charging rate that would give a sufficiently fast response and supply enough fluid to carry away the heat generated, etc. Everything seemed to be ideal for application to the Hydro-Constant system. It had a fairly linear response, a relatively uniform time constant for different levels of fill, and a reasonable charging rate producing sufficient heat dissipation. Mr Bunnelle said that, 'I did not decide that it was a good idea until I had time to think about it and go through at least a preliminary engineering analysis.'

On 2 June 1961, he submitted a second disclosure in which the results of the previous six months of analysis were presented. It was proposed to have a fluid coupling with orifices in the periphery of the coupling, a charging pump to deliver a steady flow of charging fluid, a movable nozzle which would move or direct the stream of fluid from the charging pump and a deflector which would divert or vary a portion of the fluid stream into the coupling depending on the position of the nozzle while returning the remainder to a reservoir (Exhibit 20). In the steady-state condition, the nozzle would be positioned so that fluid is added to the coupling at the same rate that centrifugal force discharges it from the orifices. If the nozzle were to be repositioned to divert more fluid to the coupling, the total amount of fluid in the coupling would increase. With this greater mass of fluid in the coupling, less slip would have to take place between the runner and impeller to transmit a given torque. This would then result in an increase in the output shaft speed. The opposite would be true if the nozzle were to be positioned to reduce the total amount of fluid in the coupling. The performance of the system could be changed by varying the charging pump discharge rate, the bleed orifice size, the geometry of the nozzle positioning system, and the configuration of the make up jet.

Phase I and the first test unit

Peerless, Indianapolis, expressed a great deal of interest in the development of a variable-speed fluid-drive unit of their own manufacture. They did want time to determine whether or not there was sufficient sales enthusiasm for the repackaged Hydro-Constant to warrant the development of an all-hydraulic system. They did not believe that they could justify the development expense to their division management in Los Angeles until they showed an improvement in sales volume of the current Hydro-Constant. However, in a letter on 26 March 1961, they asked that CEL continue gathering pertinent information so that Peerless would be able to work towards justification of an authorized development program. Peerless was willing to take a $1 500 Hydro-Constant unit off their inventory and consign it to CEL for development exploration purposes. Peerless also sent a bulletin showing a variable-speed unit made by another company using fluid coupling components from the Twin Disc Company, and suggested that CEL look into the contemplated design in the interest of reducing the tooling portion of their development cost.

In March 1962, Mr Bunnelle began to design a test unit to demonstrate his principles and develop the controls. At first he planned to build the control mechanism and apply it to the American Standard fluid coupling. However, following the suggestion made by Peerless, he decided to adapt the basic Twin Disc fluid coupling to his proposed variable-speed drive and use this in the test unit. However, while designing the fluid coupling he began to find that there were many mechanical problems associated with the movable nozzle concept. Some of the problems encountered were:

1. He was trying to move a relatively large mass.
2. The nozzle needed a flexible coupling with a positive frictionless pivot in order to be accurately positioned.
3. The mechanism used to position the nozzle would have to come out of the back of the housing where the integral pump was to be attached. The only other alternative would be to place the pressure sensing and positioning mechanism inside of the housing. However, the pressure and positioning adjustments would have to be outside of the housing. This would present problems with sealing and reading out the position of the nozzle.

In May 1962, he decided to use a fixed orifice plate to produce the jet and a movable splitter to divert the fluid, and in June 1962, he began to build and test components of the test unit. While constructing the test unit, he found that the cross-sectional area of the splitter as designed was much too small and that the angle of splitting had to be reduced. He also found that the cross-sectional area of the charging tube, which carries the fluid from the splitter into the coupling, was much too small and the angle much too flat. The problems with the splitter and the charging tube arose because, at the time of the original design work, Mr Bunnelle forgot that the fluid loses its pressure head when it discharges into the atmosphere after passing through the orifice plate. When the fluid enters the splitter, it maintains only the velocity head and the effects of gravity, and consequently slows down. He then increased the area of the splitter and put in a J-shaped charging tube with a greater cross-sectional area. The test unit was then built and testing began in July 1962.

The test results were generally favorable, but instability occurred at shut-off. He first tried to gain stability by reducing the charging rate, i.e., increasing the time constant of the system. Although this produced a slight increase in stability, it greatly increased the response time of the system. Mr Bunnelle noticed that as the unit ran for long periods of time, and the charging fluid became hotter, the unit became more stable. He reasoned that instability was due to the time lag in getting the fluid from the splitter into the coupling where the effects of a change were seen. To test this, he placed baffles in the charging tube to further slow down the fluid and the stability of the system became much worse. The tube was again redesigned so that it was almost vertical, the fluid being discharged directly into the back of the coupling.

The test results for constant-pressure control can be seen from Exhibit 21. Constant pressure could be maintained at any flow rate from shut off to maximum at any head the pump was capable of producing. The regulation from full flow to shut off required approximately 20 sec. The variation in pressure at any given flow rate was negligible.

Sales of the Peerless Hydro-Constant units were increasing very rapidly and Peerless became anxious for CEL to continue the development of the all-hydraulic unit. The test unit was completed and testing began in July 1962. At the same time the design of a production model drive was started. This model was designed using essentially the same components as were used on the test unit. In October 1962, the first phase of the development program was completed with very satisfactory results.

Phase II, III and IV

In a report to Peerless, CEL proposed a four-phase program to continue the development of the jet-controlled Hydro-Constant drive.

The first phase, which had just been completed, involved demonstrating the all-hydraulic principle and designing the first production model drive, using purchased components. In order to realize the full potential of the drive, it was proposed that Peerless enter the market with a line of jet-controlled Hydro-Constant drives in the shortest time possible, using purchased components where expedient and putting several sizes of fluid couplings in one housing to get a more complete range of coverage with the least expenditure of engineering time and tooling cost. The range of applications would be broadened by developing additional modes of control and an improved series of horizontal and vertical drives, which would be developed in the more popular horsepower ranges using fluid couplings manufactured by Peerless, and incorporating other features that would improve performance and/or reduce cost. It was felt that the field experience with the initial line of drives would provide the basis for many of the improvements.

Phase II began in January 1963 and required approximately nine months. This phase extended the horsepower range of the horizontal drives and expanded the range of applications by developing several additional modes of control such as level control, constant-flow control, remotely operated electrical control, and improved pneumatic control. Horizontal drives for 30–100 hp at 1 750 rpm using the Twin Disc 12·2 in and 14·5 in couplings were designed and work was initiated to design a vertical drive to cover the 30–100 hp range. This included testing of the charging system and some development work on a prototype. A fluid coupling to be manufactured by Peerless which would be lower in cost than the Twin Disc coupling and have improved performance, was also to be developed.

Phase III began in October 1963 and also required approximately nine months. In this phase the vertical drive was developed in the total horsepower range from 1–100 hp. A more compact gear-driven charging pump and other features to improve performance and reduce cost were installed on the units. It was proposed in this phase to complete the design of a Peerless manufactured fluid coupling and install this in the Hydro-Constant unit. However, Peerless felt that the amount of money saved per unit by manufacturing their own coupling would not justify the development cost and that it would be better to continue purchasing the basic Twin Disc coupling.

Phase IV was carried on concurrently with Phase III. This phase involved the development of a small drive for shallow-well household application. Peerless proposed a drive in which the charging fluid was drawn from the reservoir by a slinger and thrown to a tank above the coupling where gravity feed was used to produce the charging stream. A fractional to 5-hp drive utilizing this principle was subsequently developed.

EXHIBIT 17(a)   Fluid drive with ring valve control

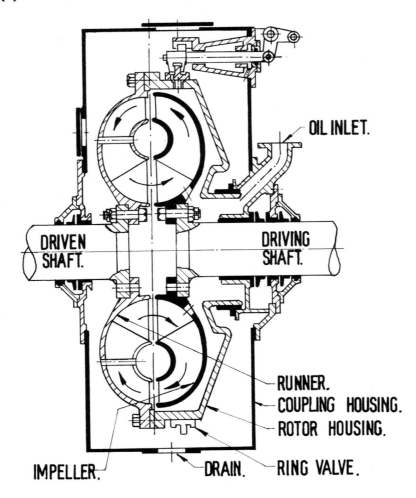

OIL INLET.

DRIVEN SHAFT.   DRIVING SHAFT.

RUNNER.
COUPLING HOUSING.
ROTOR HOUSING.
IMPELLER.   DRAIN.   RING VALVE.

EXHIBIT 17(b)   Cross section of marine-type fluid drive

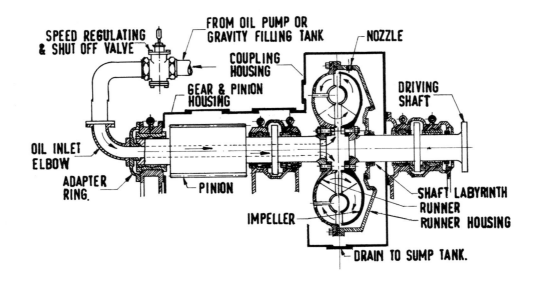

SPEED REGULATING & SHUT OFF VALVE

FROM OIL PUMP OR GRAVITY FILLING TANK

NOZZLE

COUPLING HOUSING

DRIVING SHAFT

GEAR & PINION HOUSING

OIL INLET ELBOW

ADAPTER RING.

PINION

IMPELLER

SHAFT LABYRINTH
RUNNER
RUNNER HOUSING

DRAIN TO SUMP TANK.

EXHIBIT 18   Disclosure

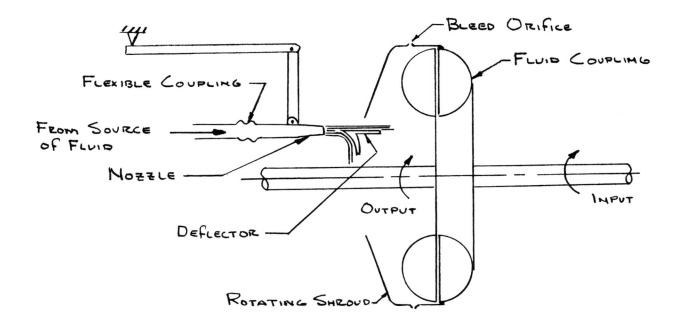

TORQUE CARRYING CAPACITY (& CONSEQUENTLY SPEED) IS A
FUNCTION OF THE AMOUNT OF FLUID IN COUPLING. WITH THIS
ARRANGEMENT, FLUID IN THE COUPLING IS CONSTANTLY BEING
DISCHARGED THRU ORIFICES. IT IS REPLENISHED FROM A PUMP
OR OTHER SOURCE THROUGH A NOZZLE. THE DISCHARGE FROM
THE NOZZLE IMPINGES UPON A SPLITTER OR DEFLECTOR PLATE.
THIS ALLOWS A CERTAIN PORTION OF THE STREAM TO ENTER
THE COUPLING & DEFLECTS THE REMAINDER. BY VARYING THE
POSITION OF THE NOZZLE OR THE DEFLECTOR PLATE, THE
LEVEL OF OIL IN THE COUPLING MAY BE VARIED

EXHIBIT 19   Hydro-Constant package automatic pressure or flow control

fmc CENTRAL
ENGINEERING
LABORATORIES

PAGE _____

TITLE *HYDRO CONSTANT PACKAGE*   PROJECT. *CE. 9615.*

AUTO PRESSURE OR FLOW CONTROL.

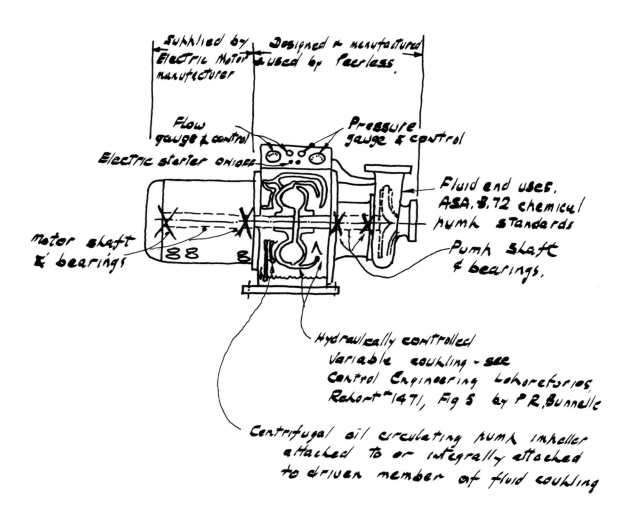

Supplied by Electric Motor manufacturer

Designed & manufactured & used by Peerless

Flow gauge & control

Pressure gauge & control

Electric starter on/off

Fluid end uses, ASA. B.72 chemical pump standards

Motor shaft & bearings

Pump shaft & bearings.

Hydraulically controlled variable coupling - see Control Engineering Laboratories, Report #1471, Fig 5 by P.R. Bunnelle

Centrifugal oil circulating pump impeller attached to or integrally attached to driven member of fluid coupling

BY _____

DATE /-7-61

EXHIBIT 20 **Disclosure**

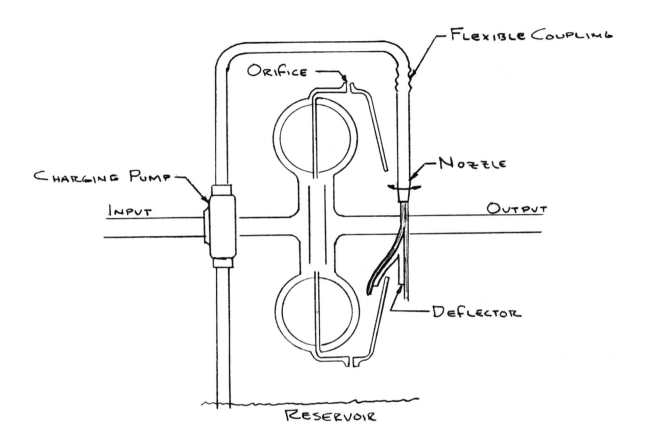

Reservoir

EXHIBIT 21   Constant pressure control

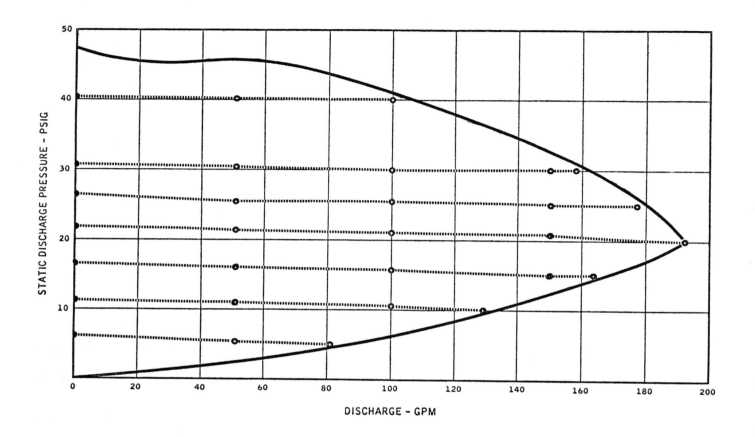

Bleed orifice size 0·082 in ; charging rate 4 gpm ; max rate of correction 20 sec from full flow to shut off

PART 6   Controls

Mr Bunnelle used a movable splitter and a stationary nozzle. In the control system, a flexible metallic bellows pushed directly against the control spring to position the arm of the splitter (see Exhibit 22). The bellows and spring were then packaged as a unit and mounted on the top of the fluid coupling housing. Since changes in discharge pressure were sensed immediately, a very high system gain could be used. This resulted in a fast response and small error. The system gain could easily be changed by changing the control spring. A stiffer control spring would result in a lower gain.

This control system was used on the First Test Unit and as can be seen from the results in the preceding section, proved to be an effective means of controlling the discharge pressure. However, the flexible metallic bellows which were supposed to operate to pressures to 60–70 psi, had a tendency to buckle when operating at this pressure. This was because the bellows were connected directly to the splitter arm and had to curve as the arm was deflected. Peerless required that the system operate to pressures of 175 psi without buckling or bulging of the bellows. It was evident that either larger bellows would have to be used or that a substitute would have to be found. Flexible metallic bellows were found to have other disadvantages, some of which were:

1. A finite fatigue life.
2. A large diameter unit with many convolutions was needed to obtain the necessary travel.
3. The large cross-sectional area meant that large forces would have to be counteracted by the control spring. This necessitated a large husky spring which not only took up too much room, but also required a large force on the adjustment screw to change the set point.
4. The bellows would be very expensive requiring $500 for a prototype and $20–30 each even in production.

In October 1962, the first production model was designed using the flexible metallic bellows, as no satisfactory substitute could be found. After designing the first model, Mr Bunnelle looked into other methods to position the splitter. While discussing the problem with some engineers at Peerless Indianapolis, it was suggested that he try to replace the bellows with a deep molded rolling diaphragm produced by the Bellofram Corp. (Exhibit 23). The Bellofram Diaphragm and piston arrangement proved to be ideal for this application. The diaphragm, cost $1 each, and piston arrangement cost approximately $20 less than the flexible metallic bellows arrangement, had an almost infinite life, and a much smaller cross sectional area, which meant that a lighter control spring could be used making a much more compact unit. The diaphragm, piston and spring arrangement were then packaged as a unit and is now being used on all the Hydro-Constant drives being produced (Exhibit 24).

EXHIBIT 22   Hydraulically controlled fluid coupling movable splitter

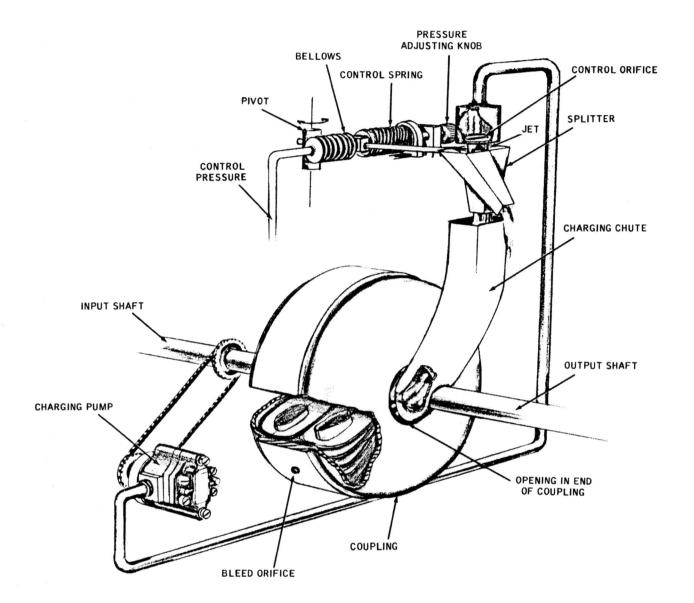

EXHIBIT 23   **Hydro-Constant control unit**

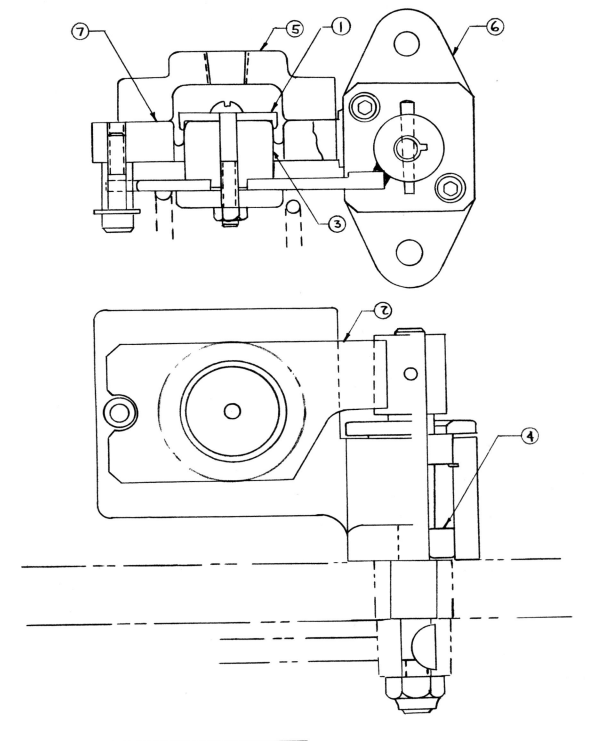

Item No.	Description
1	Cap
2	Splitter Arm Assem.
3	Piston
4	Bearing
5	Retainer
6	Bracket
7	Rolling Diaphragm

EXHIBIT 24   Production Hydro-Constant

HYDRO/constant
MP8, MP9, MP10

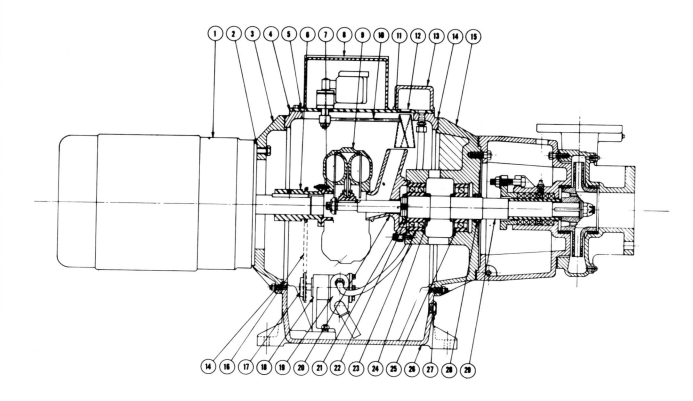

1. MOTOR
2. MOTOR GASKET
3. MOTOR MOUNTING FLANGE
4. TOP PLATE GASKET
5. TOP PLATE
6. COUPLING DRIVE HUB
7. BEARING QUILL
8. UNITIZED CONTROL COVER
9. FLUID COUPLING
10. SPLITTER
11. ORIFICE CHAMBER GASKET
12. ORIFICE PLUG
13. ORIFICE CHAMBER
14. END BELL GASKET

15. END BELL
16. ROLLER CHAIN
17. SPROCKET
18. PUMP BRACKET
19. OIL PUMP
20. CHARGING CHUTE
21. BEARING LOCKNUT
22. BEARING LOCKWASHER
23. BALL BEARING
24. SHAFT COLLAR
25. BALL BEARING
26. MAIN HOUSING
27. OIL GAUGE
28. OIL SEAL
29. OUTPUT SHAFT

EXHIBIT 25   **9·5 in Twin-Disc coupling showing section of one-fluid circuit**

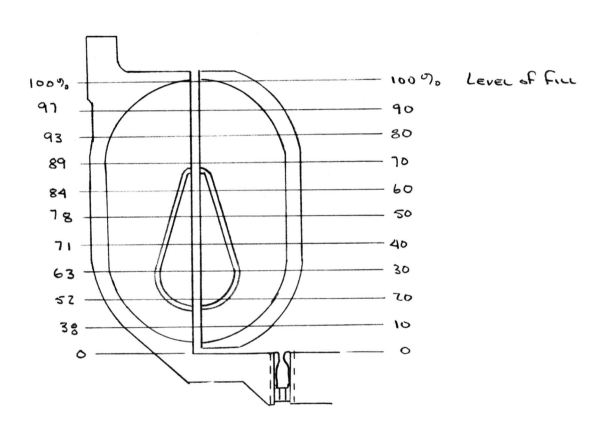

Theoretical Analysis

Mr Bunnelle began a theoretical analysis of the feedback control by deriving the equations describing the dynamic relations of the different elements of the system. The two primary elements in the drive were considered to be the coupling and the rotating elements of the drive. Therefore, equations relating the change in level of fill of the coupling with a change in charging rate, and the change in speed of the rotating elements with a change in input torque (or level of fill) had to be developed. These two equations were then combined to give the overall dynamic response of the system.

The equations predicting the dynamic response of the two primary elements in the drive were derived as follows:

1. Coupling

The coupling was viewed as a tank or reservoir with fluid entering at the top from the splitter and draining out the bottom through the bleed orifices. The rate of change of level of fill would thus be equal to the difference between the inflow and outflow divided by the area of the free surface of the liquid:

$$DL = \frac{Q_{in} - Q_{out}}{A}$$

where D $= d/dt$.

Q_{in} = charging rate, which is a function of splitter position.

Q_{out} = rate of discharge through the bleed orifices, which is a function of level of fill, and was expressed as the product of the slope of the curve, k_1, at the operating point and the level of fill (Exhibit 26).

A = area of the free surface of the liquid in the coupling at the operating point.

The transfer function for the coupling was then written as

$$\frac{L}{Q_{in}} = \frac{1/k_1}{1 = (A/k_1)D}$$

where $1/k$ = static gain

A/k_1 = time constant

2. Rotating elements

For the rotating elements, the acceleration torque would be equal to the input torque minus the load torque:

$$JDS = T_{in} - T_L$$

where J = polar amount of inertia of the coupling runner and pump impeller.

S = speed of rotating elements.

T_{in} = input torque which depends on both level of fill and slip. It was expressed as the difference between the torque at zero speed and the product of the slope, k_2 of the particular k_4 curve–question and the speed S.

T_L = load torque.

However, instead of using the torque at zero speed, an expression relating level of fill and torque was used. The term was expressed as $k_4 L$ where k_4 is rate of change of torque with change in level and is a function of k curve slope and spacing and was obtained by plotting the intercept of the tangent to the k curve at shut off with the zero speed axis v the level of fill (Exhibit 27). Load torque which for a given point of operation was assumed to be the product of the slope, k_3 of the load curve and the speed S (Exhibit 27).

The transfer function for the rotating elements was then written

$$\frac{S}{L} = \frac{k_4/(k_2+k_3)}{1+[J/(k_2+k_3)]D}$$

where $\dfrac{k_4}{k_2+k_3}$ = static gain

$\dfrac{J}{k_2+k_3}$ = time constant

The quantities k_1 k_2 k_3 and k_4 were then evaluated at shut off (i.e., zero flow), since this would be the least stable point of operation. At shut off, the gain would be the highest, i.e., a given change in pump speed would produce the greatest change in pressure. Also, the coupling k-curves and the load curve were flatter at this point, so that there would be less of a tendency to stabilize the drive of any given speed. Calculations performed to determine the frequency response characteristics of the system for shut off at 1 000 rpm are shown as follows:

Coupling

Level of fill in the coupling at 1 000 rpm zero load = 0·39 in.

Surface area of liquid = 92 in².

Slope of bleed rate curve at 0·39 in fill = 7·7 in²/sec.

Time constant for the coupling = 92/7·7 = 12 sec or 0·2 min.

Using a unity static gain to plot the frequency response diagram, the transfer function is

$$\frac{L}{Q_{\text{in}}} = \frac{1}{1+0\cdot2j\omega}$$

Rotating elements

Slope of k-curve k_2 = 0·014 ft-lb/rad/sec.

Slope of load curve k_3 = 0·036 ft-lb/rad/sec.

Slope of torque *v.* level of fill curve k_4 = 7·5 ft-lb/in.

Time constant for rotating elements = $\dfrac{0\cdot045}{0\cdot014+0\cdot036}$ = 0·9 sec or 0·015 min.

Transfer function is

$$\frac{S}{L} = \frac{1}{1+0\cdot015j\omega}$$

Dead time

In addition to the lags associated with the two time constants, there was also a transport lag or dead time, which was caused by the time required to get the charging fluid into the coupling and accelerated up to speed. By using a cam arrangement to produce a sinusoidal variation of splitter position, the experimental frequency response diagrams were obtained (see Exhibit 28 for gain comparisons). It was found that there was a large variation between the experimentally determined phase lag and the phase lag attributed to the two time constants of high frequencies. This discrepancy was presumed to be due to the dead time. By correlating the experimental and theoretical responses at high frequencies it was found that the dead time was approximately 0·3 sec (Exhibit 29).

In order to determine the open loop transfer function for the system, relationships reflecting variations in pressure into the splitter with changing rate and variations in pump discharge pressure with speed had to be obtained.

Splitter

Pressure into splitter

$$P_s = \frac{Q_{in}}{k_s}$$

where $k_s = 2\cdot5$ in^3/sec/psi

Pump

Discharge pressure

$$P_d = k_p S$$

where k_p = slope of curve relating discharge pressure to speed, obtained from pump characteristic curve by plotting pressure *v.* speed for a given flow = $0\cdot3$ psi/rad/sec at zero flow, 1 000 rpm.

EXHIBIT 26   Curve showing bleed rate *v.* level of fill

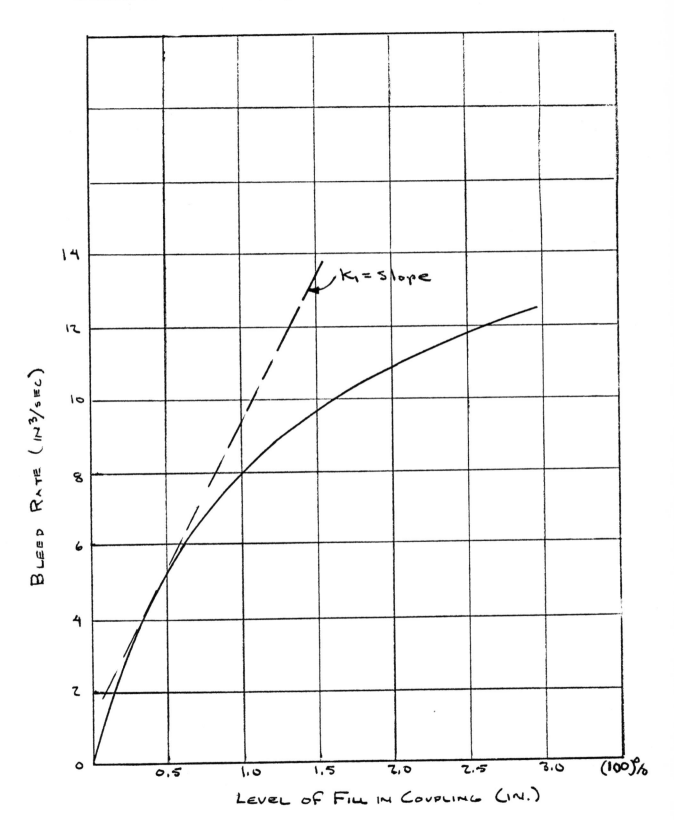

EXHIBIT 27   *K* curves for 9·5 in twin-disc coupling and torque-speed

Characteristics for pump (1½×2×10) available torque at a given speed is expressed as a function of the slope of the *K* curve, the speed and the *Y* intercept of the *K*-curve slope

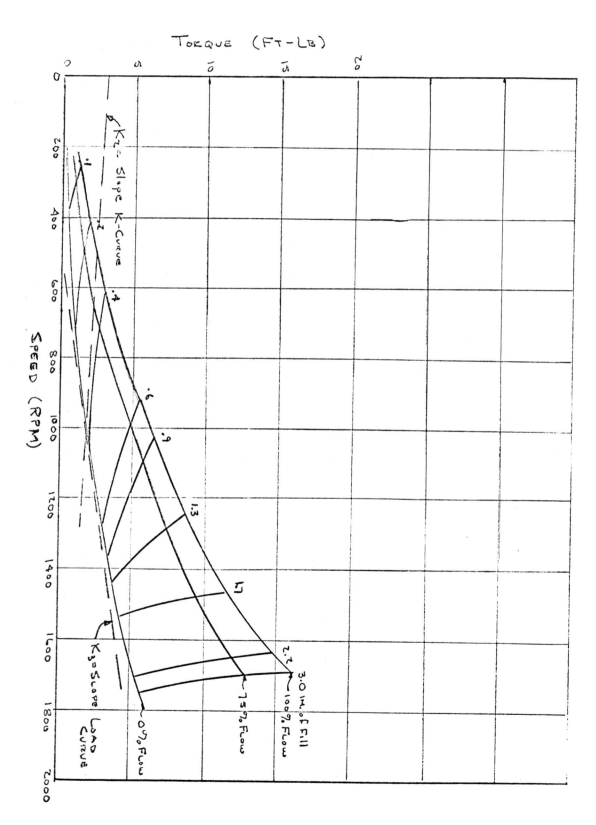

EXHIBIT 28 **Synthesized amplitude curve**

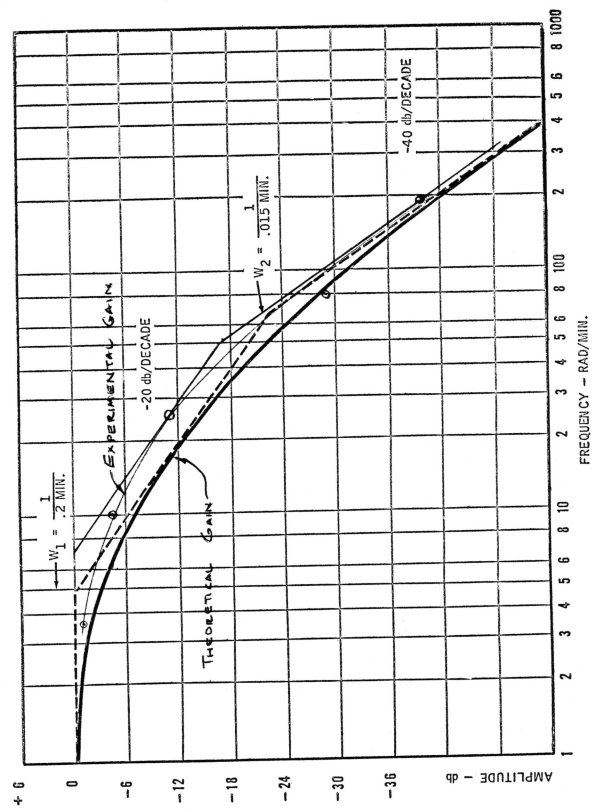

For drive at 1 000 rpm with pump at shut-off

EXHIBIT 29　　Synthesized phase angle curves

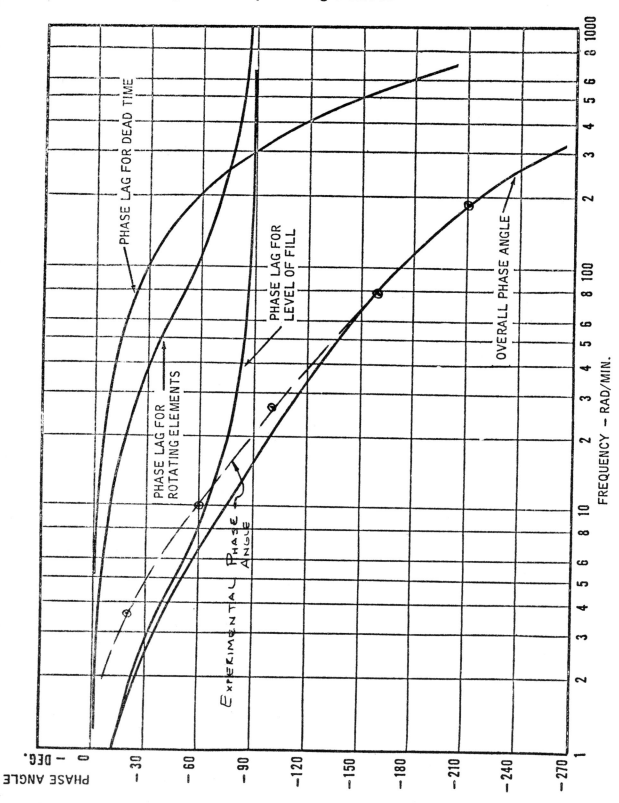

For drive at 1 000 rpm and pump at shut-off showing individual components and overall curve

PART 8

Market evaluation

Hydro-Constant development program and market study

By the end of 1962 there was an established market for the Hydro-Constant pump. The improved electromechanical system was readily sold and the newly developed jet-contolled drive would greatly reduce the costs of production.

It was anticipated that the number and range of applications for the Hydro-Constant drive could be greatly expanded if the necessary development work were undertaken. The CEL group came up with a long-term development programme presented to Peerless in a report dated 10 January 1963 (see *Part 5*).

The following figures indicate anticipated development costs and length of time of program.

Time required for Phase II – 9 months. Cost (materials, labour plus overhead) – $34 000.

Time required for Phase III – 9 months. Cost (materials, labour plus overhead) – $32 000.

The proposed development program was based on the following economic considerations.

Initial cost The Phase III (all-Peerless) line of Hydro-Constant drives would offer one of the lowest cost systems for obtaining controlled pressure or flow. The following table is a comparison of prices of two other variable-speed drives and Chicago pumps constant-pressure system. Prices shown are to user except for the uni-pressure system which represents cost of valves only. Prices adjusted to exclude motor.

Horsepower	Peerless H-C ($)	Am. St. Gyrol ($)	World Motor Vari-Drive ($)	Chicago Pump Uni Press System ($)
5	375	830	540	190
10	375	860	970	365
25	575	1 060	1 810	550
50	625	2 100	3 990	810
100	1 800	2 740	8 800	1 370

The following graph (Exhibit 30) was reproduced from an article on variable-speed drives and shows typical prices for different types of variable-speed drives. Prices of the Hydro-Constant drive, including motor, have been plotted in for comparison.

There are several factors contributing to the lower cost of the Hydro-Constant drive. One is the basic simplicity of the system. Another is that the drive is being designed specifically for operating centrifugal pumps. As a result, shafts and bearings can be made much lighter because the load is directly connected to the output shaft rather than being belt or chain driven. Also, with centrifugal machinery torque decreases with reduced speed, so that it is not necessary to provide extra torque-carrying capacity at reduced speeds.

Operating cost Regulating pump capacity by varying speed would result in lower power requirements than controlling by throttling or by-passing. Savings would depend on a number of factors, including average degree of flow reduction and head-flow characteristics of the system.

Maintenance Compared with throttling, maintenance costs would be lower. The control valve would be eliminated and because of the generally lower speeds and lower pressures associated with variable-speed drives, pump maintenance would be reduced. At the same time, maintenance and depreciation on the Hydro-Constant drive should be quite low. American Standard claims a 20-year life for their Gyrol drives with maintenance amounting to one oil change a year. The Hydro-Constant should have a comparable life and require essentially the same servicing.

Market evaluation (1963)

The Hydro-Constant development program proposed by the CEL group was given serious consideration. A market study was conducted to evaluate the proposal realistically. On 31 October 1963, Paul De Pace (Executive offices, San Jose) submitted his report which summarized the results of the market study and estimated Peerless' potential for each of the several Hydro-Constant programs.

Horizontal Hydro-Constant, sizes 3–25 hp

This programme covers the Hydro-Constant coupling sizes with the greatest immediate potential. Orders for these sizes are now at a rate of nearly $300 000 a year, and may increase to $500 000 a year or more, including couplings, motors, controls, and pumps. Couplings and control cost are now about 75% of system costs, and should decrease to about 60% of cost with the new IMC-manufactured unit.

Major markets for these sizes are for the constant-pressure control of water supply for the building trades and for industrial construction. Some market potential appears to exist for use with process pumps, although the economics of the Hydro-Constant should limit its use to selected applications. Some sales in the 3–25 hp sizes for municipal water supply can be made. However, the potential in these sizes for this and other markets studied is not of great significance.

Peerless' marketing of the American Standard Gyrol unit has strengthened its position in the sale of pumps to the building trades. It appears highly desirable, if not quite necessary, that Peerless market a constant-pressure system in order to sell building pumps. In this regard, the new unit should provide advantages of cost and simplicity as compared with the Gyrol. We encourage continued marketing effort on this program directed towards the building trades, industrial customers and process pump customers.

Hydro-Constant program 30–100 hp sizes

Major markets for the 30 hp and larger Hydro-Constant are for constant-pressure control of building and municipal water supply.

We understood that the amount of further development required on the Twin Disc couplings 12·5 in and 14·5 in diameter is not extensive, and that cost of developing and placing these sizes in production would be about $15 000. This expenditure appears justified, on the basis of the following points:

1. Future sales volume of Hydro-Constant systems in this range should exceed $100 000.
2. Margins should be significantly higher than 25%.
3. If this program is not undertaken, Peerless may not be able to be competitive in the larger sizes without its own unit, or without reducing present margins to low levels.

Vertical Hydro-Constant program

Major markets for a vertical Hydro-Constant are for variable speed pumping in sewage lift stations, in municipal water supply, and, to a more limited extent, in industrial water supply. The only one of these markets which is currently firmly established is that for sewage pumping.

The variable-speed sewage pumping system market is of such magnitude ($1 million yearly, with future sizeable growth anticipated), that Chicago Pumps furnishing

EXHIBIT 30   Price range of various basic types of varispeed drives

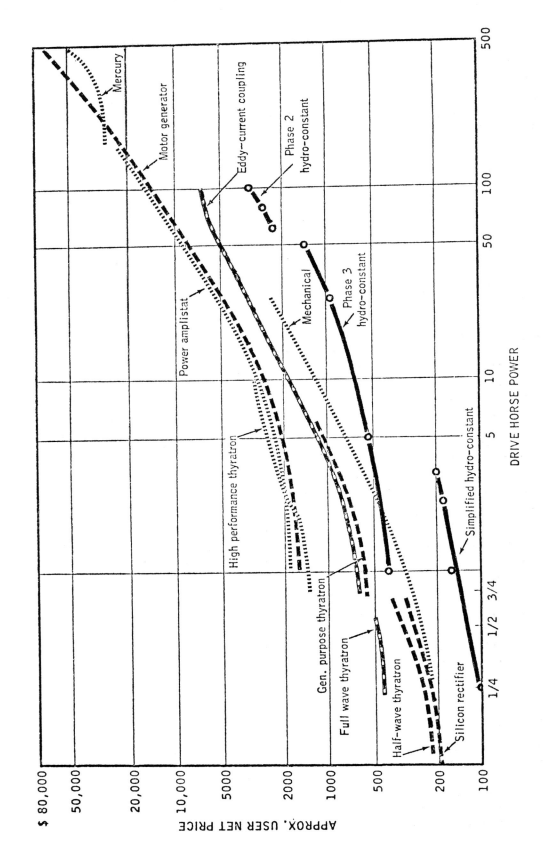

some type of variable speed drive would seem appropriate. In view of Chicago Pump's established sewage pump position, its marketing of an acceptable FMC vertical coupling would seem desirable.

If Chicago Pump should use its established distribution to promote and sell a vertical Hydro-Constant to the sewage market, then the development of vertical couplings would be desirable. Should Peerless alone be marketing a vertical coupling, we do not at this time visualize sufficient potential for the product to make an extensive development program desirable.

Peerless manufacturing costs

Peerless anticipates that manufacture of the Hydro-Constant lead to a substantial increase in profit margin. Depending upon the size and type of installation, between $400 and $800 of added margin per unit is expected in the 3–25 hp sizes.

The additional dollar margin represents an average of about 33% of the total cost of the Hydro-Constant package system, including coupling, controls, pump, base, and motor. Since Peerless margin on the system is 25%, a 33% cost reduction represents 25% of the distributor net price of a system. Therefore, Peerless could reduce distributor net by 25% and maintain the same dollar margin, or maintain selling prices and increase margin on the system to about 50%. As a result it is to be expected that Peerless will be in a position to better meet price competition, which is at least 5 to 10% below Peerless, and to show higher margin.

The American Standard Gyrol represents about 45% of the total cost of the present Hydro-Constant system. Peerless coupling manufacture would therefore represent a substantial amount of new production. For instance, at the present Hydro-Constant order rate of $454 000 a year, roughly $200 000 worth of additional product will be manufactured in the Indianapolis plant.

It is proposed by Central Engineering that Peerless manufacture its own coupling impeller and runner, rather than buy from Twin Disc Clutch Company. A comparison of quotations for clutches from Twin Disc with the total cost of the appropri-

Market summary Hydro-Constant study

	Horizontal program 5–25 hp ($)	Horizontal program 30–100 hp ($)	15–100 hp vertical Prop. ($)
Approximate cost to date for program	65 000	3 000	10 000
Estimated cost to complete program	5 000	15 000	60 000
Current yearly order rate in units	110	20	12
Current yearly order rate in dollars[1]	290 000	90 000	70 000
Potential yearly sales, in units[2]	245	50–70	70–140
Building trades market:			
Present unit order rate	90	15	—
Potential unit order rate	180	30	—
Municipal and industrial market			
Present unit order rate	20	5	12
Potential unit order rate	40	10–20	20–40
Other markets:			
Present unit order rate	—	—	—
Potential unit order rate	25[3]	10	50–100[4]

1. Includes pumps, motors, couplings, and controls.
2. Potential year dollar system sales are $1·0 to $1·5 million.
3. Primarily for process pumps.
4. Assumes Chicago Pump marketing of the vertical H-C.

Hydro-Constant program costs

Project No.	Project title	Total expenditure ($)
Peerless		
PP ED 1261–41	Gen'l dev. – H-C	654[1]
PP 4257-41	Gen'l dev. – H-C	2 537[2]
PP 1973	Fluid coup. prod.	18 386[2]
CEL 9667	Improv. on H-C	4 080[1]
CEL 9856	H-C ouput design	643[1]
		$26 300
Central Engineering		
CEL 9663	Improvement on H-C	1 153[1]
CEL 9753	H-C jet control	50 704[2,3]
CEL 9837	Fluid controls	8 589[2]
CEL 9856	H-C ouput design	474[1]
		$60 920
Totals, Peerless and CEL		$87 220

1. Project closed.
2. Through 30 September 1963.
3. Of this amount, about $10 000 is cost attributable to design of a vertical unit.

ate Hydro-Constant systems indicates that the Twin Disc clutch would be about 3% of cost. If Peerless does $450 000 of business, the value of couplings manufactured (in terms of Twin Disc's prices) would then be less than $15 000 a year. Even if Peerless should increase its Hydro-Constant sales to $1 million a year, only about $30 000 worth of clutches would be manufactured.

Design, development and production engineering expense to introduce an FMC impeller and runner would have to be quite small in order to make Peerless manufacture of this item attractive.

Further, since volume would be rather restricted, there may be a question as to whether Peerless could manufacture at a cost comparable to Twin Disc prices. On the basis of the restricted internal demand such an undertaking would not appear desirable.

SUGGESTED STUDY

1. Submit a conceptual design of at least four systems providing variable fluid flow at a constant pressure (excluding the use of a fluid coupling). Include a description of each system and its operation, a critical analysis of its features, and any necessary sketches. Determine the 'state of the art' by searching current literature, including industry journals, manufacturers' literature, etc.
2. You have been given the problem statement from Peerless to Central Engineering as stated in Part 1. Prepare a preliminary proposal listing the steps that you feel are necessary to fulfil stated requirements. How would you test the final design? Include a program for testing, and a layout of an adequate test facility. What specific information do you need to implement this program?
3. Draw the block diagram for the Hydro-Constant and identify the system parameters for all elements of the diagram.
4. Synthesize a mechanism (linkage or cam), to be driven by a constant-speed motor, which will linearize the arm rotation–discharge reiationship. Use

Mr Bunnelle's plot of 3% discharge–% tube travel. The range of tube travel is 3 in. You may want to change the range of arm rotation. Plot the results for 10° increments of arm rotation.

5. Justify, qualitatively, Mr Bunnelle's original decision to linearize the steady state flow response.

   Hint: Look at Exhibit 6 near shut-off and consider a sudden change in discharge pressure.

6. Present at least two means by which the coupling level of fill can be regulated using hydraulic controls. Include sufficient diagrams to explain your proposals.

7. You are now acquainted with the system proposed by Mr Baird in Part 4. The complete control system will consist of a pump, filter, relief valve, servo valve, and double acting power cylinder as shown on pages 169/70 in Exhibit 14.

   (a) Establish the design specifications for the system in question taking into account performance requirements, size and appearance, environment of operation.

   (b) Choose the right components (consult manufacturers' catalogues) to complete the proposed design.

8. Formulate your own conceptual design to accomplish pressure control under conditions of variable flow.

9. For constant pressure control, design a mechanism that will sense the discharge pressure of the pump and position the splitter. Show how the mechanism will be mounted on the fluid coupling.

10. Prepare a critique of Mr Bunnelle's analysis of the Hydro-Constant control. Explain how your solution is either better or worse than Mr Bunnelle's.

(a) What assumptions did he make?

(b) What would be the consequences of these assumptions?

(c) What other factors besides the transport lag would affect the stability of the system? How? etc.

Case 6   A moving map display

by Professor Geza Kardos

Prepared with support from the National Science Foundation for the Stanford Engineering Case Program. Based on a seminar given by Bob Vago at McMaster University. Assistance from Computing Devices of Canada, Ltd. is gratefully acknowledged.

This case is available in pamphlet form as ECL 166 from the Engineering Case Library, Room 500, Stanford University, Stanford, California 94305.

PART 1   Background

Bob Vago had the job to quickly produce a prototype of an improved Moving Map Display. This is a complex device which automatically shows position and heading of a plane on a map which moves as the plane flies. Optical, electrical, and mechanical subsystems are used. Bob's company originally intended to use a British license and American technology, but later decided to proceed to a novel design. Within a year a prototype was delivered. Four years later production models were sold.

In 1962 Wing Commander Lambert RAF (retired) proposed that Computing Devices of Canada Ltd. build a moving map display. At that time there existed a moving map display built by Mr Honick of the Royal Aircraft Establishment, Farnborough, with which the Wing Commander had been associated. It was proposed that Computing Devices obtain from the Ministry of Aviation a licence to build the Honick display. The intent was to ensure a ready acceptance of the Computing Devices display by the RAF. Additionally, it was reasoned that utilization of North American design techniques would lead to a much smaller and, therefore, more acceptable version of the Honick display.

Wing Commander Lambert had recently joined the British subsidiary of Computing Devices of Canada Ltd. (ComDev) as sales manager. Wing Commander Lambert had been a pilot with the RAF. He had originated the idea of a moving map display navigation aid while with the RAF and had been instrumental in having the Honick display designed to meet his operational requirements. Honick, the designer, had patented many features of the display.

When Lambert joined ComDev he proposed that they undertake the manufacture of the display under license. Throughout the development Wing Commander Lambert was in charge of marketing of the unit and had a profound influence on performance requirements. His contacts within the Ministry of Aviation were invaluable in guiding the progress of the project.

The Project was assigned to Bob Vago, an electronic design engineer: 'It seemed the programme was straightforward and plans were laid to produce a fully working model for flight trial at Boscombe Down in England.'

The license agreement with the Minister of Aviation (MOA) and a letter from the Wing Commander detailing the required operational performance was made available. Soon afterwards, it was learned that the MOA had also licensed several other companies in the UK. The most important of these licensees turned out to be Ferranti of Edinburgh. Ferranti had been working ahead of ComDev for some time with plans to provide the RAF with the Honick Displays.

Computing Devices of Canada Ltd. is a Bendix Corporation affiliate located at Bells Corner, Ontario, just outside of Ottawa. Over a third of their 1 500 staff are scientists, engineers and technologists. It does a high percentage of original design work, enabling it to license others – including competitors – to use its patents. It regularly accepts work on a crash-program schedule. In addition to supplying equipment to Canadian and UK forces, it has been successful in supplying navigational equipment to the US Airforce and the Federal Republic of Germany among others.

Unfortunately, all that was readily available to Bob relating to the Honick display were several photographs of an assembled display unit with side covers removed and schematic diagrams illustrating proposed map assemblage layout. At a planning meeting it was concluded that ComDev must be behind the competition and in addition handicapped by the absence of Honick's working drawings. Some means had to be found to catch up. The strategy decided on was to offer the RAF a different map display with many unique and added features.

The next few weeks were spent evaluating possible improvements based on several years' experience on other navigational systems. During the investigation it was determined that the large size of the Honick display could not be reduced simply by utilizing North American components. The ComDev moving map display would, therefore, bear little resemblance to Honick's and in consequence, little help could be expected from Farnborough. It was recognized that the new approach would result in a much more intensive programme than originally planned. To remain competitive the original time schedule for completion had to remain in force regardless of the expansion of the programme. The new philosophy was presented to the RAF and several weeks elapsed before they accepted it. To meet the planned schedule the project was kept active and forging ahead.

The purpose of the instrument was to provide a navigator with his aircraft's present position against the background of a topographic map. Such a facility was not new and several such devices were in existence. The limitations of these devices were small area coverage, lack of map detail, and large equipment sizes.

Honick of Farnborough and Gillfillan of California had constructed displays using a filmstrip as the basic storehouse for map data. These displays were not suitable for two-seater high performance aircraft. Honick's was too large, Gillfillan's was too dimly lit and both displays required the navigator to participate when changing from one film frame to another. Additionally, both displays were merely repeaters of positional data generated by separate navigational computers.

Computing Devices planned to make up for the short-comings in the new display by making a display with relatively small display size, high level of screen brightness, completely automatic map movement, and inclusion of a small flexible navigational computer.

The display was to be obtained by projecting miniature topographic maps set down on a 35 mm color film through an enlarging lens system onto a 5 in diameter viewing screen.

Because there is much to do in the cockpit of a fast low-flying aeroplane, the performance logistics requires navigational aids to perform virtually without attention. Most air-strikes are pre-planned and the route to target is made up of a series of legs. Each leg termination point is selected for easy recognition. To accommodate this, ComDev's display was to include a destination selector. Each one of the selector positions would correspond to the latitude and longitude of a termination point. This facility was to be available to and from the target. Turning of the selector was to cause the display pointer to orientate to the termination course and the range counter was to show the distance to termination point. To each termination the pilot would turn the craft until the course pointer was at twelve o'clock and run the range counter down to zero.

The proposed map display was to provide a continuous display of the terrain between termination points. The method for referencing the display involved flying over an identifiable earth feature and the earth feature on the display could be adjusted into the centre circle thereby providing correction.

It is not unusual to find that the inclement weather conditions or tactical events prevent usage of a pre-slected landing location. The display, therefore, was to include a means of permitting the navigator to use the map to select an alternate landing site of his choice. Range and course-to-steer to the selected site would be chosen and shown by the display. The same feature could be used in support of air strikes directed from the ground.

On the closing legs of the strike it is expected that violent evasive action on the part of the aircraft may be necessary. Under these circumstances it is desirable to provide the navigator with the means of determining whether the target will be reached. The incorporation of a 'look ahead' feature would cause the intended target to advance to the display centre circle if the pilot was on course. The same feature could also be used to look for mountain ranges or other obstacles long before they came within visible range.

As Bob Vago said, 'It is not surprising that much original system activity **was to** come out of these operational requirements and that several patents were secured.'

A schematic diagram for the overall system was used to establish the basic design philosophy. From this, component requirements and inter-relationships were established.

With the exception of servo-loops, most of the parameters shown on the schematic were established by theoretical analyses followed by breadboarding. In the case of servo-loops, it was company experience that performance could only be verified with final hardware. Breadboarding was not, therefore, undertaken.

Draftsmen were brought in to lay out the components and to provide the necessary supporting structure. At ConDev engineering approval of all drawings, layouts and detail drawings was a requirement before data is released to the workshops.

'The importance of drafting to the conceptual engineer cannot be overemphasized,' commented Bob. 'The necessary interface with the draftsman often provides the engineer with frustration. It has been common in recent years to remove the conceptual engineer from design detail carried out on the drawing board. In consequence, fine detail upon which the success of any project depends is left to a draftsman who is not always sufficiently skilled in engineering matters. The original concepts, therefore, undergo distortion, and in some instances, inadequacy in detailing. It is hoped that this state of affairs will pass with time. It will indeed be a great day for the conceptual engineer when computer-assisted design is an every day occurrence.'

At the time the programme was undertaken, ComDev was changing from a project to a functional organization, and there was a great volume of development work that had to be carried out. In consequence, Bob Vago was left largely on his own to fend with the project. While drafting assistance was available, little direct assistance of any consequence was available for several months. Much assistance was sought outside the company without which progress would have been much slower.

The first prototype was shipped to England in April 1963.

The result of Bob Vago's effort at Computing Devices of Canada was a novel aircraft navigating instrument that displays a map of the region flown over, and shows the aircraft location on the map (Exhibit 1 and Exhibit 2, Fig. A).

The display is on a 5-in screen, showing a map in colour. Present position is indicated by a small circle in the centre. The circle stays fixed and the map moves under it. The map is on a single strip of 35-mm colour film, moved automatically and continuously sideways and forward in accordance with the actual path of the aircraft. Sideways movement of the filmstrip is avoided by having it automatically wind to the next matching section and then continue to traverse.

The details of the design as abstracted from the patent application are given in the Appendix (p. 218).

The heart of the apparatus is a filmstrip having a number of frames in abutted end to end relationships spaced as shown in Exhibit 3, Fig. 3. The filmstrip is contained in a film cassette. Each frame is a transparency representing a part of a map area. A portion of a frame is positioned in the projector light path projecting an image on the screen.

The main frame of the apparatus has a light projector at one end (Exhibit 2, Fig. B) and a display screen at the other end. The light passes through the aperture in the cassette and is focused by means of a lens system onto the screen.

The film positioning is achieved by servo-mechanisms which rotate the turntable, translate the cassette, and wind the filmstrip.

The film cassette with its carriage servo components is mounted on a turntable. The turntable orients the display. The turntable is rotated by a servomotor mounted on the housing. When track orientation is called for, the table is oriented so that the aircraft heading corresponds to the fixed track line on the screen.

The film cassette is mounted on a carriage. The carriage is positioned by a motor-driven lead screw to provide north–south display movement across the width of the film. North–south position feedback is provided through a potentiometer attached to the lead screw.

The film is mounted on spools in the cassette. The spools are attached to tensioning devices. The tension devices are connected to produce a constant tension on the film. A sprocket engaging the film drives the film back and forth against the spring tension to produce the required east–west display movement. A potentiometer connected to the sprocket drive provides east–west position feedback.

The housing contains two light sources which can be interchanged in case of failure during flight. Two projecting lens systems are included so that the display can be presented at two magnifications: 1:500 000 and 1:2 000 000.

The apparatus includes a means for receiving a signal representing earth miles travelled and translating this to map displacement. This device resolves the direction and distance travelled into map grid east–west and north–south displacement. These signals and aircraft reading are applied to the respective servo drives.

Various auxiliary functions and alternate operational modes are available and are described in the Appendix.

The 35-mm filmstrip became the heart of the instrument. It was not fully appreciated in the beginning that considerable effort would be necessary to obtain a filmstrip of suitable quality. Since the company had little experience in film technology, they tried to meet the requirements by use of an outside vendor.

To start on the filmstrip it was first necessary to evaluate various map projections for suitability. The Lambert Conformal projection was selected because of its low distortion of earth features regardless of latitude. An evaluation of the projections' properties established that an area roughly 1 800 × 1 800 nautical miles could be stored on a 35-mm filmstrip 25 ft long and that the standard parallels of latitude should be 36° and 54° North.

The next requirement was to determine the type of film and its drive means. It was thought at first that the most acceptable results would be obtained by driving the film with a pressure roller. Thermal expansion of the film and inaccuracies in the drive produced errors of an unacceptable level. A film sprocket drive was chosen thus enabling use of standard sprocketed film.

Copying USAF pilotage charts onto the 35-mm film presented a formidable number of problems; corrections for Lambert scale factor between earth and chart, for chart shrinkage, for scale factor and for variation in film strip sprocket pitch. Provisions also had to be made for north–south map overlap, for enlarged markings, for high advance and for abutment accuracies.

The commercial microfilm facility used in the early phases of the project soon proved unable or unwilling to cope with the requirements.

In consequence, it became necessary for the company to purchase a microfilm facility. Before the microfilm camera could be used it was necessary to strip out the mechanism and replace it with a re-designed one which met the high standards required.

Next it was necessary to select a colour film type which would exhibit faithful colours and which would have sufficient resolution to prevent blurring of fine detail on the projection screen.

Several European and North American film types were evaluated. Resolution was determined with the assistance of a colour resolution chart provided by the United States Bureau of Standards. This chart was photographed with a high quality microfilm camera and the developed film types were compared under a high power microscope. Colour reproduction was checked by comparison of pilotage charts to full size screen blowups.

Once the film type had been selected, it was necessary to persuade the film vendor to provide the film in the required long lengths.

All servo-loops were analysed theoretically to establish response characteristics and the required component parameters. Servo testing was carried out on the final hardware. Apart from the unexpected behaviour of a critically placed antibacklash gear all servo-loops behaved as predicted.

Optical path analyses were carried out on all optical components excluding the internal elements within the compound lenses. The lenses were proprietary and had to be obtained from France.

When all the optical components were available, a thorough breadboarding programme was instituted to determine screen image resolution, screen brightness against ambient light level, and the effectiveness of heat abstraction techniques over a wide ambient temperature range. The analysis aimed at a legible display with minimum colour washout in an ambient light level of 4 000 ft lamberts. Field experiments confirmed that in some applications performance under an ambient light level of 10 000 ft lamberts was necessary.

The design objective was for quantity production not exceeding 200 fitments. This number was chosen based on typical production quantities for similar navigation equipment.

Apart from supporting structures, interconnecting linkage, including gear passes, most items were purchased from vendors. Such items included servo amplifiers, synchros, electric motors, gear boxes, transformers, saturable reactors, capacitors, semi-conductors, resistors, projection lamps, lenses, slip rings, brushes and many other small items.

Most purchased items were custom built and, therefore, required explicit specification control drawings to be generated by the engineer in order to effect their purchase. ComDev experience showed that certain vendors would over-estimate their capability of meeting the standards required.

It is usual practice on production programmes to divide a requirement between two or more vendors in order to protect the project. In development programmes, limited funds usually inhibit this practice. In addition, only one vendor can be persuaded to cut back his development costs to a reasonable level on the assurance that recovery is almost guaranteed on follow-on production programmes.

'Because development time for some of the more important components often represents a significant portion of the time allowed for the project it can prove fatal if an unsatisfactory component shows up. On this project, this occurred twice and it was most fortunate that a second vendor produced the required components in time. It was annoying to find that the only liability of the vendor under these circumstances was the cost of the component provided' Bob recalled.

Bob Vago felt that 'On short development programmes most of the skills required should be housed within the Company. The decision to make or buy should be limited only by this consideration. Such action is prudent even though the company standards may be questionable when compared with an outside vendor.'

As ideas formed they were weighed for suitability within the design constraints. If an idea did not measure up it was modified or discarded. 'The chosen concept did not evolve from a conscious permutation of known techniques, knowledge or skills, but rather by what is called intuitive thought. We were not conscious of an idea until suddenly it formed in our mind. The thought process appeared random with the time concentration' recalled Bob.

Many of the ideas evolved over a lengthy period of time. In some instances optimization occurred after breadboarding. At no time was any attempt made to employ formal analytical optimization methods.

The operational requirements for presentation of navigational data have developed over the years. Consequently there is little room for the industrial designer. Pilots and navigators are most reactionary to proposals for modifications to existing navigation presentations. There is, of course, good reason for this attitude. It is important that under the stress of operation combat, interpretation of conventional navigational quantities takes place almost as a matter of reflex.

This attitude was reflected with respect to the map display's track counter. For Bob, his display counter represented a significant advance over the conventional compass ring. Navigators' comments regarding this track display were never enthusiastic and often the request for return to the conventional compass ring was heard.

The use of the compass ring would have made the display significantly larger and this was the prime reason Bob didn't use it.

Guidance for human engineering factors in aerospace equipment is set on many programmes by the USAF document AFSCM 80-3. It is normal, therefore, in the absence of a special requirement to work to this USAF document. The characteristics of colour and contrast of the map representation on the screen was not covered by the referenced document.

Controversy existed regarding desirable contrast in the display. Mr Honick was of the view that blacks and reds were desirable. With this, clarity of the display was improved, but at the expense of eye fatigue. ComDev chose the soft greens of the USAF Pilotage Charts. The contrast was not as good as Honick's reds and blacks, but there was no problem with eye fatigue, and also there was no problem in securing wide territorial coverage.

No ready solution was available and it was generally agreed that no existing topographic chart had the desirable contrast characteristics. In the time available, it would have been out of the question to have had a cartographer create a special series of maps for the display.

A target cost for the development phase was allocated based on records of previous programmes of similar complexity.

The only portion of the programme which caused serious cost difficulty was provision of the microfilm copying facility. Fortunately, the cost involved was largely written off by the facility's ability to meet the required microfilm standards.

EXHIBIT 1

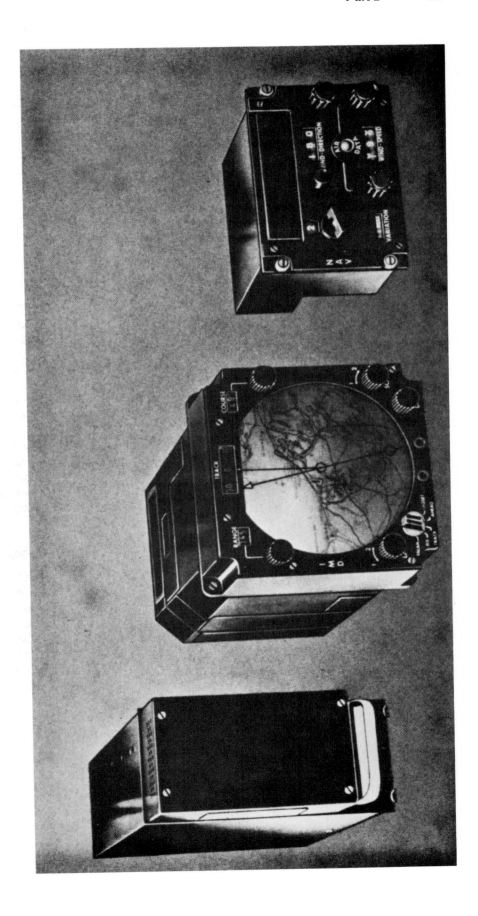

EXHIBIT 2

1. MAP DISPLAY.
2. COUNTER READOUTS.
3. MODE INDICATOR LIGHTS.
4. SLIP RING ASSEMBLY
5. FILM HOLDER (CASSETTE)
6. HEAT FILTERS.
7. STANBY LAMP ARRANGEMENT.
8. COOLING AIR DUCTING.
9. OPTICTAL SCALE CHANGER. (TYPICAL 1:2.)
10. MOTOR/GEARBOX.
11. AMPLIFIER MODULE

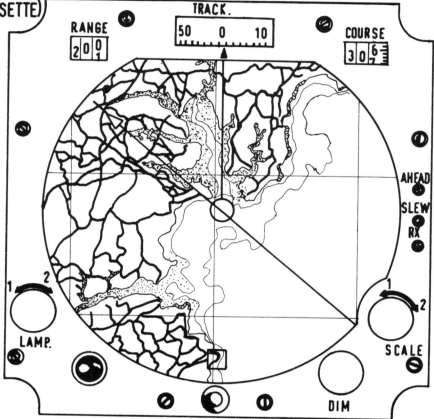

FIG. A   The appearance of the moving map display

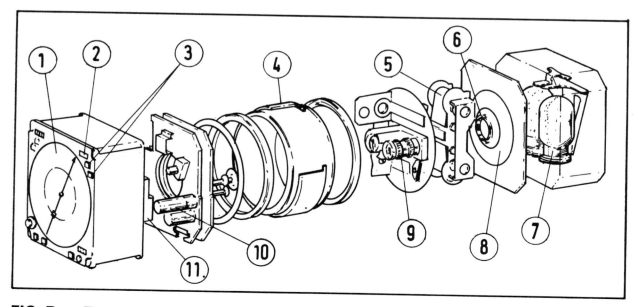

FIG. B   The mechanization of the moving map display

EXHIBIT 3

FIG. 1

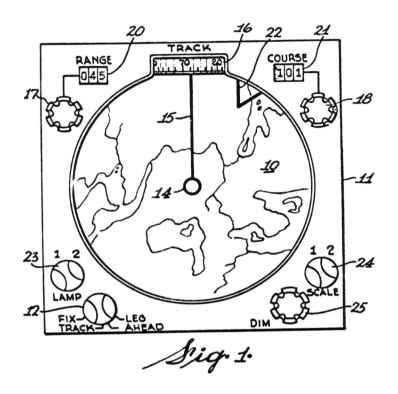

Fig. 1.

EXHIBIT 3 *(contd.)*

FIG. 3

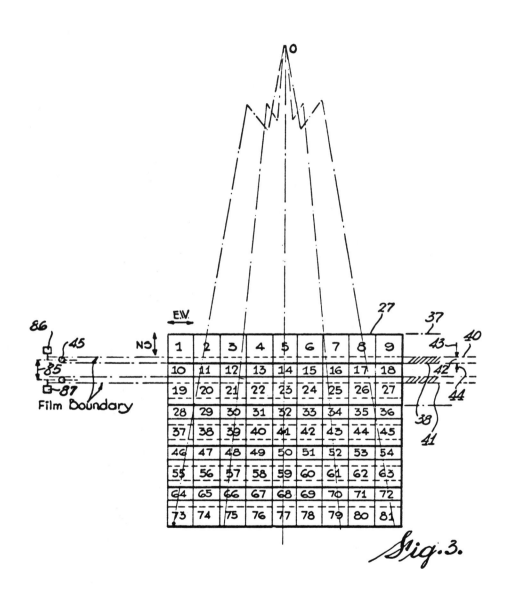

Film Boundary

Fig.3.

EXHIBIT 3 *(contd.)*

FIG. 4

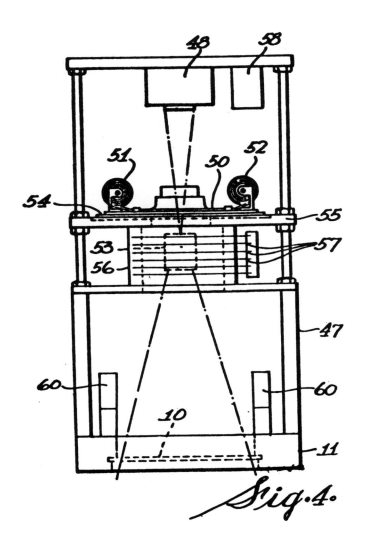

Fig.4.

EXHIBIT 3 *(contd.)*

FIG. 5

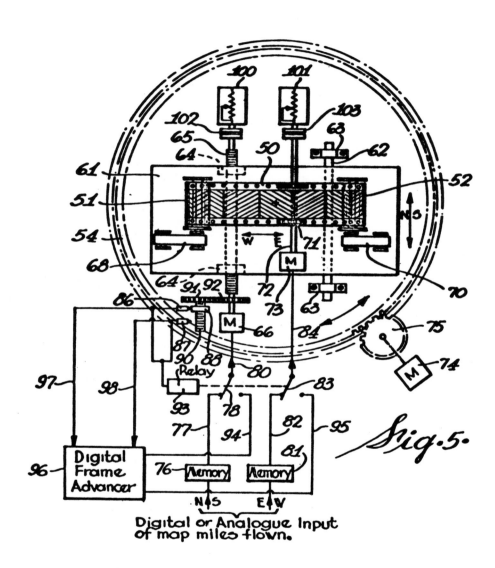

Fig. 5.

Digital or Analogue Input
of map miles flown.

EXHIBIT 3 *(contd.)*

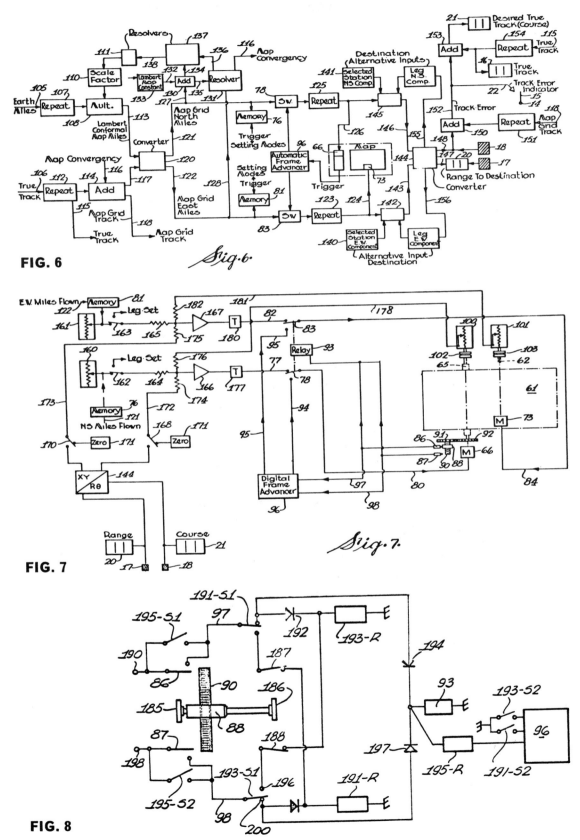

FIG. 6

$\mathscr{Fig.6.}$

FIG. 7

$\mathscr{Fig.7.}$

FIG. 8

$\mathscr{Fig.8.}$

Appendix – Detail design

This appendix describes the detail operation of the moving map display. The part numbers refer to parts shown in Figs. 1, 3, 4, 5, 6, 7, and 8 (Exhibit 3). The description follows closely that of the patent application.

Briefly, the present invention is for a navigational apparatus for use in a piloted craft and having a display unit comprising a frame, a light projector mounted at one end of the frame, a display screen mounted at the other end of the frame in the projector light path, a turntable mounted to the frame between the projector and the screen having a light passing aperture therethrough, drive means mounted to the frame and engaging the turntable for orienting the turntable in a pre-determined orientation, a film transport mounted to the turntable and having back and forth movement in a first direction, a film strip holder carried by the film transport and including a film driving sprocket rotatable in a second direction at right angles to the first direction, a first drive mounted to the turntable and connected to the film transport having a film transport driving rate responsive to a component of movement of the craft in a fourth direction at right angles to the third direction, a switch mounted to the turntable and operable upon movement of the film transport in the first direction to a forward and backward limit, respectively, means actuated by the first switch energizing the first drive for a predetermined movement at an accelerated rate, and means actuated by the first switch energizing the second drive for a predetermined movement at an accelerated rate.

The film strip holder preferably holds a strip of film having a plurality of frames in abutting end to end relationship. Each frame is a transparency representing a part of a map area. A portion of a frame is positioned in the projector light path projecting a map image on the screen. The film strip and consequently the map image are moved by the first and second drives in accordance with movement of the craft to show craft position.

The apparatus also includes the means for receiving a signal representing map miles travelled, means receiving a signal representing true track and deriving therefrom a signal representing map grid track, means converting the signals representing map miles and map grid track into signals representing north–south and east–west components of craft movement in map miles, and means applying these latter signals to the first and second drives for positioning the film strip.

In a preferred embodiment, a destination position may be set into the apparatus. Two signals are resolved from the destination position representing north–south and east–west components of craft position to energize the first and second drives causing the film strip to move to project on the screen an image showing a map representation of the desired track to destination.

Figure 1 shows the display unit as it would be seen by the operator. A screen (10) occupies the centre portion of a front panel (11) of the display unit. An image of a portion of a topographic map is projected onto the screen from a film strip. The apparatus has four operational modes which may be selected by the mode knob (12) the normal mode is the 'track' mode. When operating in the 'track' mode, a small fixed circle (14) on the screen represents the present position of the aircraft, and a fixed position of the aircraft, and a fixed line (15) represents aircraft track. As the aircraft moves over the terrain the map image moves correspondingly along track line past the present position circle. The frames in the film strip change automatically when required. The track being made is shown on track indicator scale (16). Thus, the operator always has before him a map display showing aircraft present position with aircraft track indicated above.

Another mode of operation is the 'leg' mode. When this mode is selected, a desired leg of a flight can be established to a destination at the end of the leg. The destination position may be set in by means of a separate push-button unit, or it may be set in using a range knob (17) and a course knob (18). The range or distance to the leg destination is shown on range counter (20), and the desired course or track to the destination is shown on counter (21). When this destination has been set in, the knob (12) can be turned to place the apparatus once more in the 'track' mode. Then the range and course continue to show the remaining distance and course or bearing to the destination. In addition, once a destination has been set in and the apparatus is in the 'track' mode, an indicator (22) indicates the desired track to the destination. Displacement of the indicator from track line (15) shows track error.

Another mode of operation that can be selected by knob (12) is in the 'look ahead' mode. When this mode is selected, and when a destination has previously been set into the apparatus, the map display automatically advances at an accelerated rate along the desired track to the destination. This provides the operator with an opportunity to look at the terrain, shown by the map, over which the desired track lies. If desired, knobs (17) and (18) can be rotated to slew the projected map image around to look at any particular part.

A 'fix' mode may be selected by the knob (12). This made is provided to correct positional errors in the display. If a fix is obtained and the position of the fix does not correspond to the position shown on the map display by the present position circle (14), then the 'fix' mode is selected and knobs (17) and (18) operated to position the display in accordance with the position of the fix. The apparatus is then returned to its 'track' mode, the input signals representing the aircraft movement are fed into a memory and stored. When the 'track' mode is again selected, the stored signals are taken from the memory to move the projected map image so that the display will again show the position of the aircraft.

The scale control knob (24) selects one of two optical magnifications for the projector. Thus, the magnification can be changed to meet different requirements. The scale of 1:500 000 provides a viewing radius of about 17 nautical miles on screen (10) and the 1:1 000 000 scale provides a viewing radius of about 34 nautical miles.

It will be noted that the display, as described, has the aircraft track towards the upper part of the display on Fig. 1. This is normally the most convenient display for the pilot. Provision is also made to orient the map display with North represented by the upper part of the display as seen in Fig. 1. A spring loaded switch is incorporated in knob (18) for changing the orientation.

The apparatus could be readily adapted for use with various map projections. It was, however, preferred to use a Lambert conformal conic projection. Such projections are well known.

Film layout

Referring to Fig. 3 there is shown a schematic view of the layout of the frames representing portions of the map area of interest. The map is divided into a series of smaller parts designated by the numbers (1) to (31). The parts are arranged in rows and columns. Each of the designated 81 areas represents a part of a topographic map photographed as one frame on a film strip. The frames are positioned so that the frames (1–9) represent a continuous strip, frames (10–18) represent a continuous strip, and so on, and there is no overlap or separation on the film strip between frames.

Because of the interruption when a frame boundary is reached in the display resulting from N.S. aircraft movement, it is convenient to have an overlap between areas represented by adjacent N.S. frames. This N.S. overlap is shown in Fig. 3. The amount of overlap is chosen so that on any frame it will provide a full display when projected onto the display screen, and will just fill the screen. Thus, when the film strip is being moved to change the projected image from one frame to another representing an adjacent N.S. map portion, just before and just after the change the same image will be displayed and each image will fill the display without showing the edge of the frame.

Detail operation

In Fig. 4 there is shown a general layout of the main parts of the apparatus. The light system (48) directs light through a portion of a strip of film (50) extending between spools (51) and (52). These are on a turntable (54) rotatably mounted in a bracket (55) on frame (47). A slip ring assembly (56) is mounted to move with the turntable. A set of brushes (57) mounted to frame engage the slip rings. A blower (58) is provided for cooling the projection system and servo mechanisms.

The turntable (54) is shown from the top in Fig. 5 with its associated parts. Mounted to the turntable (54) is a film transport (61) which moves back and forth in the direction marked N.S. Towards one end of transport (61) a shaft (62) is fixed to the transport, driven by shaft (65) which in turn is driven by a reversible motor (66). When motor (66) is energized it moves the film transport in the direction marked N.S. The shaft (62) supports and guides the transport to ensure only movement in the N.S. direction.

The spools (51) and (52) are mounted to the film transport (61) with the film (50) extending between the spools. Tensioning devices (68) and (70) are mounted to film transport and connected to the spools. The tensioning devices are designed to keep a substantially constant tension on the film strip regardless of the amount of film on each spool. A sprocket (71) engages the edge holes in film strip (5). A reversible motor (73) is connected to the sprocket shaft (72) for driving the film from one spool to the other in an E.W. direction.

A turntable drive motor (74) is provided for driving the turntable (54). The orientation of the turntable in Fig. 5 is in a N.S. direction. The motor (74) is driven in response to a signal representing true track.

Light from the projector passes through a portion of film strip (5) to provide a map display image. As the aircraft moves the motors (66) and (73) are driven to provide movement of the film in accordance with aircraft movement, and the motor (74) is driven to maintain the required orientation of the turntable and the film strip. The transport drive motor receives its control drive signal from a N.S. input through a N.S. memory (76). The film drive motor (73) receives its control drive signal from an E.W. input through an E.W. memory (81).

When a N.S. boundary of a frame in the film strip is approached, the film sprocket drive is energized to move the film strip nine frames at an accelerated rate while at the same time the film transport drive is energized to move the film transport to its opposite side at an accelerated rate. This action is initiated by two micro switches (86) and (87) mounted to the turntable. Thus switches (86) and (87) define limits of movement of the film transport.

Referring now to Fig. 5, the micro switches (86) and (87) are shown mounted to the turntable (54). The micro switches (86) and (87) are actuated by a travelling nut (88) which is on a threaded shaft (90). As shaft (90) rotates, the nut (88) is moved along it. The shaft (90) has a gear (91) meshing with gear (92) mounted on shaft (65) driven by motor (66). Thus, the position of the travelling nut (88) always has a direct relationship with the position of the film transport.

When the film transport has reached a position that is its limit of movement north, the switch (86) is actuated by nut (88). A relay (93) is operated to move switches (78) and (83). When the switches (78) and (83) are in this position, the signals representing N.S. and E.W. map miles are stored in the memories (76) and (81). At the same time, micro switch (86) initiates the operation of a digital frame advancer (96). The digital frame advancer (96) comprises a pulse generator that provides a discrete predetermined number of pulses to motors (66) and (73). These pulses drive the motors at a substantially faster rate than they are normally driven and they drive them by an exact predetermined amount. Thus, the motor (66) moves the film carriage to its opposite limit, and the motor (73) moves film (50) by exactly nine frames. When the required number of pulses have been applied to the drives the relay (93) operates to return switches (78) and (83) to their normal position. The memories (76) and (81) then feed out the signals stored to bring the position of the film strip up to date.

It is desirable in practice to include means for preventing random operation of the frame advancer (96) during unusual operational manoeuvres. For example, if the aircraft should fly into the region represented by a boundary between two adjacent map strips, say the boundary between frames (1) and (10), and should then turn and fly in an E.W. direction parallel to this boundary, the apparatus should provide the correct display. As another example, if the aircraft should fly in a N.S. direction to a boundary and initiate a frame change and should then immediately reverse its direction, the apparatus should recognize this and provide the frame for the correct display. Figure 8 shows a means for preventing random operation under such unusual manoeuvres. To aid in identifying the relays with the switches they control, they are given the same designation number – the relay coil being followed by the letter R and the switch being followed by the letter S.

The travelling nut (88) has switch operating arms (185) and (186), which are shown in the drawing as extended from the nut to indicate engagement of switches (86) and (87) on one side and switches (187) and (188) on the other side with movement of the nut (88). With the display at a N.S. boundary, the travelling nut is in a position where arm (185) engages switch (86) and closes it.

The closing of switch (86) energizes relay (193-R) moves switch (193-S1) to ground to complete the circuit through the frame advancer. The relay (93) operates switches (78) and (83) as described in connection with Fig. 5. The relay (195-R) functions as a locking relay and when energized closes switches (195-S1) and (S2). This serves to maintain a closed circuit between terminal (190) and conductor (97) after nut (83) with arm (1985) moves away from switch (86) during the frame advance.

When switch (86) has just been closed and the circuit has been established as described above, the frame advancer (96) is placed in an operative condition. A short delay, of the order of 250 milliseconds, is built into the frame advancer to ensure that all relays have completed their operation. After this delay, the frame advancer (96) generates pulses to drive motors (66) and (73). As the pulses are generated, the shaft (90) is rotated and nut (88) with arm (185) moves away from switch (86) allowing it to open. The switch (195-S1) is, however, closed and the opening of switch (86) does not affect the circuit. At the moment the frame advance is complete, the travelling nut (88) is positioned with arm (185) just closing switch (87). There is a circuit established from a voltage source at terminal (198) through switch (87) [and also through switch (195-S2)], conductor (98), switch (193-S1), contact (196), switch (188) and relay (193-R) to ground.

When the frame advancer (96) finishes the frame advance it momentarily provides an open circuit and thereby stops the current flow through relay (195-R) and permits switches (195-S1) and (195-S2) to open. The opening of switch (195-S1) stops the flow of current from terminal (190) and de-energizes relay (93). However, there is still a circuit from terminals (198) through switch (87), switch (193-S1), switch (188) and relay (193-R) to ground. Thus, switches (193-S1) and (193-S2) do not move. Before frame advancer (96) is permitted to operate again, the switch (193-S2) must open and then be re-closed.

It will be seen that if the aircraft should turn and fly parallel to a boundary at this time, that is fly in an E.W. direction, the arm (185) will not move and switch (87) will remain closed. This will keep relay (193-R) energized and switch (193-S2) closed, preventing operation of the frame advancer.

Now if the aircraft reverses its direction during a frame advance period initiated by switch (86), the frame advance will be complete with arm (185) just closing switch (87) and then the movement of the aircraft in the reverse direction will cause the travelling nut (88) to move farther towards switch (87). Switch (188) is set to open with a short over-travel of nut (88) past the position where switch (87) is just closed. When the additional one-half mile has been travelled, the arm (186) opens switch (188). This de-energizes relay (193-R) so that switch (193-S1) moves to contact (200) and at the same time switch (193-S2) opens. The opening of switch (193-S2) resets the frame advancer (96), and the switching of (193-S1) completes a circuit from terminal (198) through switch (87) to energize relay (191-R), relay (93) and relay (195-R). Thus, a frame advance is initiated through switch (87) in a manner similar to that initiated by switch (86) previously described.

In normal flying, the aircraft would continue on a course across a boundary between adjacent strips. A frame advance would be initiated for example by switch (86) and would be completed with nut (88) positioned to just close switch (87). As the aircraft continued its normal flight, the nut (88) would be moved away from switch (87) permitting it to open. This would de-energize relay (193-R), move switch (193-S1) to its other contact (200) and open switch (193-S2). The circuit is then ready for the next frame advance.

Referring once more to Fig. 5 it will be noted that two potentiometers (100) and (101) are connected through clutches (102) and (103) and disengage during the frame advance operation so that the change in frames does not affect the potentiometer record. The manner in which these potentiometers (100) and (101) are used will be described later.

Referring to Fig. 6, the two main inputs to the apparatus are indicated at (105) and (106) representing earth miles and true track. The earth miles signal at (105) must be transformed to a signal representing amp miles before it can be used in moving the map display. The earth miles signal at (105) is passed by a repeater (107) to a multiplier (108), where it is multiplied by a scale factor from scale factor unit (110). The scale factor in a conic projection varies with latitude in some manner.

For any given conformal projection used for preparing the film strip, the upper and lower standard parallels are known and are set into the apparatus prior to using it. The map latitude in degrees is derived in the apparatus to a very close approximation from a map miles to degrees converter or resolver (111). Thus, the output from multiplier (108) is a signal representing map miles.

The true track signal at (106) is passed by a repeater (112) to an adder (114). The repeater (112) also supplies a true track signal at (115) to be used in another portion of the apparatus. The adder (114) also receives a signal at (116) representing map convergency C from a source that will be described later. The output signal from adder (114) represents map grid track and is available at (117) and (118).

The signal representing map miles is input to converter (120), and the map grid track is applied on another input. The converter (120) is a ball resolver or sin cosine resolver, and is designed to convert from terms of measurement by angle and distance to terms of distance in rectangular co-ordinates. The output from converter comprises of a signal representing map grid miles in a N.S. direction (ΔN.S.), and a signal representing map grid miles in an E.W. direction (ΔE.W.).

The signal ΔE.W. is applied through switch (83), repeater (123) and to motor (73) to control the sprocket drive to move the film strip longitudinally. The signal ΔN.S. is applied through switch (78), repeater (125) and to motor (66) to drive the film transport. Memories (76) and (81) store the ΔN.S. and ΔE.W. signals when the switches (78) and (83) are operated for setting or changing frames or whenever the apparatus is not operating in the 'track' mode.

Signals are available for driving motors (66) and (73) to position the film strip and accurately depict aircraft present position. When a N.S. boundary of the film is reached the automatic frame advancer (96) automatically changes frames as described. An automatic and substantially continuous display of present position is provided. The only adjustment required may be to correct the position when necessary.

The signals representing ΔN.S. and ΔE.W. map miles are also applied to adder

(130) and resolver (131) where they are processed to determine the appropriate scale factor by resolver (137) and (111).

'Leg' mode operation

Provision is made to set a destination into the apparatus by a separate push button control or by the controls provided on the front panel. X and Y values of any given position may be applied to the apparatus by a push button unit indicated in Fig. 6 by blocks (140) and (141). Thus adder (142) receives a signal representing E.W. from repeater (123) and a signal representing the X or E.W. component of a selected station, and provides a summed output to a converter (144). Similarly, adder (145) receives a signal representing ΔN.S. from repeater (125) and a signal representing Y or the N.S. component of a selected station, and provides an output to converter (144). These signals represent the E.W. and N.S. component distances in map miles from a reference to the destination, and are converted to a range and bearing. Converter (144) is a double purpose converter which performs xy to $R\theta$ conversion or performs $R\theta$ to xy conversion. The distance part of the output is used to drive a range counter (20) which shows range to destination. The bearing portion of the output is applied to an adder (150). The other input is a map grid track signal obtained from adder (114) through repeater (151). The adder provides a signal which is the difference of two input signals and represents track error. The track error signal is used to position pointer (22) (Fig. 1) and is also applied as one input to an adder (153). A true track signal from repeater (112) and repeater (154), is used to drive the true track indicator (16), (Fig. 1) and is also applied as a second input to adder (153). The output from adder (153) is the desired true track to destination, and is used to drive counter (21) (Fig. 1).

When the apparatus is in the 'leg' mode, the range knob (17) and the course knob (18) may be manipulated to set in a desired destination. The knobs (17) and (18) adjust signals applied to converter (144) which then acts as an $R\theta$ to xy converter supplying signals representing N.S. and E.W. components of the change. These components are applied to leg component blocks (157) and (158) to alter them to provide the required output signal to adders (145) and (142). The operation is then the same as when selected station components blocks (141) and (140) were used. The 'leg' set mode is completely independent of map motion.

PART 3

In the development of the moving map display and its subsequent field trials, Bob Vago ran into several major problem areas.

The first had to do with the film sprocket drive system (71, 72, 73, Fig. 5, Exhibit 3, Part 2). The original design had the motor driving the sprocket through an antibacklash gear. This gear also drove the feedback potentiometer. The antibacklash gear used was of the spring loaded two-piece type.

During the initial assembly phase, unexpected hunting of the associate servo-loop occurred. It was found that since the motor had to drive the film against the film tensioning spring, high torque had to be transmitted by the antibacklash gears. The original backlash springs in the gear were not sufficient to carry the torques. Consequently, the gears distorted and set up vibrations in the feedback loop.

The fault was corrected by first placing heavier springs in the antibacklash gear. Subsequently fitting a synchro in place of the gear system did away with the problem all together.

A second problem appeared while the equipment was on trial in England. Its presence was masked at first by other faults in electronic sub-assembly. A large

EXHIBIT 4

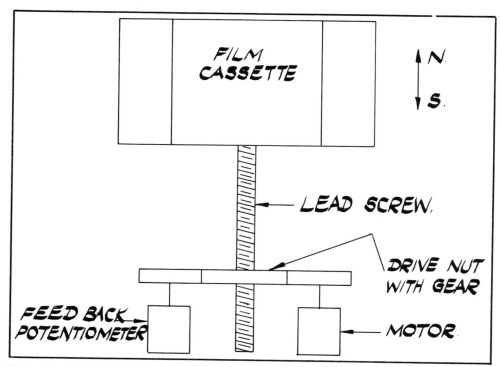

FIG. 1   **North–south cassette drive**

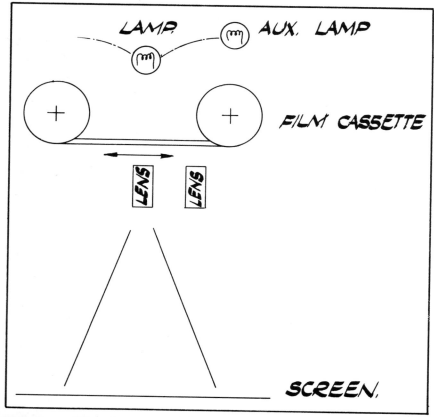

FIG. 2   **Film, lamp, and lens system**

standoff error occurred in the aircraft's position indication after automatic film advance. A simple circuit modification involving wire re-routing took care of the fault.

A more serious fault was the failure of an electronic sub-assembly which continued to plague the design for many months. The Automatic Frame Advancer (96, Fig. 6, Exhibit 3), was a digital unit which indicated and recorded the relationship between frames when the edge of the film was reached. This was the only digital unit in an entirely analog system. Because the power requirements were excessive for the digital unit, the circuit was susceptible to unwanted triggering from transient voltages emanating either from relays or associate power supplies.

The heavy involvement of the electronic circuit designer on other high-priority programmes prevented this fault from being rectified sooner. The re-occurrence of this fault caused a most unfavourable impression during field trials.

Another problem that continued to plague the project was the north–south cassette drive. A sketch of the system is shown in Exhibit 4, Fig. 1.

The north–south drive motor drives the nut through a gear integral with the nut. This same gear was used to drive the feedback potentiometers. Consequently, the positioning of the film cassette was dependent on the dimensional accuracy of the lead screw and the associated hardware. Any backlash in the system showed up as position error. This fault showed up during shakedown of the system in England. It took an inordinately long time to correct because a suitable machine shop to make a more accurate lead screw was not available at Boscombe Down.

A final problem that came up was brightness of the display. Although the unit was considerably brighter than existing display systems, it still was not considered optimum. The field trials did lead to criticism of insufficient brightness. The limiting factor on brightness was heating of the film. The film, lamp and lens system are shown in Exhibit 4, Fig. 2.

For reliability, two lamps were necessary and could be swung in position as shown. The two magnifications necessary required the use of two lens systems. The geometric relations of all these components with associated cooling made it impossible to increase the brightness of the screen without excessive heating of the film.

The requirement for two magnifications and the resultant two lens system caused considerable argument between Bob Vago and Wing Commander Lambert. Bob questioned the need and utility of the two magnifications. He felt that it added unnecessarily to the complexity of the device. He also was of the opinion that with a single lens system the components could be rearranged to increase the brightness. Wing Commander Lambert was insistent that the two map scales were absolutely necessary from an operational view-point and consequently from the sales viewpoint.

When asked about these problems, Bob Vago commented, 'It probably seems unreasonable to have the continuous failure of one electronic sub-system jeopardize all the work carried out to that date. It is, therefore, worth mentioning the fact that situations apparently unconnected to engineering can have an effect upon a project's success.

'The map display project was running in the order of six to eight weeks late with respect to the original schedule. Information available at the time led the Company to believe that Ferranti was going to make their model available to the RAF if ComDev did not show up on time. In addition, it was understood that the RAF had an aircraft standing by awaiting installation on the ComDev equipment.'

On the day that the map display was first brought together as a system, an urgent telegram was received which strongly recommended shipment of the system to Boscombe Down. The equipment was summarily packed into crates and shipped by air to the United Kingdom. Bob Vago and a field service representative were sent off on the same flight.

While enroute over the Atlantic, Bob was still contemplating a fix for the electronic sub-assembly. It was on the following day that the equipment was unpacked on a bed in the George Hotel in Boscombe Down. The following morning the equipment was to be presented to the RAF. An assistant from the London Office of ComDev was sent out to purchase a soldering iron from a local store so that wiring modifications could be made.

'Promptly at 9.00 on the following morning the equipment and supporting team were made available to the RAF. It soon became apparent that the aircraft was occupied for another six weeks on other equipment trials. During the next several months no sign of the Ferranti display was seen,' continued Bob.

The RAF made room for the Computing Devices equipment in a small corrugated Nissen hut in which radar equipment was being serviced. Unfortunately, the skills, materials, techniques, and components necessary to the map display were nowhere to be found. 'On occasion, it became necessary to have machine shop work done. The nearest usable facility was 60 winding miles away in London and the supply of electronic components was limited to those available at a TV repair shop 15 miles away in Salisbury.'

During the next four weeks the project team worked 16 hours a day, 7 days a week, to obtain satisfactory bench operation of the map display. 'Frequently during this time period the electronic sub-assembly acted up and finally we had the circuit designer flown in from Ottawa to do what was possible. Once again the primitive facilities were against the project and only a limited fix could be implemented.' By the time the RAF was ready for flight trial of the display everything was working well except for the occasional failure of the electronic sub-assembly. It was with much trepidation that Bob and the project team looked forward to the first flight trial.

After about one week's installation work the display was ready for trial. 'The first flight went off without a hitch,' recalled Bob, 'and the display did all that had been hoped for. It was almost worth all the trouble to be able to recognize earth features from the aircraft by means of the map display. The accuracy was very good and the RAF navigator was favourably impressed. On the journey back to base the set was switched on and almost exact correspondence existed between the map display and the TACAN set display all the way back to Boscombe Down.'

After the first flight trial Bob Vago was rotated back to Ottawa. Subsequent trials were marred by occasional failure of the electronic sub-assembly. Several months later, due to low priority, the fault was cleared back at ComDev. 'Unfortunately, the damage was done.' The RAF flight trial report issued approximately one year late commented on the fault. This same report criticized the uneven steps of the display. 'The size of these steps was $\frac{1}{64}$, which gives some measure of how subjective a display of this type can be,' was Bob's response. The uneven motion was caused by the finite resolution of a wire-wound servo potentiometer. No attempt was made to find a potentiometer with better resolution while the equipment was on trial. Later trials in a Javelin aircraft involved placing the display at eye-level in the cockpit. In this position, washout of the display occurred in direct sunlight.

The trials were complete with only one major fault, the solution of which was just a matter of time.

The company was not in a favourable position for tender. Unfortunately, the Ministry of Aviation found it necessary to cancel the Hawker Hunter for lack of funds. Company's sales activity then went into low-gear and eventually Ferranti managed to persuade the Ministry to purchase their version of the Honick display for another aircraft fitment.

PART 4   Epilogue

Shortly after completion of the project Bob Vago left Computing Devices of Canada. The cancellation of the aircraft by the Ministry of Aviation, competition from Ferranti and an indifferent flight trial were factors which played a part in the unsuccessful attempt to get a production contract.

Bob felt, 'This demonstrates that the success of an engineering project can depend upon circumstances largely beyond the control of the engineer in charge.'

Computing Devices did not give up on the display. They continued a vigorous sales effort and field trials.

Three of the original development models produced for flight trials were tested by the Royal Air Force, the French Army and by the Royal Canadian Air Force.

Royal Air Force trials

The Royal Air Force trials were conducted between April 1963 and May 1964. For the first phase of the trials, the Projected-Map Display was installed in a Hastings transport aircraft. The system was provided with true heading and drift angle and ground speed from a Doppler radar. Principal objectives of this phase were to assess the accuracy of the display and to evaluate the human factors of readability and handling, principally as a navigator's function.

For the second phase of the flight trials the Projected-Map Display was installed in a Javelin aircraft. Sensor inputs to the system were magnetic heading and true airspeed. Magnetic variation and wind velocity were manually set. The principal objective of this phase of the trials was to assess the display as a pilot navigation aid in low flying, high-speed strike and reconnaissance aircraft roles.

The flight trials were conducted with some night flying at an average height of 250 ft and an average air speed of 450 knots.

The final phase of the RAF trials was a prolonged bench test of equipment which included 24 simulated flights of 300 nautical miles and the cumulative results (for a sub-system of Display Unit and Analog Computer) of these simulated flights showed mean errors of 0·33% along track, 0·05% across track and a standard radial error of 0·26.

The first trials of a projected colour film map display provided a valuable assessment of display performance and human factors. Low altitude track-keeping ability was excellent, as was navigational orientation. Fixing capability was continuous and simple because sensor errors which caused the map to drift out-of-phase with the ground were readily detected and corrected. Map display controls have been greatly improved in design as a result of the human factors assessment in flight.

French Army trials

The trials were sponsored by the French Army Light Aviation Group. Bench trials were conducted at SUD Aviation in 1964, and flight trials in an Alouette III helicopter during 1965.

No trials report was available: however, an operational requirement for a projected map display was introduced after the trials. Pilot debriefing following trials established the general acceptance and enthusiasm for operational ease of handling and the actual display of information, as presented.

Two significant statements made by Captain Chaillet, the project pilot, placed numerical values on the advantages of the projected map over hand-held maps. These were:

(a) the projected map enabled a reduction of crew from two pilots to one.

(*b*) that mission navigation success rate (i.e. maintaining prescribed course with one kilometre at 'nap-of-earth' altitude) rose from 20–25% to 95%.

During the above trials development continued. Updating and improvements included reduced step size on the film drive, joystick slewing and position update, relocation of mode controls, mode lights and improvement in film frame advance.

RCAF flight trials

The trials vehicle was a CF100 aircraft and tests were run in 1965 and 1966.

Evaluation included Phase 1 using heading, true airspeed and hand set wind velocity and Phase 2, using excellent sensors of stable platform heading and K band roll stabilized doppler.

Although an official report of the flight trials has not been issued, liaison has established that the display was extremely well received. Pilot comment confirmed that the development of improved brightness under direct sunlight, on which work was being done, was a definite requirement.

These flight trials showed that the film presentation concept, the basic outline of the equipment and capabilities of the design were well established. Accordingly, further development work proceeded along the original lines incorporating anticipated improvements.

Design updating included improved cooling, modified film drive, frame advance speed up and accuracy, packaging, film-type change. optical components and brightness level.

A-7 aircraft flight trials

Flight trials were carried out by the United States Naval Ordnance Test Station, China Lake, in mid-August 1967 on an improved system shown in Exhibit 5.

Flight testing and evaluation was conducted by VX-5 Air Development Squadron in an A-7A aircraft.

The purpose of the trials was to determine the advantages of the Projected Map Display over the paper roller map presently installed in this aircraft type.

The results of these efforts was a successful bid to equip the Corsair II A-7E with the system. George Wilaneous, Vice-President of Marketing for ComDev remarked, 'Our success in this case was largely due to proven capability of the system, as a result of many field trials.'

In reviewing the case history, Gordon M. Mount who acted as programme manager at ComDev during the A-7 competition stated:

'It is my opinion, that the "non-success" of the early RAF trials was not due particularly to ComDev ineptness, which was Vago's original implication. Vago attempted to maximize his diligence, etc., but felt he was the victim of circumstance. To me, the major cause of early customer reluctance was a combination of "the market wasn't ready," and the engineering design clearly had severe functional limitations.

'It should be made clear that only the broadest similarity of concept exists between the current design and the original prototype; the present-day Projected Map System is one of the more successful items of avionics hardware on the A-7 airplane.'

EXHIBIT 5   Design improvements

PROBLEM	SOLUTION
1. Inaccuracy in present position after automatic frame change operation attributed to lost or spuriously gained stepper motor pulses during frame change.	Redesign of film drive system to replace stepper motor drives with servo motors and replacement of potentiometers with synchro transmitters for feedback of film position.
2. Obvious coarseness of map incremental step motion.	
3. Lengthy frame advance time.	
4. Insufficient display illumination for satisfactory viewing in direct sunlight. Note: Increasing projection lamp excitation would shorten lamp life and cause film damage by overheating.	New projection lamp operating at a higher excitation voltage, improved cooling and modified optical elements.
5. Insufficient heat removal at high altitude and high ambient air temperature accelerates film fading. (Revealed during MIL-E-5400H Class 1 environmental testing).	Redesign of cooling system to provide two blowers, one using unfiltered ambient air to cool projection lamp and the other using well filtered air to cool the film-gate heat glass.
6. Film cassette subject to up/down and forward/backward vibration modes ranging from mild to severe with sufficient displacement to cause loss of focus.	Redesign of complete film cassette and turntable to provide maximum rigidity.

EXHIBIT 5 *(contd.)*

PROBLEM	SOLUTION
7. Position resolution, repeatability of position indication and some-time erratic operation of film drive, all attributed to film position feed-back derived from potenti-ometers.	Redesign of film drive system utiliz-ing servo motors and 3 speed servo feedback provides high resolution and accurate repeatability.
8. Film damage caused by dust particles	Redesigned film cassette with housing and use of filtered air for film gate cooling.
9. 'Hot Spot' or non-uniform illumination of screen as a result of increased illumination.	Reselection of projection lens.

EXHIBIT 6   Functional block diagram

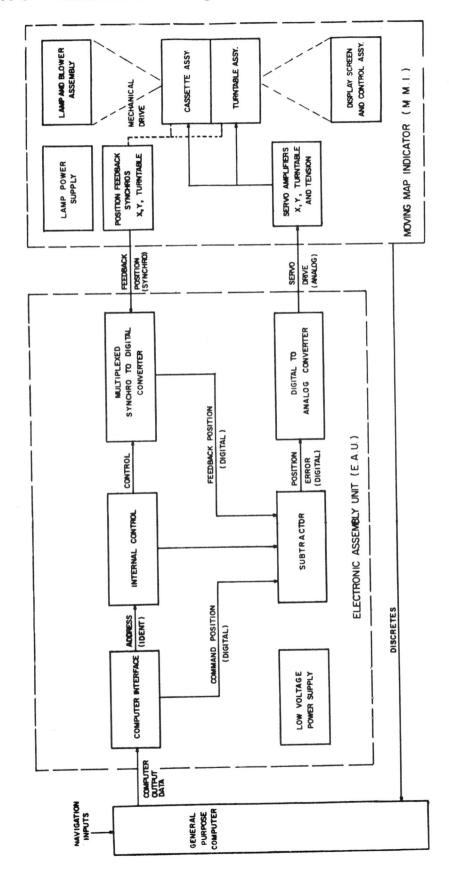

EXHIBIT 6 *(contd.)*   **Exploded view of the moving map indicator**

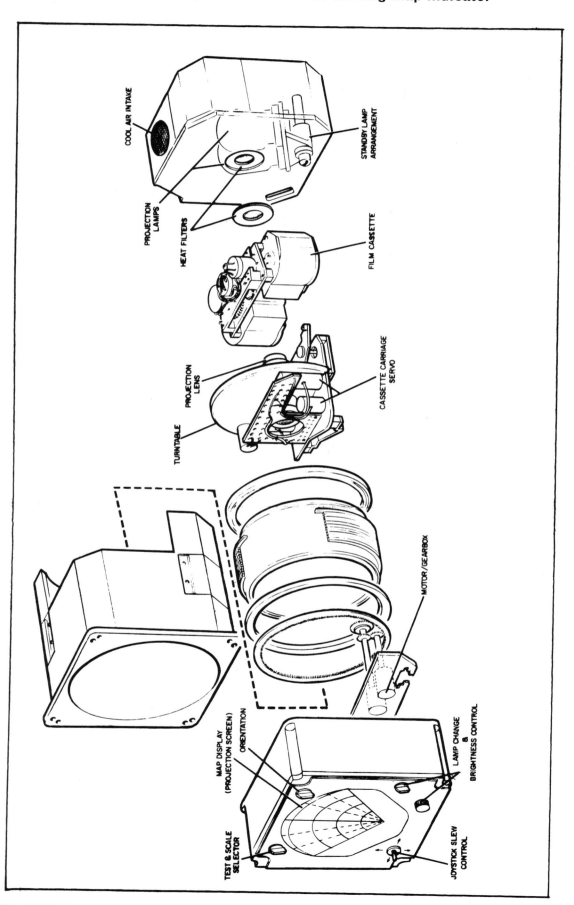

COOL AIR INTAKE

STANDBY LAMP
ARRANGEMENT

PROJECTION
LAMPS

HEAT FILTERS

FILM CASSETTE

PROJECTION
LENS

CASSETTE CARRIAGE
SERVO

TURNTABLE

MOTOR/GEARBOX

MAP DISPLAY
(PROJECTION SCREEN)
ORIENTATION

LAMP CHANGE
&
BRIGHTNESS CONTROL

TEST & SCALE
SELECTOR

JOYSTICK SLEW
CONTROL

EXHIBIT 6 *(contd.)*   **Signal flow diagram of the moving map indicator**

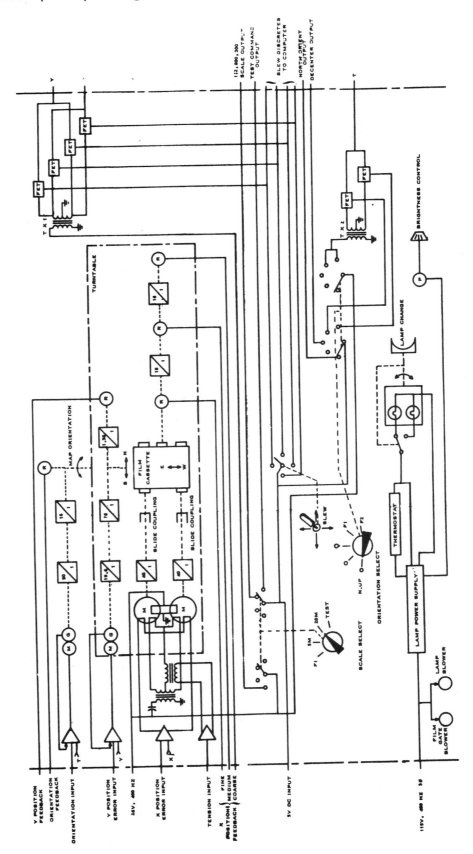

EXHIBIT 6 *(contd.)*   **Centre section with components assembled**

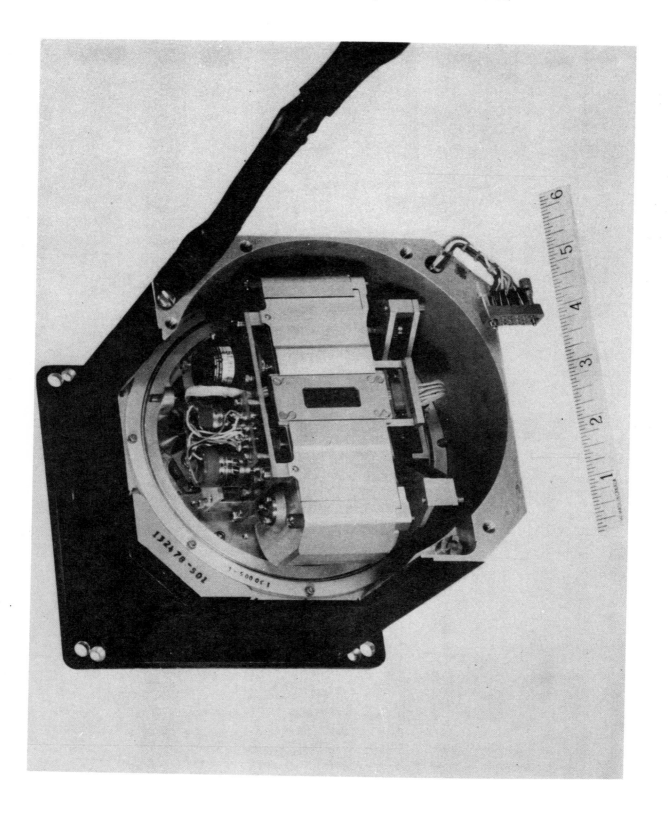

EXHIBIT 6 *(contd.)*   **Method of inserting cassette**

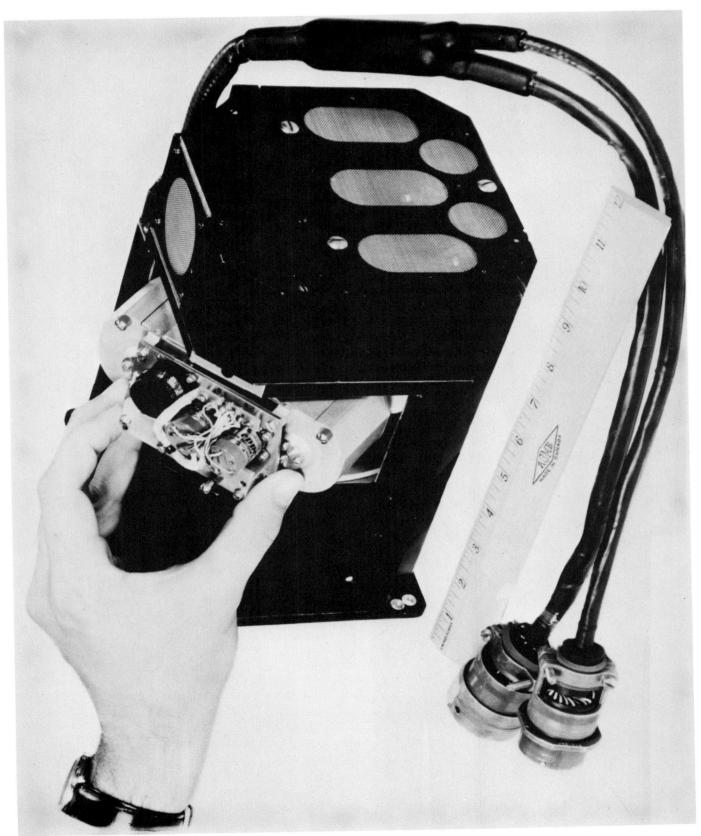

EXHIBIT 6 *(contd.)*   **View of cassette showing film path**

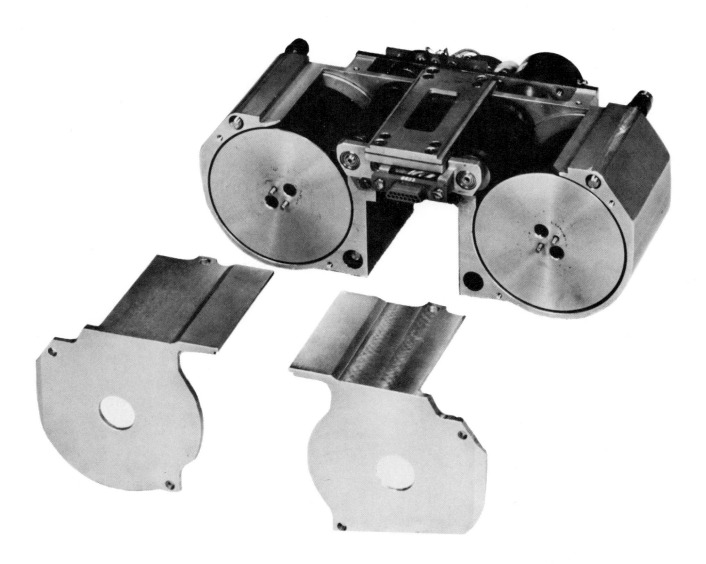

SUGGESTED STUDY

The case presents the development of a piece of avionics hardware. The case includes three complete iterations in the development. The first three segments of the case are presented from the point of view of Bob Vago, the project engineer. Bob Vago's comments reflect some of the disillusionment and frustrations that faced him. The final segment demonstrates success of the item after Vago leaves the firm and shows that the project has an independent identity.

Part 1

Describes the origin of the project, the equipment performance requirements are established. Class discussion may be focused on some of the following:

1.1. The merits of changing the company strategy.
1.2. Specification of performance requirement of the moving map display.
1.3. Preparation of a schematic diagram for the moving map display to achieve the specified functions.
1.4. Bob Vago's comment 'It will indeed be a great day for the conceptual engineer when computer assisted design is an every-day occurrence.'
1.5. Scheduling and planning of the engineering activity necessary to meet objectives.

Part 2

Discusses the development of the prototype. Details are given in the exhibits and in the appendixes. Class discussion may be focused on the following:

2.1. How the unit operates.
2.2. Does the device meet the requirements set out in Part 1?
2.3. Human factors requirements.
2.4. Merits of the north–south drive system.

Part 3

Discusses the shakedown problems encountered making the equipment functional. The problems in preparing the equipment for field trials are also covered. Class discussion can be focused on:

3.1. The problems and their correction.
3.2. Solutions to insufficient brightness.
3.3. Bob Vago's statement 'Situations apparently unconnected to engineering can have effect upon a project's success.'
3.4. Was the project successful?
3.5. What should the company do?

Part 4

Is presented as an epilogue. It tells what happened to the project after Bob Vago left the company. Discussions may be concerned with the following:

4.1. The various field trials and their contribution to the revised design.
4.2. Comparison of final product with objectives set out in Part A.
4.3. Gordon M. Mount's statement 'Only the broadest similarity of concepts exist between the current design and the original prototype.'

Case 7   Containing a large nuclear reactor

by Maheshkumar Desai, Robert McKechnie, and
Bruce Shawver

Prepared originally in partial fulfilment of their requirements for the Master of Engineering in Mechanical Engineering at the University of California, Berkeley. It has been edited and shortened by R. F. Steidel, Jr. We are indebted to Mr Douglas Kelly and the Department of Mechanical Engineering of the Pacific Gas and Electric Company for this project; also to Mr J. O. Schnyler, Mr C. Ashworth and Mr D. Barton who have generously given of their time. The cooperation of Mr C. H. Robbins of the General Electric Company is also acknowledged.

This case study has been prepared and written with the aid of a grant from the Ford Foundation for the demonstration and development of a new Master of Engineering programme in design.

This case is available as a separate pamphlet from the Department of Mechanical Engineering, University of California, Berkeley, California 94720.

Foreword

This case study describes the development of a new concept in nuclear reactor containment, from the point when it first was suggested as a means of containment to the point when the Atomic Energy Commission issued a license for construction of such containment. As the story unfolds, we shall see how the Pacific Gas and Electric Company plotted the course for the development of this concept into a workable system, how obstacles along the way were overcome, and how the various decisions were made.

PART 1   Introduction

Need for containment

Nuclear reactors, whether used for power production or for purely experimental purposes, often use a system of circulating water to remove heat from the fissioning core of fuel elements. Steam, of course, is a useful by-product of this cooling process. If a rupture should occur in this high-pressure system, there would be a sudden release of steam and super-heated water to the surrounding area. Because these fluids contain radioactive materials in a nuclear reactor, and because radioactivity presents dangers to life of all kinds, precautions must be taken to prevent the fluids from leaving the immediate reactor area in the event of an accident. The means by which these radioactive materials are confined to a specific safe area is commonly called *containment*.

A far stronger reason for containing the reactor products is the protection of the reactor against a rupture in the pressure system. If, because of a rupture, the core of fuel elements loses its supply of cooling water, and if it is not possible to slow the reaction quickly enough with the appropriate control measures, the heat generated from the decaying of fission by-products could melt the fuel elements and possibly part of the reactor. If this occurred, large amounts of highly radioactive material would be released from the reactor, and, in the absence of a containment system, these would be distributed over the plant area and into the atmosphere.

The conventional method of containment

Prior to 1960, the problem of containing the mixture of water, steam, and radioactive materials which would be released in case of an accident was handled by completely enclosing the reactor in a large, leakproof vessel. In the event of an accident, this vessel would be sealed from the rest of the plant by double valves installed on each pipe leading in or out, and the radioactive mixture would be contained.

Because the water in the reactor during operation is saturated and pressurized, its release in the event of a rupture would result in the liberation of large quantities of flash steam as the pressure is reduced to atmospheric. This flash steam, plus the steam already in the system, would cause a pressure rise in the containment vessel. The containment vessel not only had to be leakproof, but it had to be able to withstand a positive pressure as well.

The final design of the containment vessel was largely a matter of economics. A small vessel, because it would incur higher pressures, demanded heavier construction; a large vessel, although it could be of lighter construction, required more material and occupied more space. A trade-off among these considerations typically resulted in the reactor being enclosed by a steel sphere or cylinder around 100 ft in diameter, designed for a pressure of 40–50 psig, which came to be a familiar containment vessel.

The spherical or cylindrical containment vessel had several disadvantages. For one thing, it was expensive – both in materials and in construction costs. For another, its large size forced the reactor to be located above ground, which in turn forced the installation of heavy radiation shields around the vessel. A reactor located underground uses the earth for a shield in all directions except upward. Furthermore, one of the desires of nuclear plant designers is to have their plants resemble conventional plants as nearly as possible. This is an attempt to gain public acceptance of nuclear power by making the differences less obvious. The large sphere or cylinder around the reactor was a very noticeable distinction from a conventional plant. It tended to impress the public with the idea that reactors were dangerous.

In contrast to the method to be described, this first containment system has come to be known as *dry containment*. A nuclear plant using a typical dry containment system is shown in Fig. A. The plant, located in Puerto Rico, was designed by the Bechtel Corporation.

Pressure suppression containment

To devise a containment system better than the steel sphere or cylinder has been a standing problem in the nuclear power industry. One idea which has been considered is to release the first rush of flash steam to atmosphere, and, after a few minutes, to isolate the containment vessel by means of valves. Because the bulk of the flash steam would have been released before the vessel is sealed, it could be designed for a much lower pressure. Proponents of this idea justify the release of radioactive steam to the atmosphere by the fact that the radioactivity at this stage is almost entirely due to the Nitrogen 16 and it would therefore quickly become negligible (N_{16} has a half-life of 7 seconds). Another idea which had been considered was to build a light vessel around the reactor with a vent pipe leading off the top and into a large inflatable rubber or plastic bag. Again, a lighter structure would be required, and the radioactive materials would be safely contained in the bag. However, for one reason or another, nothing was developed from these ideas.

The origin of pressure suppression containment

During the last months of 1957, an entirely new method of reactor containment was suggested. This method, which later came to be known as *pressure suppression containment*, is based on the idea that the pressure rise in the containment vessel can be limited by quenching the steam released during an accident.

Figure B is a schematic illustration of this basic concept. The reactor vessel (V-1) is enclosed and supported within a chamber (V-2) which is the dry well. Several large diameter pipes lead out of the dry well, and terminate below the surface of a large pool of water. The pool is enclosed in a vapor-tight container (V-3), forming the suppression chamber.

In the event of a pipe break within the dry well, it was anticipated that the steam or water–steam mixture would flow through the vent pipes into the water pool and be quickly condensed. Because of the condensation of the steam, pressure build-up in the chamber would be relatively small, and hence the term, 'pressure suppression'. Furthermore, it was hoped that those radioactive materials which accompanied the steam as far as the pool would be almost completely absorbed in the water.

When it was first considered as a means of containment, it was anticipated that pressure suppression would offer the following advantages:

1. Because of the lower design pressure and the relative simplicity of the design, it would be less expensive than the conventional dry containment. A preliminary

Fig. A. A nuclear plant using dry containment

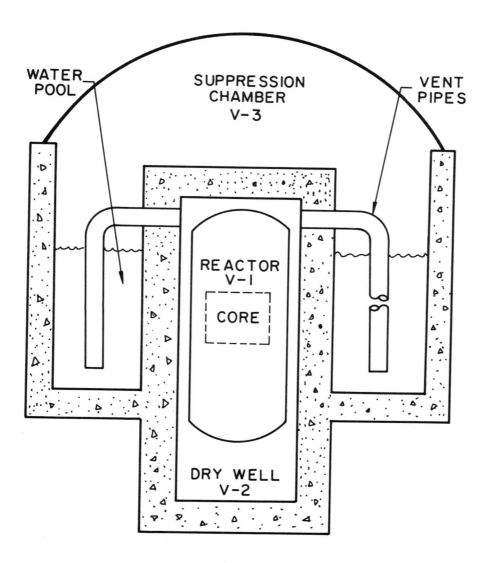

Fig. B. Simplified arrangement of pressure suppression containment system

estimate indicated cost savings in the order of $500 000 (exclusive of development) over a dry containment structure.

2. Its relatively small physical size would make it possible to locate either a portion or the entire containment structure below ground. Aside from saving space and improving plant appearance, this offered the advantage that additional radiation shielding would be provided by the earth, in all directions except directly overhead.

3. In the event of an accident, fission products would remain largely in the small dry well volume, thus limiting the spread of radioactive materials, permitting the radiation shielding to be localized and highly effective, and permitting subsequent plant cleanup with a minimum of accidental release or exposure to personnel.

4. Any fission products escaping from the dry well through the vent piping would pass into the water in the suppression chamber, and would be largely retained. Those which did find their way into the suppression chamber would have less tendency to leak out, since the difference in pressure between the inside and outside of the chamber is relatively small.

5. If it became necessary to flood the reactor core in an emergency, the pressure suppression pool would provide a quick and convenient supply of water.

It should be pointed out that the pressure suppression method of containment is applicable primarily to single-cycle boiling water reactors. It has been investigated for possible application to the pressurized water reactor and the dual-cycle boiling water reactor, but it appears to offer no cost advantage over the dry containment in these cases because of the more complicated systems involved.

The status in 1957

With the basic concept of pressure suppression containment established, we must now provide some background information on the situation at the Pacific Gas and Electric Company of San Francisco (PG & E) towards the end of 1957. At this time, PG & E and the Atomic Power Division of the General Electric Company (GE) at San Jose, were jointly studying the feasibility of building a nuclear power plant. It was first thought that a relatively large 200 megawatt plant might be built. As the study progressed, it became apparent that it would be more economical to build a smaller plant, and, in the last few days of 1957, it was tentatively decided to go ahead with plans for a 50-megawatt boiling water reactor, to be located about 300 miles north of San Francisco at Humboldt Bay. This decision was affirmed in February 1958.

In order to obtain a fixed cost, PG & E contracted with the San Francisco office of Bechtel Corporation, which is a large consulting and construction engineering corporation located in San Francisco, to have the latter perform the detailed design work, purchasing, scheduling, and construction supervision. While sub-contracting to Bechtel did relieve them of much of the design work, PG & E was still responsible for all the major decisions to be made in the project.

The organization of the Pacific Gas and Electric Company

Since this case is largely concerned with decision making, it is important to understand how PG & E was organized at the time. All major decisions were officially made by the company president, Mr N. R. Sutherland, and his Board of Directors. In making engineering decisions, Mr Sutherland relied heavily on the Engineering Department, as he was not an engineer. The responsibility of co-ordinating the work of the Engineering Department and providing Mr Sutherland with the information he required to make his decisions was delegated to the Vice-President of Engineering,

Mr Walter Dreyer. The Chief Mechanical Engineer, Mr C. C. Whelchel, and the Humboldt Bay Project Manager, Mr J. O. Schuyler, were principally concerned with the Humboldt project. Mr James O. Schuyler was a Supervising Mechanical Engineer in PG & E's Department of Mechanical Engineering. He received his B.S degree in Electrical Engineering in 1948 from the University of California at Berkeley. He joined PG & E in 1950, and has since been engaged in the electrical and mechanical design of steam power plants. He is a registered professional engineer in California, and a member of the Institute of Electrical and Electronic Engineers and the American Nuclear Society.

Other engineers in the Engineering Department will be introduced later in the case study.

The pressure suppression proposal   PART 2

Development of the concept of pressure suppression at General Electric

In the early part of 1957, a special unit was set up in the Atomic Power Equipment Division of the General Electric Company to suggest and study means of reducing the cost of atomic power. Mr Leo Schantz was in charge of the unit, and Mr C. H. Robbins was one of the twelve engineers in the unit.

Among the several possible means of reducing costs suggested by this unit was the idea of using pressure suppression as a means of containment. It was felt that such a containment, if it worked, would lower construction costs because of the lower pressure requirements, and possibly offer several safety advantages over dry containment.

The concept of pressure suppression did not originate at General Electric. The idea of using a water pool to condense steam and thus prevent pressure rise had been suggested in the industry before, but, as far as was known, no one had ever proved the feasibility of the idea. Thus, pressure suppression might be a means of reducing the cost of nuclear power, if it could be proven workable.

Some quick tests were made to check on the fundamental principle underlying the pressure suppression concept. No one knew if and how fast steam passing through water would condense. The tests, which took place at the Vallecitos Atomic Power Laboratory, near Pleasanton, California, consisted of blowing steam through a small diameter pipe into a large tank of water. The steam was completely condensed in these preliminary tests, and thus the engineers concluded that the underlying principle of pressure suppression had some credibility.

All that remained was to try and design a pressure suppression containment system for a nuclear reactor, and to show that this system would work in practice. How to proceed bothered Mr Schantz, for he realized, first, that such a development program was going to be expensive, and second, that a great deal of other research was being carried out at the time at General Electric.

Mr C. C. Whelchel, Chief Mechanical Engineer of the Pacific Gas and Electric Company, and a General Electric official were at a conference together in Washington. When a heavy snowstorm caused their stay to be extended a day, the two men spent some time informally chatting about atomic power, and among other things, about how money could be saved. It was at this point that the General Electric official (who remains anonymous) suggested the concept of pressure suppression. He also pointed out that safety could probably be improved as well, with this type of containment. Mr Whelchel, who is characterized as being very open-minded, liked the idea immediately, and asked that PG & E be told more about it. Thus, along with the proposal GE gave PG & E on the original idea of a 200-megawatt plant, an alternative scheme was included using pressure suppression containment.

Unfortunately, a sketch of this containment scheme, as originally presented by GE, is not available. However, it was essentially the same as the scheme later presented for the 50-megawatt plant (but on a larger scale), the details of which are presented in the section on PG & E's decision to proceed (p. 252).

PG & E's reaction to the pressure suppression idea

The reaction of PG & E to the pressure suppression concept was that it did look interesting. Since the company hadn't yet decided that a plant was to be built, no real consideration was given to a concept or design.

When a decision to build a 50-megawatt plant at Humboldt Bay was reached in February of 1958, interest in using pressure suppression as a means of containment was reborn. Mr J. O. Schuyler, project engineer for the Humboldt plant for PG & E, liked the idea very much. He and Mr Whelchel decided to spend a part of the next few weeks looking into the possibility of using pressure suppression and what would be involved in developing it. Their objective was to make a recommendation to PG & E management as to whether or not the company should pursue the idea. It would then be up to management to make the final decision. Their findings, and some thoughts on the matter, were summarized in a memorandum written by Mr Whelchel to Mr Dreyer on 2 April 1958.

The memorandum emphasized some of the potential advantages of pressure suppression. It pointed out that with dry containment, it is possible for molten radioactive fuel released in an accident to be plastered on the inside of the capsule, thus giving rise to the following undesirable effects:

1. Damage inside the capsule could be extensive.
2. Access to the reactor would be denied for a long period of time.
3. A dry containment capsule required a radiation shield, where nearby areas are occupied and equipment was to be kept in service, since the capsule protrudes high above ground (thus a large area surrounding it is exposed to harmful radiation).

One of the important advantages of pressure suppression containment would be that the suppression pool could be used to flood the reactor core in the case of an accident, preventing the formation of molten fuel and avoiding undesirable effects and eliminating the need for a radiation shield. Other advantages outlined in the memorandum were that because of the lower capsule pressure, construction costs could be reduced, and by eliminating the large dry containment structure the plant would more closely resemble a conventional plant.

Next, the memorandum suggested that, although the pressure suppression system proposed by GE for the 200-megawatt plant retained the capsule shell designed to a lower pressure, an improvement would be to eliminate the capsule altogether, substituting a concrete enclosure with a gas-tight deck over the reactor. Although no details were given at the time, it was felt that this would further reduce the cost of the containment structure, and give a neater appearance.

The memorandum suggested that a reasonable plan of attack would be to reopen the subject with GE by having GE make a proposal to PG & E regarding the cost and details of the development program in sufficient detail so that PG & E could judge its success. At the same time, Bechtel would be asked for an estimate of the cost savings that would result from this type of containment. Such an estimate would be difficult before the development work was completed, but it was felt that some idea could be obtained.

Finally, it was suggested that it would be desirable to freely discuss the matter with the AEC and exchange information with them during the research and development program. This would obtain the benefit of their thinking and experience and make more likely a containment design of which the AEC Reactor Safeguards Committee would approve.

This memorandum was passed on to Mr Dreyer for the purpose of obtaining Mr Sutherland's approval to ask GE for a proposal. Both Mr Whelchel's memorandum and Mr Dreyer's covering letter to Mr Sutherland are enclosed as Exhibits 1 and 2.

It should be pointed out that while it would seem that the decision to ask GE for a proposal would be fairly automatic, this was not the case. The cost of the research work would be borne by PG & E, or at least shared between the companies. It is important to realize that a research and development program such as that under consideration was not standard practice for G P& E, whose research work was generally on a much smaller scale, i.e., developing and testing small pieces of equipment such as transmission line insulators, solving operating problems, etc.

The General Electric proposals

When GE received the invitation to submit a proposal on the development of pressure suppression containment, Mr C. H. Robbins, one of the twelve engineers in the original cost-reduction unit, was put in charge of preparing the proposal. While he did most of the actual writing himself, he had two other engineers working with him on this project.

In drawing up the proposal, Mr Robbin's basic philosophy was to have the development program start off slowly and relatively inexpensively, and to proceed one step at a time. After each step, the results of the previous one would be carefully evaluated, and the next one would be taken only if the results appeared favorable.

In a little less than a month, Mr Robbins, and his staff had written the required proposal, entitled, 'Pressure Suppression System Development Programme Proposed for Pacific Gas & Electric Company', and derived the necessary time and cost estimates. The objectives of the proposed development program, as stated in the proposal, were as follows:

To design a pressure suppression system of reactor containment and to provide proof to all who are concerned with reactor safety that the system will work well if ever needed. The work was necessary to prove the following essential features:

1. What is the maximum credible accident which the pressure suppression system should protect against?
2. How should the dry well be designed to make sure the steam and water released do not bypass the water pool? What were the requirements so the dry well can withstand maximum static pressure, blasts, and missiles?
3. How could the steam be injected into the water pool so all of it was condensed?
4. How effective was the water pool in trapping fission products?
5. How could water from the pool be used to keep the core from melting following an incident?

The remainder of the 14-page report described in detail the work considered necessary to accomplish these objectives. It was proposed that the investigation be carried out in three sequential phases. Phase I, to be carried out by GE, was to include a literature search, some tests to establish a suitable means of injecting the steam into the water pool, and the development of methods of analysis for predicting the performance of a pressure suppression system. It was suggested that if this phase should prove encouraging, Phase III, consisting of extensive testing and experimentation would be carried out jointly by GE and PG & E. Phase II, to be conducted by GE, was to interpret the test results into the form of a workable design.

EXHIBIT 1   Letter — Dreyer to Sutherland, 2 April 1958

PACIFIC GAS AND ELECTRIC COMPANY   **COPY**

ENGINEERING

Proposed R&D Program for
Improved Reactor Containment

April 2, 1958

MR. N. R. SUTHERLAND:

Mr. Whelchel's attached memorandum suggests
that we ask GE for a proposal on a joint PG&E-GE research
and development program for an improved pressure suppression
type of reactor containment as contrasted with the capsule
type used at Vallecitos.

A recommendation to proceed with such a program
would not be made until after a thorough investigation of
the GE proposal and a belief that the desired results would
be obtained.

I believe we should ask GE for the proposal.

WALTER DREYER

CCW:ca
Attach.

EXHIBIT 2   Proposal, 1 April 1958

519177 9-56-100 PDS.

PACIFIC GAS AND ELECTRIC COMPANY
DEPARTMENT OF ENGINEERING

PROPOSED RESEARCH AND DEVELOPMENT PROGRAM
FOR
IMPROVED REACTOR CONTAINMENT
APRIL 1, 1958

This memorandum outlines some of the considerations involved in the problem of reactor containment, and suggests that we consider at this time a Research and Development program with General Electric on an improved type of containment as a continuation of our work to reduce the cost of atomic power.

Vallecitos Type of Containment

The primary purpose of the Vallecitos capsule is to prevent the escape, following a reactor rupture, of radioactive (1) steam, (2) noncondensable gases, and (3) molten fuel which may be ejected from the core.

The size of the capsule and the thickness of its steel plate is a function of the volume and pressure of released steam. The additional pressure caused by the radioactive noncondensable gases is small although their containment is important. Since, in the worst probable accident molten radioactive fuel may be plastered on the inside of the capsule, a shine shield is required in those instances where nearby areas are occupied and equipment is to be kept in service.

If an accident should occur which makes use of the capsule for the above purposes, the damage inside the capsule may be great, and access may not be possible for a long period.

Pressure Suppression Type of Containment

Containment which will limit or prevent the formation of molten fuel or release of steam should reduce the cost of containment and limit the damage to the equipment and the housing. The "Pressure Suppression System" which was included in GE's November 6, 1957 proposal for the 200 MW unit was designed to condense most of the steam and flood the core to prevent the fuel elements from melting. They, however, retained the capsule type shell, but designed it for much lower pressure. Improvement on their concept would eliminate the low pressure shell and substitute a gas-tight deck over the reactor, with its probable further reduction in cost resulting from the complete elimination of the capsule and the capsule shine shield, because the latter could be made a part of the deck over the reactor.

The information becoming available from AEC containment studies leads us to believe that the subject should be reopened with GE by having them make a proposal to PG&E, with its estimated cost including an outline of the proposed work in sufficient detail so that we, and they, can judge its probable success. Such an R&D program might cost between $200,000 and $300,000.

Attached are pages 161 and 162 from the 200 MW November 6, GE proposal, which outlines the "Pressure Suppression System" that GE had in mind at the time of the proposal. From these sheets it appears that a steam supply system will be required. We suggest that whatever power plant engineering is needed be done with PG&E forces, and perhaps the models for testing could best be constructed by us at Pittsburg or other plants where steam would be available.

EXHIBIT 2 (page 2)

619177 9-56-100 PDS.

PACIFIC GAS AND ELECTRIC COMPANY

DEPARTMENT OF ENGINEERING

Proposed R&D Program for
Improved Reactor Containment   #2   April 1, 1958

We would also propose to obtain from Bechtel an estimated cost
saving that would result if this type of containment could be used at
Humboldt. This may be difficult at this time and before the new containment
work is completed, nevertheless some idea could be obtained. The net saving
after paying for the R&D program might be as much as $500,000 if the capsule
could be eliminated and the shine shield incorporated in the deck. GE estimated
the saving to be $2,500,000 in the case of the 200 MW unit.

If the R&D program is successful, its cost could be considered a part
of Humboldt because it should lower the overall Unit No. 3 cost. If the R&D
program is not successful, its cost could be considered as an operating expense.

It may be desirable to freely discuss and exchange information with
the AEC during the R&D containment program to obtain the benefit of their
thinking and experience, and thereby be surer of working toward a design of
reactor containment that the AEC Reactor Safeguards Committee would accept for
Humboldt.

C. C. WHELCHEL

CCW:ca
Attach.

The final page of the report, reproduced here as Exhibit 3, showed the times estimated for the various phases of the development programme. With reference to this figure, we see that GE estimated Phase I would take a little less than three months, Phase II about four months, Phase III two months, and the entire program eight months.

Three copies of the report were given PG & E on 30 April 1958, along with a letter quoting the prices and terms of the various phases of the development program. This letter from Mr Ridgway to Mr Whelchel is included as Exhibit 4.

In addition to a verbal description of the program, several drawings were included in the proposal describing equipment and the required flows of steam and air. Since these details will be presented later when the tests are described, they will not be included here.

While a major decision such as this was ultimately up to Mr Sutherland and his Board of Directors, it became the task of Mr Whelchel and Mr Schuyler to gather all information pertinent to the decision, evaluate it, and make a recommendation. It would be largely on Mr Whelchel and Mr Schuyler's recommendation that Mr Sutherland and his board would make a final decision.

Before describing the steps Mr Whelchel and Mr Schuyler followed, however, it is necessary to supply some background information on how the pressure suppression idea had been developing.

PG & E decides to proceed with Phase I

With the description of the proposed development work, times estimates, and price quotations thus in hand, PG & E management faced a major decision. Did the development program as described by GE look promising enough to spend $38 500 on Phase I? Some of the specific questions that needed answers before making this decision were:

1. Would the proposed development program be sufficient to produce a suitable containment design, and prove it would work in practice as well?
2. What cost savings could be expected if this type of containment were to be used at Humboldt, and would these savings outweigh the cost of the development programme?
3. Would Bechtel have any difficulty incorporating such a containment system into the Humboldt design?
4. Would the development program be finished in time for the results to be used at Humboldt?
5. The AEC would have the final say as to whether or not the pressure suppression system could be built and operated; what would be its reaction to the idea?

Evolution of the pressure suppression concept

PG & E had also been thinking about the design of the containment system, and they had come up with the idea shown by the sketches in Exhibits 5 and 6. This scheme envisioned the enclosed reactor with the suppression pool filling the space between the caisson wall and the reactor dry well. A significant feature of this design was that there was no steel capsule enclosure. Furthermore, its proponents felt that there would be no need for a radiation shield, since the water pool surrounding the reactor would serve this purpose.

A few days after GE presented their proposal to PG & E, they submitted a drawing outlining how they thought the pressure suppression idea should be incorporated into the Humboldt plant (see Exhibit 7). This design was patterned after that proposed earlier for the 200-megawatt plant, only on a smaller scale. GE proposed to enclose the reactor in a steel capsule, as in a dry containment system except the

EXHIBIT 3   Schedule–Pressure suppression system development program

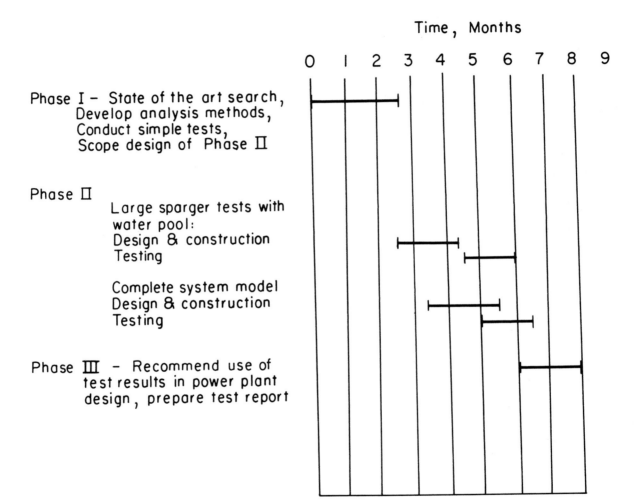

Time, Months

Phase I – State of the art search,
Develop analysis methods,
Conduct simple tests,
Scope design of Phase II

Phase II
Large sparger tests with water pool:
Design & construction
Testing

Complete system model
Design & construction
Testing

Phase III – Recommend use of test results in power plant design, prepare test report

EXHIBIT 4   Letter, Ridgway to Whelchel, 30 April 1958

GENERAL ⊛ ELECTRIC

COMPANY   APPARATUS SALES DIVISION

PACIFIC DISTRICT

235 MONTGOMERY STREET, SAN FRANCISCO 6, CALIFORNIA . . TELEPHONE DOUGLAS 2-3740

April 30, 1958

Mr. C. C. Whelchel
Chief Mechanical Engineer
Pacific Gas & Electric Company
245 Market Street
San Francisco, California

Subject: Development Program for
          Nuclear Plant Pressure
          Suppression System

Dear Mr. Whelchel:

This will confirm and amplify our recent conversation in your office
concerning the proposed pressure suppression system development
program. The basis of our discussion was the brochure entitled
"Pressure Suppression System Development Program Proposed for Pacific
Gas & Electric Company by the Atomic Power Equipment Department",
together with accompanying drawings. Several copies of this material
were left with you following the meeting. The following comments
concerning costs refer to this brochure.

The program is divided into three phases:

Phase I

    Phase I consists of the initial work to determine the analytical
    methods to be used in the study of the preliminary air and steam
    tests to confirm hydrodynamic theory. This work will cover a
    period of approximately two and one-half months and provide a
    firm basis for the conduct of the tests to be made in Phase II.
    The firm price of Phase I is $38,500, to be paid net cash in
    full 30 days after completion of Phase I.

Phase II

    Phase II is broken into two parts: a) a large scale test of
    the sparger system and b) a small scale test of the complete
    suppression system.

EXHIBIT 4 *(contd.)*

GENERAL ⊛ ELECTRIC

Mr. C. C. Whelchel   -2-   April 30, 1958

a. Large Scale Test of Sparger System

We propose to conduct this test at a PG&E steam electric
plant where steam is available at a rate of approximately
50,000 lbs/hour at approximately 100 psig. The price quoted
assumes that steam will be delivered to the test tank at no
charge and that a crane will be available at the test tank
for the lifting of the cover, which weighs approximately
15,000 lbs. The price includes the design, construction,
and operation of the facility. It is entirely conceivable
that you may elect to construct the test facility to our
design and with our supervision of construction, in which
case the price will be as noted.

b. Small Scale Test of Completed Suppression System

The equipment to be used for this part of Phase II will be
assembled at APED's San Jose plant and tests will be conducted at the Vallecitos Atomic Laboratory.

The estimating price for Phase II, in the event General Electric
has complete responsibility, is $274,000.

In the event that you elect to construct the sparger test facility
(which we would prefer), as described in (a) above, the estimating
price will be $131,000. For your information, an equipment list,
showing what you would supply and what General Electric would
supply for this test facility, is attached.

Phase III

Phase III consists of the analysis required to extrapolate the
results of Phase II into design criteria for application to the
construction of nuclear power plants of 50 MW and larger. A
complete report will be prepared which will include design specifications for pressure suppression systems for a 50 MW and a 200
MW boiling water reactor power plant. The estimating price for
Phase III is $6,950.

The proposed schedule for the complete program envisions the
completion of Phase I in approximately two and one-half months.
Phase II would require approximately four months and Phase III
approximately two months. A small overlap between phases would
result in completion of the entire program in eight months. It
is important that this program be started promptly if the results
are to be available for the design of your 50 MW power plant to
be constructed at Humboldt Bay.

EXHIBIT 4 *(contd.)*

GENERAL ⊕ ELECTRIC

Mr. C. C. Whelchel   -3-   April 30, 1958

The development program described will provide you with rights
to use all information, designs and patents obtained in the
course of the study. These rights include the right to have
manufacturers other than General Electric manufacture for your
use equipment based on the information, designs, and patents
obtained in the course of the study. The General Electric
Company will retain all other manufacturing and licensing
rights.

We would appreciate your careful review of this proposal and if
it meets with your approval we will undertake the development
program immediately upon receipt of your authority to proceed.
Should questions arise, please let me know.

                    Very truly yours,

WR:lj
Enc.   W. RIDGWAY
                    MANAGER, ELECTRIC UTILITY SALES

EXHIBIT 5   PG & E's original containment design idea (A)

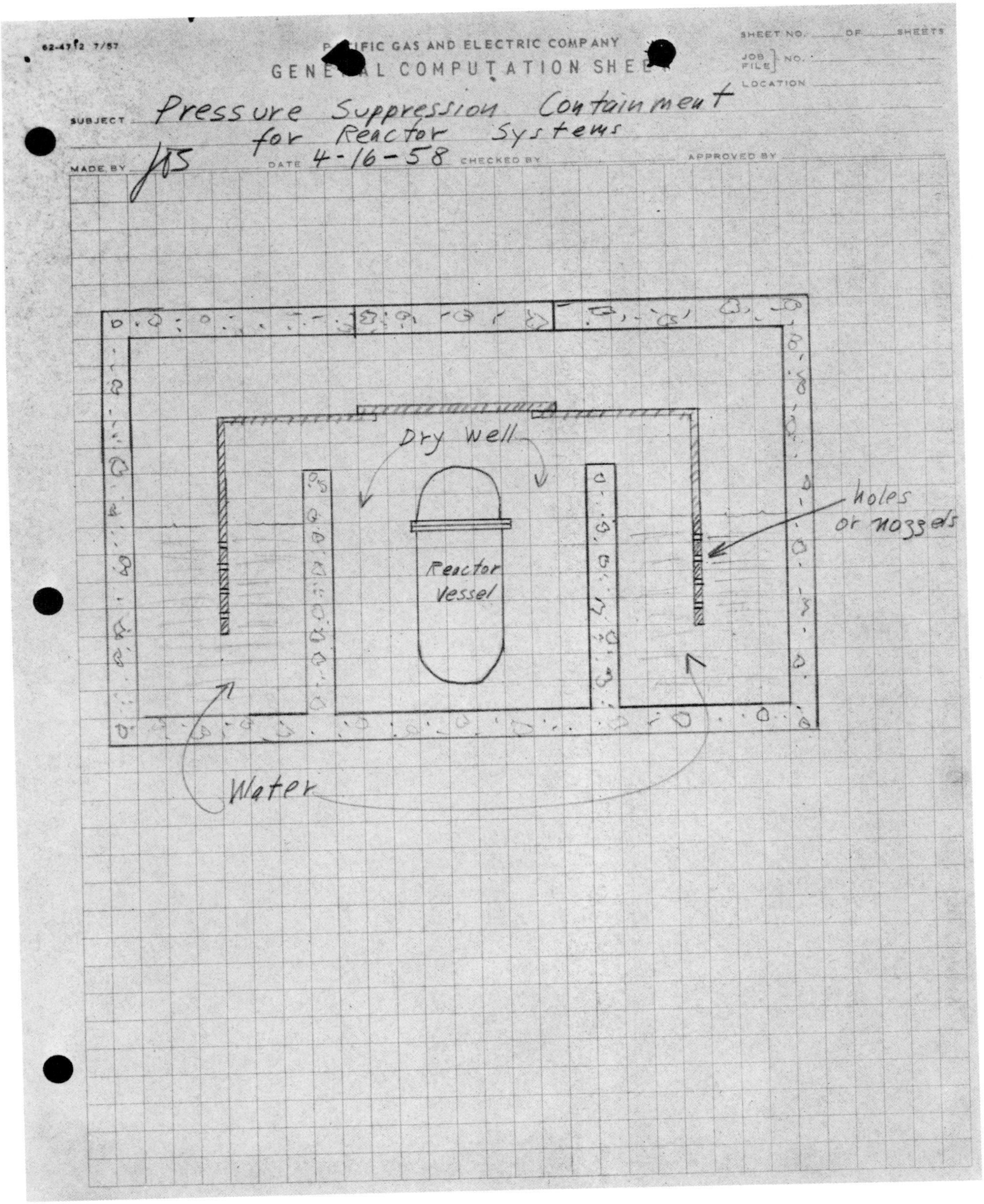

EXHIBIT 6   PG & E's original containment design idea (B)

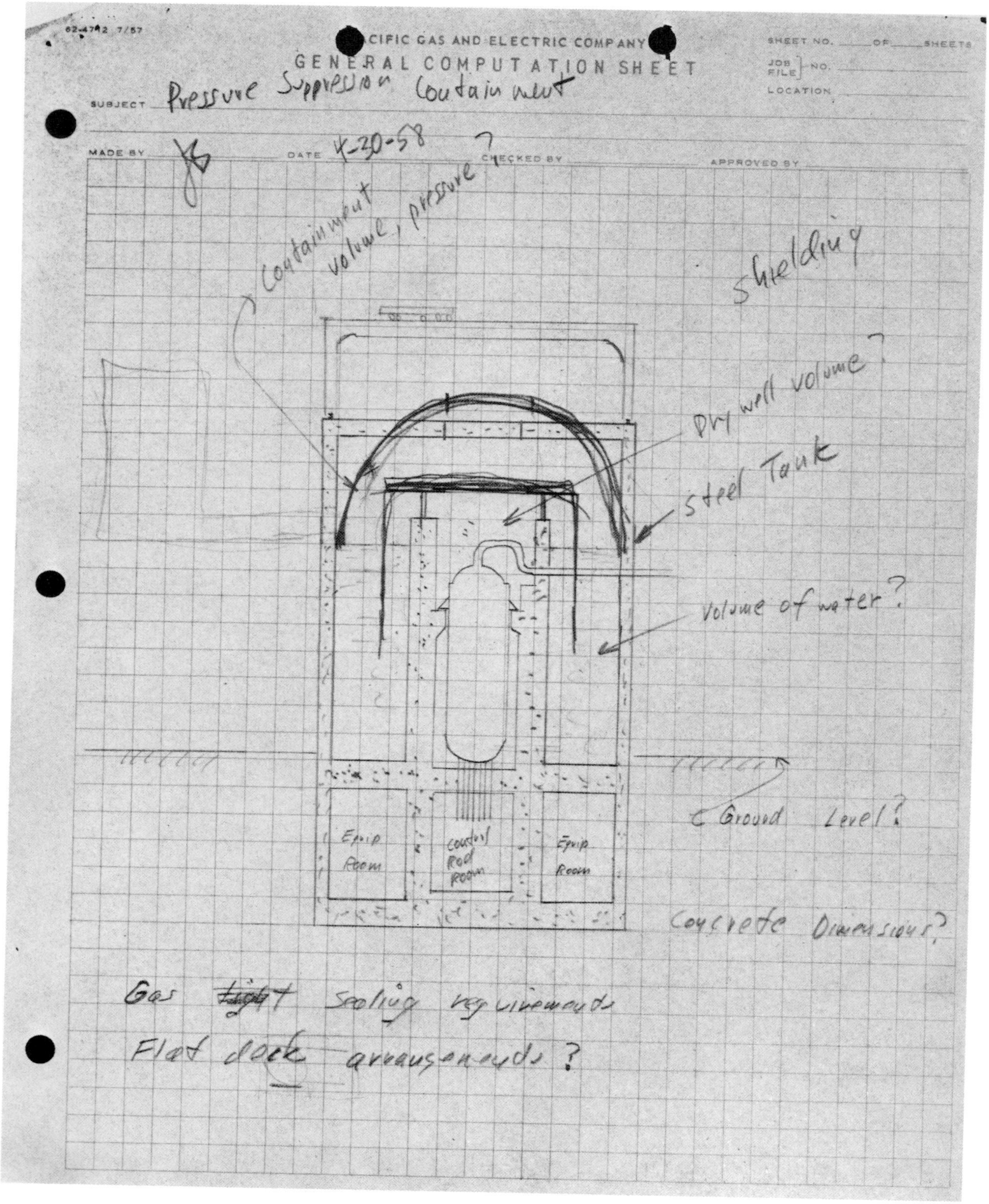

EXHIBIT 7   General Electric reactor containment study

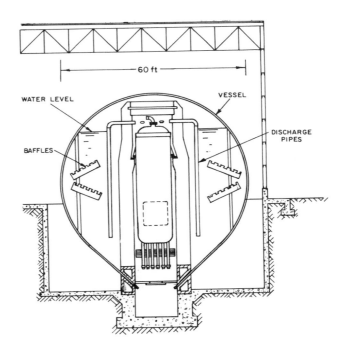

capsule would be of much lighter construction due to the reduced pressure requirements anticipated. The reactor would sit in a concrete enclosure in the middle of the
capsule, and the space between the enclosure and the capsule would be filled with
water but for an air space near the top. Vent pipes would lead out of the top of the
enclosure to the bottom of the pool, and baffle plates would be located within the
pool to help dissipate the steam and make condensation easier.

Up until this point, Bechtel had not been consulted formally about the pressure
suppression idea. They had been working hard on the dry containment design for
Humboldt, intending to present it to PG & E sometime in September 1958. Of interest
here is the containment system they had been proceeding with up to this point, and
it is shown in a Bechtel sketch in Exhibit 8. This is a typical dry containment system –
the reactor sitting in a concrete structure surrounded by a 90-ft diameter steel cylinder with a spherical top.

Reaction to the pressure suppression proposal

Mr Schuyler first made a very rough estimate of the cost savings that could be
realized with a pressure suppression system. To do this, he considered the $2·5 mil-
lion figure estimated by GE as being the possible saving using pressure suppression
in the original 200-megawatt plant proposal. By comparing the amount of steel which
would be saved in building a 50-megawatt plant, rather than a 200-megawatt plant,
he thus estimated a saving of at least $500 000 would be possible using pressure
suppression. PG & E would gain $180 000 after paying $320 000 for the Research and
Development Program.

It was next decided to ask for the opinion of the Bechtel Corporation of the practicability of the pressure suppression scheme, what cost savings could be expected,
and how it could be incorporated in the Humboldt Bay design schedule. For this purpose, Bechtel was given a drawing showing the light steel capsule enclosure of GE,
and the PG & E sketches showing the concrete structure. Bechtel immediately indicated they were disturbed by the possible change in the containment system. They
did not know how to proceed, whether they should continue to the original design, or
whether they should delay the portions of it which would be affected by the enclosure change, and they asked for guidance. PG & E advised them to hold back on the
containment design until it was known which would be used.

Bechtel's reaction to the pressure suppression idea was obtained in a series of
meetings Bechtel called between themselves, PG & E, and GE during the next few
days. First of all, Bechtel found several things wrong with the GE containment design
including the following:

1. There was no provision for an enclosure protection during refueling, at which
   time the capsule, as proposed, would have to be opened. Thus, a building would
   have to be constructed around the capsule to contain the radioactive products
   which could be released in a refueling accident.
2. There would be a problem in getting at the reactor control rods from the bottom
   of the capsule.
3. The construction of such a vessel would require an awkward and expensive construction sequence, including lowering the reactor through the top of the partially
   completed capsule shell.
4. The vessel was not designed to withstand an earthquake, and there was an over-
   turning moment about the base ring.

Bechtel pointed out that they could not properly make an estimate on the design without correcting for these factors. GE agreed that the criticisms were valid but pointed
out that it was primarily a conceptual design and it was decided to let Bechtel revise
the design before making a cost estimate.

EXHIBIT 8   Bechtel reactor containment study

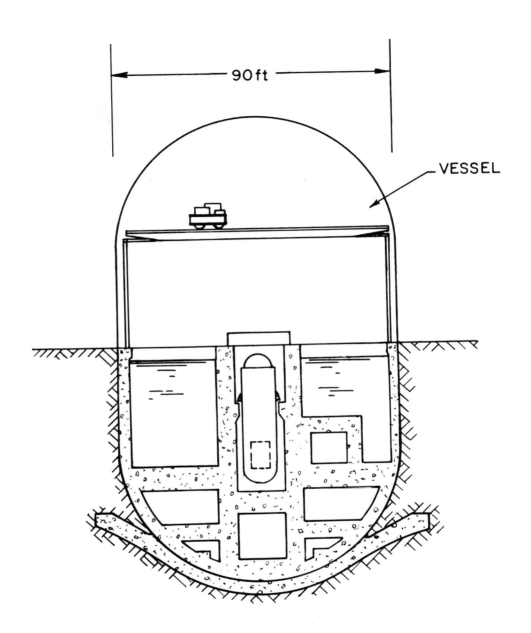

Bechtel also found several things wrong with the PG & E containment design:

1. The reactor was completely surrounded by the water pool, which meant there was no way of getting at the lower part of the reactor.
2. The top of the reactor was too high for refueling purposes.
3. The amount of concrete in the structure would make the cost prohibitive.

Because of these considerations, it was decided not to make a cost estimate.

After a few days, Bechtel announced they had estimated the savings possible using the revised GE capsule enclosure to be $280 000. This upset Mr Dreyer, the Vice-President of Engineering, because it was considerably less than the $500 000 that the PG & E staff had estimated could be saved with a pressure suppression system. He suggested that the PG & E engineers meet with the Bechtel engineers to resolve their differences. At this point, Bechtel produced what they claimed was a more sensible containment system. This suggestion, shown in Exhibit 9, had the reactor enclosed in a concrete caisson below grade level. The suppression pool was located separately from the reactor, and was connected by relatively long vent pipes. In the course of the meeting, Bechtel managed to convince PG & E that their mutual estimating efforts should be concentrated on this separate pool design.

The Atomic Energy Commission Reactor Safeguards Committee

While waiting for the estimating results, PG & E arranged for a meeting with GE and the AEC Reactor Safeguards Committee. The meeting took place on 28 May 1958, in Washington, D.C. Present were Messrs Whelchel and Schuyler of PG & E, Messrs Robbins and Sege* of GE, and several members of the AEC representing Hazards and Evaluation, Licensing and Regulation, Technical Operations, and Reactor Safety.

The purpose of the meeting was, first, to acquaint the AEC with the improved reactor containment under consideration, second, to obtain their opinion on the proposed system and development program, and third, to find out if the AEC had made any similar development work, or knew of any. The results of the meeting are summarized from its minutes as follows:

1. The AEC saw nothing wrong with the basic pressure suppression concept, but indicated that their final approval would depend on PG & E's ability to convince them that the system would operate satisfactorily under all credible accident conditions.
2. The AEC indicated that they had not performed any research and development that would have any appreciable bearing on the idea under consideration, nor were they contemplating any.
3. Various members of the AEC made comments throughout the meeting, including the following:

   (a) Blast effects, if any, should be properly analyzed and included in the design.
   (b) Water in the pool following an accident would be radioactive, and therefore precautions should be taken to prevent its leakage from the pool.
4. The AEC expressed the hope that PG & E would proceed with the pressure suppression system, and, in general, presented a favorable reaction to the idea.

With the AEC's stand thus determined, Mr Whelchel and Mr Schuyler next considered the time element. They felt that if the development scheme could be completed within the time estimated by GE, Humboldt would be able to start up sometime in 1962 as originally intended. However, what if unforeseen factors arise, and the development program fell behind schedule? The predicted demand for electricity could be met by the existing facilities at least into the early part of 1963. If a development program meant a few months' postponement in start-up date, there would

* Mr Sege was an engineer in GE's Reactor Safeguard Evaluation Department.

EXHIBIT 9   **Bechtel separate pool containment design**

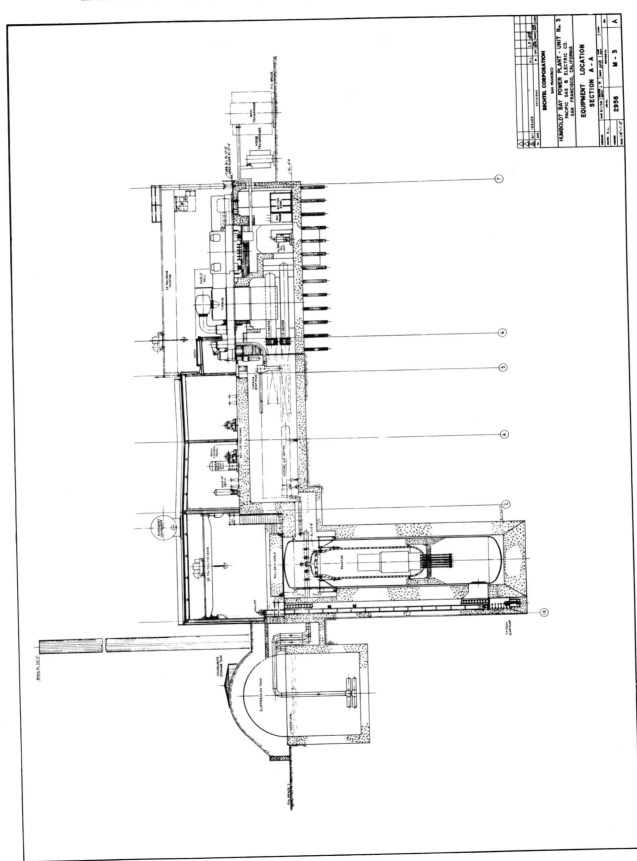

EXHIBIT 10   Letter — Dreyer to Sutherland, 17 June 1958

62-6224

PACIFIC GAS AND ELECTRIC COMPANY   COPY

ENGINEERING

GENERAL FILE
DEPARTMENT OF ENGINEERING

Proposed R&D Program for
Improved Reactor Containment

243-6060
"Water - R+D Program

GENERAL FILE
DEPARTMENT OF ENGINEERING

June 17, 1958

MR. N. R. SUTHERLAND:

    Herewith Mr. Schuyler and Mr. Whelchel's
memorandum of June 17. I recommend authorizing
this R&D program estimated to cost $275,000. I
also recommend that the cost of this program be
held in suspense against an Engineering suspense
number until such time as it is decided how the
charges shall be properly accounted for.

    Authority is requested to proceed with
the R&D program as recommended.

WALTER DREYER

CCW:eon

Attach.

cc: CCW
     FEB
     SLS

    TRS

APPROVED R. Sutherland   6/17/58
PRESIDENT

EXHIBIT 11   **Proposal**

619177 9-56-100 PDS.

PACIFIC GAS AND ELECTRIC COMPANY

DEPARTMENT OF ENGINEERING

PROPOSED RESEARCH AND DEVELOPMENT PROGRAM
FOR
IMPROVED REACTOR CONTAINMENT
JUNE 17, 1958

MR. WALTER DREYER:

This memorandum summarizes the considerations and findings to date concerning the proposed research and development program for improved reactor containment, and recommends a course of action.

Atomic Energy Commission

The pressure suppression concept was discussed with the AEC in Washington and described in our letter to Mr. Sutherland of June 3, 1958. AEC's reaction was favorable.

Application to Humboldt

PG&E has estimated the cost of several designs using the pressure suppression system for Humboldt. These estimates show a capital saving of between $600,000 and $850,000, depending on the design selected. Bechtel also estimated the cost saving; and their estimate, based on one of the designs, substantiates ours very closely. The final design and cost saving cannot be made until the R&D program has been successfully completed.

We have reviewed with General Electric the R&D time schedule and believe, with them, that it is a tight schedule, but can be completed in nine months. Based on this time, Bechtel are of the opinion that the Humboldt completion date will be unchanged. They estimate the extra engineering expense to be about $50,000.

Probability of Success of R&D Program

In the Vallecitos type of reactor containment, any benefit from the condensation of steam resulting from an accident is neglected in designing the capsule. The R&D pressure suppression program has as one of its main objectives the rapid condensation of steam in a water pool, thereby greatly decreasing the volume of the containment vessel. The proposed R&D program will determine to what degree the pressure suppression system will function. It is expected to obtain nearly 100% condensation of the steam; but if it is less, the program could still be considered successful. The General Electric Co. believes the chances of success are very good.

General Electric Quotation

A proposal from GE dated April 30, 1958, was forwarded to Mr. Sutherland with our letter of May 21. We also have received the attached revised General Electric proposal dated June 11, 1958. The principal change being the inclusion in Phase I of the design of equipment required in Phase II, insofar as this is

EXHIBIT 11 *(contd.)*

619177 9-56-100 PDS.

PACIFIC GAS AND ELECTRIC COMPANY

DEPARTMENT OF ENGINEERING

Proposed R&D Program for
Improved Reactor Containment #2 June 17, 1958

practicable. This is desirable to improve the time schedule and help insure completion of the program in eight months.

Phase I Consists of preliminary tests, development of methods of analysis, literature survey, and design of part of the equipment required in Phase II. General Electric has quoted a firm price of $43,000 for this work.

Phase II Consists of designing, procuring, constructing and testing:

(1) A large scale model of the water pool and the system of introducing steam. (Sparger Facilities)

(2) A small scale test of a complete pressure suppression system.

It is thought that the sparger test would be conducted at Moss Landing Power Plant.

General Electric has estimated their cost for doing all of the Phase II work to be $269,500. If PG&E were to procure and construct the sparger test facilities estimated to cost $60,000, then General Electric estimates their cost would be only $126,500. If handled in this manner, as we believe it should be, the Phase II cost to PG&E is estimated to be $186,500 instead of $269,500.

Phase III Consists of evaluating the results of the R&D program so that a design for a reactor plant can be made. General Electric has estimated this cost to be $6,950.

Estimated PG&E Expenditure

Phase I	GE Price	$43,000
Phase II	Estimated direct cost with PG&E furnishing the sparger test facilities.	186,500
Phase III	Estimated GE cost	6,950
	Total	$236,450
PG&E Engineering, Superintendence & Accounting		15,000
Distributable Construction Items		2,000
Total		$253,450
Overhead Construction Cost		21,550
TOTAL ESTIMATED COST		$275,000

60,000 - ML material
126,500 - Vallecito + GE manpower

EXHIBIT 11 *(contd.)*

619177 9-56-100 PDS.

PACIFIC GAS AND ELECTRIC COMPANY

DEPARTMENT OF ENGINEERING

Proposed R&D Program for
Improved Reactor Containment   #3   June 17, 1958

Recommendation

        We recommend that authorization be requested for this R&D program,
with the understanding that we, and GE, will re-appraise it at the end of
Phase I, since at that time the probability of its success and effect on
plant capital saving can be more accurately evaluated.

James O. Schuyler
J. O. SCHUYLER

C. C. Whelchel
C. C. WHELCHEL

JOS:CCW:ca

Attach.

APPROVED
....
                    PRESIDENT

be no drastic consequences. A long delay was undesirable, and PG & E did want to start as close to the original date as possible. The eight months' delay predicted by GE to complete the development program seemed long, and GE was asked if it would be possible to save some time by performing the detailed design and procurement of the large-scale test equipment in Phase I instead of in Phase II, as proposed. By making this change, it was hoped some time could be saved by accomplishing the design and procurement work in parallel with the analytical work of Phase I. GE replied, saying this would be possible but that it would mean re-allocating the cost of this work from Phase II to Phase I, a matter of $4 500.

On 17 June 1958, a meeting took place in which Bechtel presented their estimate for the cost saving which could be expected using the separate pool type of pressure suppression containment. It was shown to be about $500 000 and the PG & E engineers had agreed. It was pointed out that this was not the final design, and that the estimate was only intended to establish an approximate order of magnitude of the possible savings. The final design could be established only after completion of the Research and Development program. Bechtel agreed that if the Research and Development program went according to schedule, the Humboldt completion date would be unchanged. They also estimated the extra engineering expense, because of the change in containment, would be approximately $50 000.

The results of the investigation during the preceding few weeks were summarized by Messrs Whelchel and Schuyler in a memorandum to Mr Dreyer on 17 June 1958. The memorandum is included as Exhibit 11.

Messrs Whelchel and Schuyler recommended that authorization be requested for the Research and Development program with the understanding that PG & E and GE would re-appraise it at the end of Phase I before proceeding further. Mr Dreyer sent this memorandum to Mr Sutherland with a covering letter (Exhibit 10) suggesting the cost of the program be held in a suspense account until it was known how the charges should be properly accounted for. It had been decided that if the Research and Development program were successful, its cost should be considered a part of the Humboldt expense because it would be lowering the overall cost of the unit. If the program were not successful, its cost would be considered an operating expense.

Mr Sutherland gave the authorization to proceed with the Research and Development program the same day.

Phase 1   PART 3

General

When the General Electric Company received authorization to proceed with Phase I, Mr Robbins was placed in charge of the pressure suppression development program. He was assigned three men for the Phase I work:

Mr Fiock: an engineer who had had experience in steam turbine design and whose job it was to follow the design of the model test facility in the program.

Dr E. Janssen: who was to perform the necessary analytical work.

Mr Nelson: who was to be on general assignment, perform the drafting, etc.

How these tasks were accomplished, and the resulting conclusions, were presented in a report entitled, 'Pressure Suppression System Development Program – Report of Phase I Work' dated 8 September 1958.

PART 3

Review of work by others

A search revealed that the use of water to suppress pressure buildup following a reactor accident had been proposed by several different organizations, including the Bechtel Corporation, Sargent and Lundy, and American Machine and Foundry. However, it was also found that no extensive Research and Development program had been carried out to develop these ideas. It was further found that no one else was carrying out a development programme similar to that proposed by GE.

In addition, a considerable amount of literature pertaining to the formation and behavior of discrete bubbles, bubble stability, steam condensation, and behavior of jets was reviewed. Notes were made on pertinent bits of information for later use in the design of the steam injection system. Similarly, pertinent literature was obtained and reviewed in preparation for the study of the effects on the dry well of blast, shock, and missiles.

Preliminary tests

It was originally thought that for complete condensation, the steam would have to be injected into the water in the form of small bubbles in the water pool. To aid condensation, sparger boxes, which are boxes with holes drilled in the top surface, were to be used to break up the steam into small bubbles as it entered the water.

The idea behind the preliminary tests was to observe the effect of passing steam through a sparger box into a pool of water using different flow rates, box sizes, hole sizes, and hole spacings, and to incorporate these effects in the sparger design. GE engineers were particularly worried about the possibility of the bubbles coalescing into one large bubble just after leaving the sparger surface and that these large bubbles would be able to rise to the pool surface without condensing, causing the pressure in the vessel to build up.

To find the answer to these questions, a wooden box roughly $4 \times 4 \times 4$ ft was constructed and moved to the Vallecitos Atomic Laboratory. The box was filled with water, and steam was admitted to a small sparger in the bottom at flow rates varying up to 4 000 lb/hr.

In the first test, a sparger plate having 173 holes $\frac{5}{16}$ in diameter spaced equally over a 9×9 in area was used. In this case, it was not possible to achieve a steady flow of steam through the holes – all condensation of the steam took place within the box. In a second test, a plate with only 36 holes was used, and in this case, a steady stream of bubbles was produced. Although air in the steam caused some difficulty in observing the steam condensation, it was concluded that most of the steam had condensed within about 2 in of the top of the sparger plate.

Thus, it appeared that effective steam condensation could be achieved by means of discrete bubble formation, and that there would be no tendency for these bubbles to coalesce. However, about this time, someone tried jetting steam into the tank through a pipe directed vertically downwards, and it was seen that the steam was still completely condensed. This led to another series of tests in which 1 in, and $1\frac{1}{2}$ in pipes discharged up to 4 000 lb/hr of steam into a large water storage tank at the Vallecitos site. The conclusions from these tests were that:

1. Steam is condensed in cold or warm water before it is very far from the end of the pipe. At the highest flows tested, the condensation length is about one pipe diameter.
2. The steam jet provides excellent mixing of the water, except at very low sustained flows. At low flows, a layer of hot water can be created at the top of the pool which does not condense effectively.
3. The jets must be submerged 8 to 10 pipe diameters to prevent large amounts of air from being sucked in through a circulation vortex.

4. The steam flow rate limit was about 900 lb/hr through a 1 in pipe into a 3·5 × 12 × 48 in box; this is a condensing capacity of about 770 lb/hr ft³.

Development of methods of analysis

Hydrodynamic theory A thorough investigation was conducted of the applicability of theories developed by Zuber and Tribus of UCLA, initiated by them to predict analytically the maximum heat transfer rates for pool boiling. In the course of the investigation the literature of Zuber and Tribus was reviewed, both men were consulted, and references used in their work was pursued. Sparger boxes were adapted to the pressure suppression system to provide an analytical base for design of steam discharge surfaces involving many holes in flat plates. The theory also predicted conditions under which steam tends to lift the water and coalesce into a big bubble. The finishing touches were just being put on the adaptation of this theory when the preliminary testing just described indicated that large, high-velocity jets would be a superior means of injecting steam into the water pool. Since the behavior of such jets was not covered in the Zuber–Tribus work, the design using sparger boxes was abandoned.

Transient analysis Dr Earl Janssen was responsible for developing the method of analysis for determining the pressure in the dry well and reactor vessel as a function of time. With reference to this, it is seen that pressure versus time could be predicted under variation of the following system parameters:

> volume of the reactor vessel
> initial state of the steam/water in the reactor
> size of the break
> volume of the dry well
> vent areas
> vent submergence.

The method was checked by performing hand calculations for one set of conditions, and the results appeared reasonable. Because the calculation process was tedious, however, it was decided to program the analysis for the IBM-650 computer for more intensive use later on.

Missile and blast effects To establish the problem here, let us postulate a break in that portion of the primary system located in the dry well. The flow of steam and/or water through the break is established immediately out of the reactor vessel and into the dry well. A shock wave may be expected to proceed in advance of the interface between the steam/water and the air in the dry well. The steam/water flowing into the dry well will attain a very high velocity (in the order of hundreds of miles per hour), and its potential for destruction may be likened to that of a small tornado. Any pipes, structural members, regions of the dry well wall, etc., upon which this flow impinges will experience very high dynamic pressures. Furthermore, any parts torn loose may be accelerated by the high velocity flow and become missiles capable of further damage to the dry well and its contents.

An analysis of these effects was made by Dr Janssen. Without going into the details of the analysis, his results showed that the maximum pressure to be expected as a result of a shock wave would be in the order of 750 psi or less, a pressure which could easily be withstood by the dry well. As for missile effects, what was termed a relatively gross and conservative analysis produced the conclusion that a steel lining of less than 0·4 in thickness would be adequate to protect the dry well under the worst possible circumstances. The basis of this analysis was in calculating the maximum velocities which could be attained by either an 850 lb missile from the 8 in diameter feedwater line (116 ft/sec), or a $\frac{1}{2}$ in piece of shrapnel from the reactor

(597 ft/sec) and then using known relations between armor plate and armor-piercing projectiles to determine the required thickness of plate.

Conclusions and recommendations

1. Results obtained in Phase I indicated that pressure suppression systems can very probably be used to advantage with boiling water reactors. What is believed to be the maximum credible accident can be contained by a pressure suppression system very similar to those already proposed by GE. In other words, the results to date substantiate previous estimates of system performance and cost.
2. Phase I work, though encouraging, was not sufficient to design a pressure suppression system, or to prove to the AEC and others concerned that pressure suppression can be used safely.
3. The balance of the pressure suppression development program proposed by GE to PG & E should be carried out. Major parts of this additional work were:

   (a) Development of steam injection equipment using large, steady steam flow into a large water tank to be built at the Moss Landing Power Plant of PG & E.
   (b) Simulation of the transient operation of a pressure suppression system in a small scale model to be built at the Vallecitos Atomic Laboratory. Data would be obtained to confirm or improve the present methods of transient analysis. This test operation would help provide convincing proof of satisfactory operation by mocking up a complete system.
   (c) Utilize test results and analysis to prepare design specifications for pressure suppression systems.

PG & E decides to proceed with Phases II and III

Four copies of GE's 'Pressure Suppression System Development Program – Report of Phase I Work' were presented to PG & E on 18 September 1958, at precisely the time predicted by GE in their original time estimate. Along with the reports was a letter pointing out the results of Phase I were most favorable, and urging PG & E to proceed with Phases II and III. The letter also said the following:

'As a result of the much more complete understanding of the scope and magnitude of Phases II and III, we regret to advise that we find it necessary to revise our original quotation (of $133 450) upward. We now quote a maximum firm price, not subject to escalation, of $158 000, to complete the work of Phases II and III as outlined in the Phase I report. This is on the basis that Pacific Gas and Electric Company will be responsible for construction of all of the test equipment at Moss Landing Steam Plant. Should the cost of doing the work be appreciably less than this maximum price, we will be pleased to make appropriate downward adjustment of this price.

It was necessary for us to increase our previous rough estimate of the cost of Phases II and III because our detailed study of the problem in the past three months has indicated a substantial increase in the engineering time required to accomplish the desired results and to cover increased work which was indicated to be desirable. For instance, we now intend to prepare detailed hazards evaluation for the AEC reactor safeguards report. Also, we find that the mathematical transient analysis is considerably more complicated than was anticipated. This results in a substantial increase in the computer time and cost required for the theoretical analysis.'

At this point, Messrs Whelchel and Schuyler prepared a memorandum to Mr Dreyer summarizing the results of the Phase I program and recommended that authorization be requested to proceed with Phases II and III of the Research and

EXHIBIT 12 Letter, Dreyer to Sutherland, 18 September 1958

62-6224

① ~~SS~~ ② File 115.14

PACIFIC GAS AND ELECTRIC COMPANY

COPY

ENGINEERING

R&D Program for
Improved Reactor Containment
Phases II and III

September 18, 1958

MR. N. R. SUTHERLAND:

On June 17, 1958, Phase I of the Research and Development Program on Improved Reactor Containment was authorized. This phase of the work has been satisfactorily completed; and a summation, along with the planning for Phase II, is given in the attached memorandum by Mr. Whelchel and Mr. Schuyler. A separate sheet gives General Electric's "Conclusions and Recommendations" as taken from their report to us dated September 8, 1958.

The program originally estimated at $275,000 is now estimated to cost $300,000.

Authority is requested to proceed with Phases II and III of this program.

WALTER DREYER

CCW:ca
Attach-2
cc:SLS
CCW
FEB
LWC
TRS
EVN
6 x

EXHIBIT 13   Research and Development Program

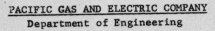

PACIFIC GAS AND ELECTRIC COMPANY
Department of Engineering

RESEARCH AND DEVELOPMENT PROGRAM
FOR
IMPROVED REACTOR CONTAINMENT
SEPTEMBER 18, 1958

MR. WALTER DREYER:

This memorandum summarizes the results of the Phase I program and makes a recommendation regarding the Phase II and III programs.

I. RESULTS OF PHASE I

A. Preliminary Tests

Tests were run discharging air then steam into a four foot cubic test tank to determine the effect on jets in water. These tests showed that condensation occurred very rapidly and that the superior method of injecting steam into water was by large high velocity jets.

B. Development of Methods of Analysis

A method of analysis for determining the pressures in a pressure suppression system allowing for the variation of all major parameters has been developed. Typical calculations have been run using an IBM-650 computer.

C. Review of Work by Others Including Literature Search

The literature search has (1) provided information on the action of steam in water, (2) shown some applicable work on blast and shock effects has been done by others, and (3) shown that no one has conducted a similar R&D program.

D. Design and Planning of Phase II Tests

(1) Moss Landing Sparger Facilities

The design of the facility has been essentially completed and arrangements made to procure the material and construct the facility upon the approval of Phases II and III.

(2) Small Scale Pressure Suppression System

The design of the test set-up is almost completed. This design had been originally scheduled for Phase II.

EXHIBIT 13 *(contd.)*

Research & Development Program for
Improved Reactor Containment   #2   September 18, 1958

II.   <u>CONCLUSION</u>

     The results obtained so far promise ultimate success of the development
program and the basic approach originally outlined still appears sound.

III.   <u>GENERAL ELECTRIC QUOTATION</u>

     The attached General Electric proposal dated September 18, 1958 quotes
a maximum firm price of $158,000 for conducting Phases II and III of the program.
Should the cost of doing the work be appreciably less than this maximum price,
General Electric will make a downward adjustment in the price. The work is
generally as originally outlined, but it has been expanded to cover a study of
the "maximum credible accident" which the pressure suppression system must be
designed to protect against. The inclusion of this study, in addition to a better
understanding of the work to be done because of the Phase I investigations, has
shown that the work for Phases II and III will require more engineering effort
than originally anticipated. For this reason the General Electric quotation for
Phases II and III is $24,500 higher than the original estimate.

IV.   <u>REVISED ESTIMATED PG&E EXPENDITURE</u>

Phase I	GE Price		$43,000
Phases II and III	GE Price	$158,000	
	PG&E Furnished Equipment	60,000	
	Total		218,000
			$261,000
PG&E Engineering, Superintendence & Accounting			15,000
Distributable Construction Items			2,000
Total			$278,000
Overhead Construction Cost			22,000
TOTAL ESTIMATED COST			$300,000

V.   <u>RECOMMENDATION</u>

     We recommend that authorization be requested to proceed with Phases II
and III of this R&D program.

                                             _____
                                             J. O. SCHUYLER

                                             _____
                                             C. C. WHELCHEL

JOS:CCW:ca
Attach-GE ltr. dtd 9-18-58

Development program. Mr Dreyer passed the memorandum on to Mr Sutherland, along with a covering letter in which he pointed out that the program, originally estimated to cost $275 000, was then estimated to cost $300 000. Mr Sutherland subsequently gave his authorization to proceed with Phases II and III.

The memorandum requesting approval and Mr Dreyer's covering letter are included as Exhibits 12 and 13.

PART 4   Phase II

Test facilities

Phase II of the pressure suppression development program was a testing and evaluation phase conducted in two sections. The Atomic Power Equipment Division of GE was to build and operate a test facility to determine transient behaviour of the system. The PG & E was to build a steady state condensing test facility to be operated under GE's supervision. This unit would be used to determine the effect of changes in the relative dimensions of nozzles, baffles, and other parts of the system on the steady state condensation of steam in a water pool. An illustration of the Transient Unit follows (Exhibit 14).

By the middle of September 1958, all of the bids were in for the condensing test unit. Seven bids were submitted ranging from $13 825 to $27 795 with the average being $20 000. On the basis of Chicago Bridge and Iron Company's lowest bid and a satisfactory completion date, they were awarded the contract.

The initiation of Phase II of the development program formally began on 30 September 1958 with Mr Sutherland's official approval.

It was decided that the condensing tests should be located at the PG & E thermal power plant at Moss Landing, California. This was because of the large steam flow that was necessary in the testing process. During the month of August, the construction department of PG & E planned the foundations necessary for the test tank. Chicago Bridge and Iron Company advised PG & E that their schedule allowed erection of the tank to begin on 17 November 1958.

Communications with General Electric

Although PG & E was responsible for design and construction on the condensing test tank, engineers from GE were to use the equipment to gather the required data. Consequently it was imperative that good communication be established between GE and PG & E engineers. Mr D. Barton was the engineer at PG & E in charge of the Phase II design efforts at PG & E, and he was the logical one to act as liaison officer with GE. He therefore spent much of his time with the GE engineers during this period to establish an understanding. Most of the conversations and exchanges of information were casual. Most exchanges were personal meetings or telephone conversations, although there were occasional formal meetings.

Credibility of a pressure vessel failure

As decided earlier, a part of Phase II was to be a report by GE on the credibility of a nuclear power reactor pressure vessel failure. This report was completed on 17 November 1958 and submitted to PG & E. Covered in this report were the possibility of failure of the pressure vessel and also the nature of the types of failure. Because of known catastrophic brittle failures of large steel structures in the past, brittle failure

was of prime importance. The conclusion reached was that if the pressure vessel was kept above its critical temperature (a property of each steel but usually in the range of $-10°$ to $+50°$ F ($-23°$ to $10°$C) as expected in operation there would be no problem. The effects of welding and radiation damage were also discussed. Based on this report, previous GE experience, and current design practice in the field of nuclear power reactors, the maximum credible operating accident was ultimately defined for the Humboldt Bay reactor as the instantaneous severance of the largest pipe leading from the reactor pressure vessel.

Transient tests

Throughout this period, the engineers at GE had been working on the design of a transient test facility, which was to be built at their plant in San Jose. The objective of the transient tests was to obtain data and information on pressure transients that might be encountered in the actual pressure suppression system. Phase I had yielded some analytical techniques, and confirmation or modification of this analysis was needed.

The transient test apparatus was built with three distinct volume segments. One segment was a pressure vessel containing water. This vessel had a rupture disc separating it from the simulated dry well. When pressure and temperature were built up in the first volume to the desired level, the rupture disc would be breached and the flashing steam would flood the simulated dry well and then be carried over into the water in the third volume or pressure suppression chamber. Such parameters were to be varied as the initial pressure and temperature, rupture disc size, dry well volume, and vent depth of submergence in the suppression chamber. Determination of test parameters was partly by approximating the expected operating conditions and partly by '. . . guessing as to what to test', quoting Mr Robbins, the responsible engineer for GE.

These transient tests yielded an unexpected result. In some of the tests, water was forced back into the dry well volume. The water in the dry well condensed some of the steam left in the dry well, creating a partial vacuum. To allow for this, vacuum breakers were later designed into the system. These breakers were placed in the dry well wall and were designed to withstand the expected internal pressure. If the pressure differential occurred in the opposite sense, these units would collapse inward breaking the vacuum. Exhibit 14 is a sketch of the transient pressure test facility.

The mathematical analysis undertaken in Phase I assumed essentially no flow resistance in the path from the dry well to the containment volume. This condition was closely met in the transient tests. As the design of the system progressed however, it was apparent that a measurable flow resistance might be encountered. This was a problem which later had to be completely answered before AEC approval of the pressure suppression concept was obtained.

Here is a situation that in retrospect appears to be an oversight. It must be remembered, though, that the design of the complete system was ill-defined at this early stage of testing. The thought at the time of the early designs and the initial tests was that very little flow resistance would be encountered because of the assumed dimensions that were involved. It became evident later during the detailed design work, that appreciable resistance was indeed present. To have revised the analytical computations at this point and re-tested them would have meant considerable effort. This effort was not thought to be justified. The engineers felt that their intuitive understanding of the system led them to sufficiently accurate predictions of its behavior with this new parameter. Consequently more analytical work and testing were not considered.

In addition to determining pressure transients, the transient tests were to measure the effectiveness of the water pool as a barrier to fission products. This was done by introducing samples of krypton and xenon into the water in the pressure

EXHIBIT 14   **Transient pressure test facility**

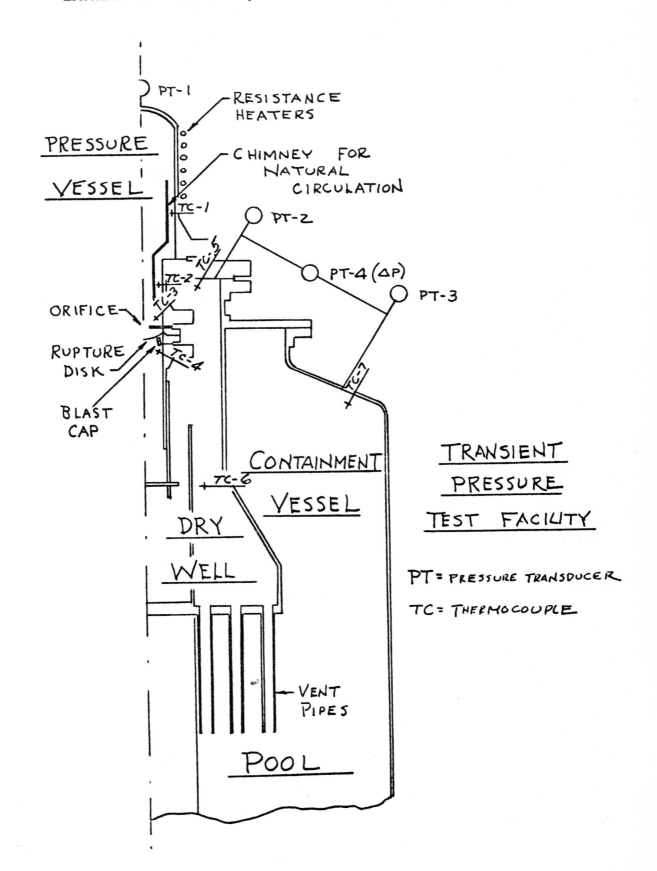

PT-1

RESISTANCE HEATERS

PRESSURE

VESSEL

CHIMNEY FOR NATURAL CIRCULATION

TC-1

PT-2

TC-5

TC-2

PT-4 (ΔP)

PT-3

ORIFICE

TC-3

RUPTURE DISK

TC-4

BLAST CAP

TC-7

CONTAINMENT

TC-6

VESSEL

DRY

WELL

TRANSIENT

PRESSURE

TEST FACILITY

PT = PRESSURE TRANSDUCER

TC = THERMOCOUPLE

VENT PIPES

POOL

vessel just at the point of rupture of the disc. Samples were then taken at various points within the test apparatus and measured for content of these elements by use of a mass spectrometer. Soluble salts were also placed in the pressure vessel and similarly detected. Corresponding tests were made with solid particles. The results were encouraging, but such factors as scale, geometry, and chemical species could play an undetermined role. This again was a problem of what to test and the engineers handled it primarily by simply choosing materials that were likely to be present in the actual system. This was more or less a guessing game due to the lack of exact information.

Safety analysis and the preliminary hazard summary report

Future plans in the area of safeguard included an evaluation, to be completed by mid-January 1959, of all significant credible causes of failure of the pressure vessel. This work had been preceded by the 'Credibility of a Nuclear Power Reactor Pressure Vessel Failure' study mentioned earlier. The continued interest in possible accidents points out the lack of exact knowledge about this crucial area and the effort that was expended to define it. A clearer understanding of this problem area was perpetually sought by PG & E.

A meeting with PG & E, GE, and Bechtel engineers was called for 12 January 1959, to discuss the safety analysis section of the Preliminary Hazard Report that eventually had to be filed with the AEC. This report was to discuss the hazards to be expected and the method of dealing with them. It was decided that GE would write the section dealing with transient behavior, and PG & E would be responsible for the steady state considerations. This division was quite a natural, since GE had made the transient tests whereas PG & E was limited to the steady state condensing tests. To gain additional information and properly organize it for presentation to the AEC, the Hazards Summary was to be submitted in March or April 1959.

Interim progress reports

By the middle of December 1958, progress on Phases II and III had been significant enough for an interim report to be compiled. A progress report was written by Mr Fiock of GE. His report noted the work accomplished and outlined the anticipated future courses of action. Achievements to date had consisted primarily of procurement and erection of the condensing test tank as well as detailed planning for experiments to be conducted. Delivery of all equipment for the transient tests was scheduled within a few days and checkout and testing to begin immediately. Some expected data points had been calculated with an IBM-650 computer.

The second interim progress report on Phases II and III was completed on 5 February 1959. The condensing tests were encouraging. With any submergence of more than one inch of the vent pipe in the water pool, steam condensation was virtually complete. Also noted was the close agreement of the analytical model with the transient test data. Less severe pressure peaks were encountered than were expected, although the time scales were in accord. This lower pressure was attributed to the water above the orifice in the pressure vessel being somewhat subcooled (and therefore not saturated) at the time of the rupturing of the disc. This led to a lower initial flashing rate and consequently to a lower peak pressure. Some assumptions about the ideal nature of the flow in the analytical model had been made. These assumptions yielded pressures that were somewhat conservative for design purposes. The results were quite promising, and it was felt that these data would provide valuable information for design. In short, everyone was pleased with the outcome.

Pool cooling

For some time there had been concern about whether the suppression pool would need cooling. Would the reactor produce sufficient heat (direct thermal as well as gamma radiation effects), and would enough of this heat be transferred to significantly increase the pool temperature? Would the conduction to the earth be adequate to keep the pool cool? What was 'cool enough'? Some tests had been run with the steady state apparatus to determine at what water pool temperature the pressure suppression concept became unusable. Pool temperatures were tested as high as 120°F (48°C) and satisfactory results were obtained. The pool was not expected to reach this temperature in actual operation. Conduction to the earth, however, was a more complicated subject. Mr Barton made a study of the problem and outlined the temperature considerations discussed above. He made a few calculations as to what magnitudes of temperature rise could be expected. His report was completed on 13 February 1959. On the basis of these rough calculations and the desire to maintain the pool relatively cool even after an accident, a cooler was ultimately incorporated into the design.

Exploratory meetings with the AEC

By now, PG & E felt that significant progress had been made and that a meeting with the AEC was in order. On 18–19 February 1959, a meeting with the AEC was held. Both the General Electric Company and PG & E were present. A progress report was made to the AEC. The meeting was fairly informal, with the major purpose being an exchange of information and an attempt to understand the opinion the AEC had on pressure suppression as a safety concept.

The AEC wanted the results of the tests submitted to enable a better review of the idea. Timing of filing of various documents with the AEC was discussed, and AEC regulations on various operating procedures were reviewed. The AEC made a few minor suggestions and reiterated its request for certain test information. The tone of the AEC was quote, 'optimistic', and PG & E was very encouraged.

PART 5   Phase III

Relation of GE and PG & E in Phase III

General Electric was to conduct Phase III, the application of data gained during the testing to the design of the actual system. Additionally, the engineers at PG & E thought that it would be well for them to work on the same design. Consequently, a study was undertaken to examine the several arrangements that had been considered and to investigate the costs and feasibilities of each. Mr C. Ashworth conducted this study.

The attempt of PG & E to anticipate the moves of their subcontractors was typical of company philosophy throughout the development program. It stemmed from their desire not to be caught unaware of contractor design and development and always to be cognizant of all possibilities. They were in the position of hiring experts to do the analysis and wanted to be able to understand and constructively criticize the results.

As a result of this philosophy, many engineers at PG & E, notably Mr Ashworth, had thought a great deal about the configuration that might be the final design.

Mr Ashworth's conceptual design (See Appendix)

By the first week of March 1959, Mr Ashworth had made several sketches of his own of various possibilities and had listed advantages and disadvantages of each design. These are shown in Exhibit 15. He had considered four distinct system types:

15(A): With the pool separate from the dry well and some distance away
15(B): With the pool concentric with the dry well
15(C): With the pool adjacent but to one side of the dry well
15(D, E, F): With the reactor vessel completely submerged in the pool.

The main disadvantage with (A) and (B), according to Mr Ashworth, was the possibility of differential settling of the earth around the structure. Scheme (C) would make it easy to obtain the required flow area and total enclosure was feasible, but the turbine would have to be farther away from the reactor than in the first two types. Schemes (D), (E), and (F) offered maximum blast and missile protection with very compact physical size, but the problem of pool water activation was present, and there were maintenance and operating problems. These ideas later proved to be valuable background material because they pointed out several design problems. This policy of encouraging a free influx of ideas was a part of the engineering philosophy, and many aspects of the problems were considered by several people.

It was felt that these concepts well covered the variations likely to be considered later. All of them presumed a reactor vessel and dry well located in a concrete caisson which furnished structural and shielding functions. The pool was to be concentric about the dry well. The difference between the concepts was primarily one of arrangement of the exterior structure or the containment building. The numerous advantages and disadvantages were listed and estimates were made by the Civil Engineering Department of the costs of each. Costs ranged from $660 000 to $710 000 – not a significant difference in light of the cursory and approximate nature of the estimates. No conclusions were drawn in the study as to which was best. This was merely an attempt to better prepare for future decisions and anticipate Bechtel's work.

By this time the bounds of the several variations of containment vessels were becoming well defined. On 10 June, a memo was written at PG & E summarizing basic possibilities to be considered in determining the best enclosure. The six variations listed were:

1. A separate pool with gas-tight building around the refueling area
2. A separate pool with a deck refueling enclosure around the immediate reactor in time of refueling – this would be in an ordinary building
3. The same as 1 but with a concentric pool
4. The same as 2 but with a concentric pool
5. A concentric pool with a circular containment building designed for low pressures – those expected in the event of a refueling accident but insufficient for a major accident
6. The same as 5 but designed to be easily adapted to dry containment in the event that pressure suppression containment was not acceptable.

This summary was to be used as a basis for comparison with whatever configurations Bechtel proposed at the conclusion of their design effort. In the end, Bechtel was to submit a few alternatives, one of which was chosen by PG & E for the final application. All during this time, Bechtel was, of course, busy working on various proposals of its own.

Refueling

In an attempt to save money in the construction of the power plant, it was hoped that open refueling could be used. This would mean that no refueling building would be

EXHIBIT 15   Pressure suppression–arrangements A and B

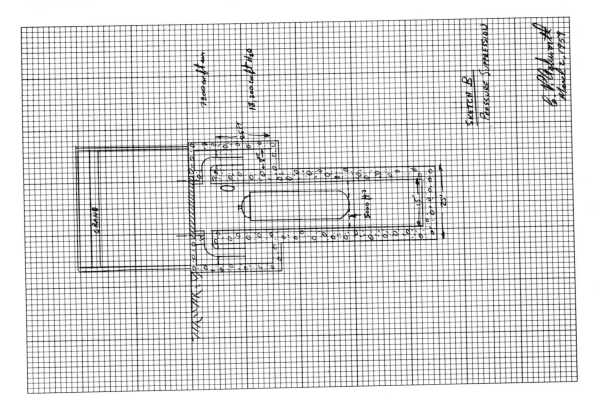

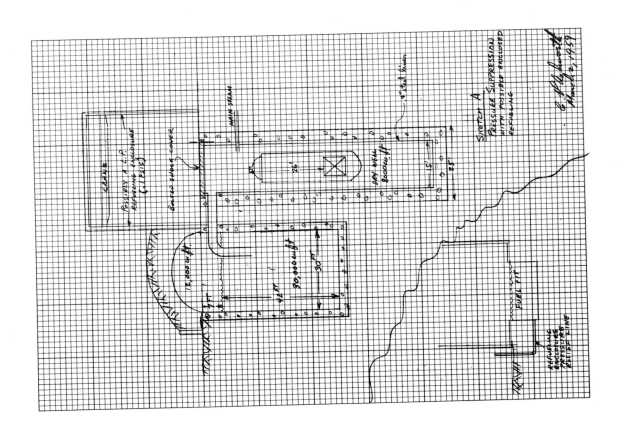

EXHIBIT 15 *(contd.)* **Arrangements C and D**

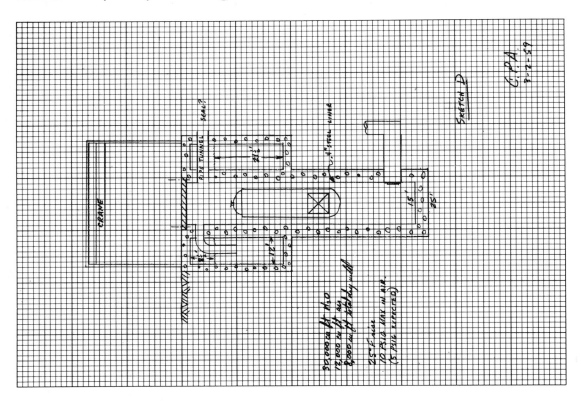

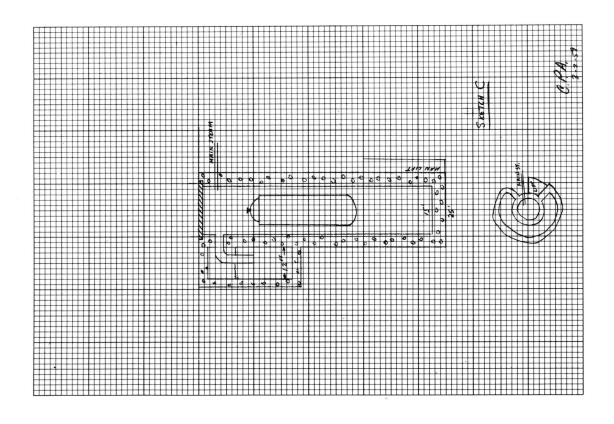

EXHIBIT 15 *(contd.)*   Arrangements E and F

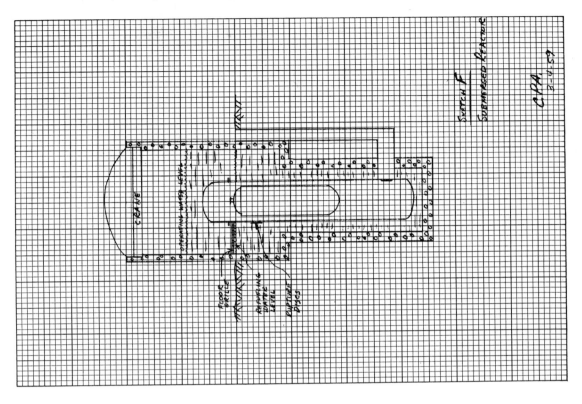

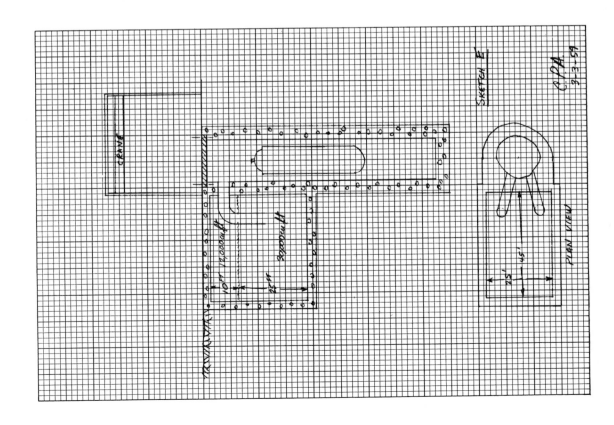

necessary over the reactor at all, except possibly a roof to give rain protection. If this idea didn't prove feasible, a sealed deck system was to be tried. In this case, a sealed container would be placed around the open reactor when refueling was to be done. Should this scheme also not be granted approval by the AEC, a 'gas tight' containment structure was to be provided. This containment structure would not meet the requirements to contain a maximum credible accident, but only serve to contain the leakage from a refueling accident. Eventually this was the refueling containment that was adopted because it offered the most safety.

Explanation of data curves

Mr Barton sent out a memorandum on 19 May 1959, concerning the expansion of understanding and definition of design problems, and explaining the pressure suppression data curves. He made a few pertinent comments about some of the values obtained and reviewed the requirements of the system. He noted that pressure suppression could not very well accommodate large vessel breaks (assumed to be 'incredible'). According to the data received, he calculated that if a '...vessel break should have an area on the order of one-tenth of the vessel cross section or larger, dry well pressures of 250–500 psig would be achieved for vent areas less than 50 ft². Vent areas of 150 ft² or more must be used to reduce dry well pressure below 100 psig, and such vent areas are probably impractical.' Since 50 ft² was the reasonable upper limit for practical vent areas, he concluded that 'it appears that the practicability of pressure suppression depends on the lack of any credible major vessel breaks.'

The General Electric report on Phase II

The first test report for the pressure suppression development program was submitted to PG & E on 2 April 1959. This report contained detailed results and some conclusions, but no design information. The report primarily reiterated the progress reports, tied all the results together, and gave the data that was obtained.

The final design report from GE was dated 21 May and contained explicit conclusions as to methods of design for a pressure suppression system. This report was intended to provide '...a designer with the necessary system response information for detailed design work', as it stated in its introduction. Dry well design, blast and dynamic pressure, static pressure, missile effects, steam injector equipment, the water pool, and the containment vessel were all subjects of this document.

The designers and engineers at PG & E were somewhat disappointed with this report and felt that it was rather incomplete. It didn't have the depth that was hoped for. Due to PG & E dissatisfaction, GE prepared a much more detailed report some months later for PG & E's future use. Of course by the time the next report was completed, PG & E engineers had given considerable thought to the system design and Bechtel had essentially completed their designs.

After GE interpreted the results of Phase II and wrote the design report, Bechtel was to do the detailed design of the system. The engineers at PG & E felt, however, that it would be to their advantage to anticipate Bechtel's operations and do some work themselves. This work was not to be in great depth but rather background and preliminary design work. As a result of this philosophy, Mr Ashworth, Mr Barton and Mr Schuyler independently prepared lists of pressure suppression design requirements. These lists contained some requirements that the design had to meet and some parameters that had yet to be determined. These were used as comparisons and references later when Bechtel had accomplished the same thing and had progressed further in their design work. Revision and addition to these lists continued through the first half of May but in a less formal manner.

Design limitations

On 1 July Mr Whelchel commented on the pressure suppression containment system. He made the point that it was very desirable to keep the design of the Humboldt plant as much like the test conditions as possible to obviate the need for further tests. He stressed the need to utilize all information gained at Moss Landing, avoiding further tests. To eliminate more expense and time, the engineers were limited to a configuration that would conform to the test results.

Mr Whelchel had long been a proponent of completely burying the reactor underground. He viewed this as a sound approach for safety reasons besides being a good public relations move. Reactors buried underground give rise to psychological assurances. Because of Mr Whelchel's conviction that the system should be underground, much thought was given to this possibility. On 20 July, Mr Barton wrote a memo with some comments on below grade containment. He expressed the view that completely below ground level containment was unnecessary as only slight safety improvements were gained and those at great cost. He also noted that there were higher construction risks and that erection would be more complicated and therefore most costly. He claimed that building safety feature on top of safety feature was excessively redundant and expensive. He implied that pressure suppression should be adequately designed so that additional safety features were not required. He also counseled against becoming involved with a particular type of construction and being forced to continue with it in the future. In effect, he warned against getting committed to a costly approach that wasn't necessary. Mr Barton didn't advocate building the power plant completely above grade because there were some economic advantages to using earth as a structural material. His opinion was that entire burial might, in spite of its advantages, prove unreasonable in the long run.

In a meeting with Bechtel, Mr Whelchel pointed out the problems of having Bechtel proceed with the detailed design when there was the chance that the AEC might veto the system. In view of this dilemma, Bechtel was asked to consider various methods of procedure in designing the pressure suppression system in such a way that it could be converted to conventional containment. They were asked to approach the problem in a manner to expose PG & E to the least possible financial loss. Bechtel later presented their estimate of costs incurred in converting to dry containment by 1 December as $300 000 to build an 80-ft sphere and an additional $200 000 for necessary concrete shielding. In conjunction there would be a further charge of $80 000 to $90 000 for design work.

The consensus at Bechtel was that alternate methods of procedure did not have to be considered until and if pressure suppression was turned down. Bechtel was confident that the experience gained on this job plus their general capabilities from earlier jobs would allow them to complete the design of a dry containment system in a matter of a few weeks.

The Bechtel proposals

Bechtel had spent a great deal of effort to arrive at an acceptable design. Much had been gained in the frequent exchanges of information and opinion that had taken place between the engineers concerned at the different companies. As a consequence of this effort Bechtel had proposed several schemes to PG & E. On 22 July, Mr Schuyler wrote a memo outlining these ideas and giving the engineering consensus of their advantages and disadvantages. These are shown in Exhibit 16.

In his discussion of the options, Mr Schuyler dismissed 1A and 1B because of the high costs and because they were conventional dry containment systems. The object was to develop pressure suppression containment as a cost-saving alternative. The inclusion of these as alternatives served as a basis of comparison. Schemes 2A and 3A were of little value since a rectangular building could not be easily constructed to meet specifications of leakage rates. Plan 2B was eliminated because it

EXHIBIT 16   Alternative designs for pressure suppression

	Scheme	Cost ($)
1A	A dry capsule similar to the Vallicitos installation plus a radiation shield.	1 733 000
2A	Pressure suppression with an offset pool and with an operating floor at grade level (i.e., reactor below grade) and surround with a 1 ft-thick rectangular building.	1 186 000
3A	Pressure suppression with a concentric pool and operating floor at grade level and with a conventional reactor building.	1 218 000
4A	Pressure suppression with a concentric pool and the entire structure buried below grade and with a 25-ton polar crane and a 60-ton gantry.	1 634 000
1B	The same as 1A but with a somewhat different crane.	2 023 000
2B	The same as 2A except with a dry well and reactor lowered 12 feet for refueling dome.	1 231 000
3B	The same as 3A except with the building of 1-foot thick concrete and gas tight.	1 288 000
4B	The same as 4A except with an operating floor at grade and a circular building provided over the reactor.	1 361 000
2C	The same as 2A except with a gas tight building.	1 256 000

was cumbersome and didn't provide for refueling, which was now conceded to be a necessity. In comparing 2C with 3B, Mr Schuyler noted that the essential difference was the price differential of $32 000. He felt that due to the approximation of estimates here, these figures were substantially the same. He considered the advantages of 3B to be:

1. GE believed the results of development were most applicable to a concentric pool
2. The concentric pool gave a structure which enclosed the dry well and therefore served as a good shell
3. There were a large number of vent pipes from the dry well and therefore high reliability. In spite of the disadvantage that 3B was a more complicated structure, the engineering consensus was that 3B was to be preferred over 2C.
Schemes 3B and 4A were similar because they both had concentric water pools.

The advantages of 4A, were:

1. Minimum possibility of radiation 'shine'
2. Maximum convenience of adaption to dry containment in case pressure suppression was ultimately turned down by the AEC
3. Greatest psychological advantage.

The main disadvantages of 4A were:

1. Highest cost of all the pressure suppression schemes
2. Very cramped working quarters
3. Required auxiliary buildings for fuel storage.

The engineers believed that 3B had ample working space and was easily adaptable to dry containment. The 1-ft thick walls were believed quite adequate for radiation shielding. General Electric had previously expressed its opinion that an accident was so remote a possibility that there was no need to shield the operators. The walls could easily be changed to 2 ft thick, with only minor design changes if more shielding were required. From these comparisons, it was concluded that scheme 4A was not worth the additional cost of $346 000. Scheme 4B was summarily eliminated because it had very little advantage over 4A while containing most of its disadvantages. Mr Schuyler concluded that 3B should be the plan to be constructed.

This summation marked the point of commitment of PG & E to a particular path although official approval had not yet been given. The engineers at PG & E were each personally convinced that 3B was the most desirable alternative. It is interesting to note that no one really knows the origin of the various Bechtel schemes. They seem to have just evolved. Certainly, scheme 3B arose at Bechtel, but to what extent outside sources influenced it is not known. The idea of locating the pool concentrically around the reactor was an old one. The earliest conceptual sketches from GE and PG & E showed such a pool. It was a logical move to place the pool there because this concept had great simplicity – just put the reactor in the pool. Unfortunately some construction and access problems accompanied this scheme. For this reason modifications were made.

Meeting with GE, Bechtel, and PG & E

On 28 July, PG & E met with GE and Bechtel to discuss the selection of a pressure suppression arrangement. PG & E wanted GE and Bechtel to concur with their selection of plan 3B. PG & E presented three Bechtel schemes and commented on the advantages and disadvantages of each. GE agreed with PG & E on the 3B plan as the most acceptable and a week later sent a letter confirming this opinion. The possibility of a reactor failure was brought up again, and GE was to look into it. Even though several previous reports had been written, new considerations on safety continually arose.

Approval and commitment

On 29 July, the day after this meeting, Mr Whelchel indicated to Mr Sutherland that the development program was essentially complete and that the results were very favorable. He also noted the tests had shown pressure transients could be predicted, and that the water pool was effective in preventing release of fission products. He went on to list the conceptual designs that had been considered and outlined the reasons given previously for the preference of 3B.

Mr Whelchel said in his letter that a good indication of the AEC's intention toward pressure suppression should be obtained after the Advisory Committee on Reactor Safeguards meeting in November 1959. He then made some estimates of when certain documents should be filed and when meetings with certain AEC subcommittees should be held. He hoped that the construction permit would be granted in March or April 1960, although he pointed out that this might be optimistic. Mr Whelchel concluded with a request for Mr Sutherland's approval, which was given.

Although Mr Sutherland's approval was the official origin of the construction process, the real point of commitment had been the engineering consensus, as evidenced by the memo from Mr Schuyler on 22 July, that scheme 3B was the best. From that point on, the engineers had been committed to convincing management and the AEC of the correctness of their choice.

AEC acceptance

PART 6

Preliminary hazards report

With the development program completed and a final design configuration chosen, it remained now for PG & E to convince the AEC that the pressure suppression containment system would provide the high degree of safety required of a nuclear plant.

When a power company proposes to build a nuclear plant, it is common practice to submit a Hazards Report to the AEC describing in detail the construction and operation of the plant, and what precautions have been taken to assure safety. The AEC then carefully scrutinizes this report, and asks the proposer to explain certain features, correct weaknesses, etc. The company responds to the AEC's criticisms and suggestions in a series of amendments to its original report, and continues to do so until the AEC is completely satisfied. Only after the AEC is completely satisfied with the Hazards Report is a license issued.

P G & E submitted the Preliminary Hazards Summary Report on Humboldt Bay to the AEC in April 1959. This report covered reactor vessel breaks, nuclear excursions, chemical reactions, release of fission products, backup core cooling systems, maximum credible accident, and safety margins in design. Included in the report was a study of the possibility of brittle failure of the reactor vessel. This event had been covered before and was of low probability, but because of the catastrophic consequences, it was important and needed an investigation. The maximum credible accident was postulated here and defined as the 'worst coolant-loss accident which could result from near-instantaneous complete severance of any pipe penetrating the reactor vessel at any location'. The accident was assumed to occur under the condition of maximum energy content at the normal operating pressure, and no energy contributions from other sources were to be considered coincident with the coolant loss. As it hadn't been established that pressure suppression containment would be used at this time, the original report did not discuss containment at all. It was pointed out that a discussion would follow when it had been decided which type of containment would be used.

Following the submission of the Preliminary Hazards Report, a series of exchanges took place in which the AEC would ask for clarification on various points of the Humboldt Hazards Report and PG & E would reply, making revisions where necessary. PG & E consulted GE in making the replies. The various amendments to the original Hazards Report covered all aspects of the Humboldt plant. Only those aspects pertaining to pressure suppression will be discussed here.

Six amendments

Amendments No. 1 and 2 follow the Preliminary Hazards Report, answering some questions the AEC had on other aspects of the plant. In September, PG & E submitted Amendment No. 3, in which they stated that pressure suppression had been chosen as a means of containment and that a conceptual design had been selected for the Humboldt plant. This amendment also described the general features and design requirements of the proposed containment system, and gave results of the development program. The legal document is Exhibit 17. To it were attached 39 pages of technical explanation, drawings and diagrams. These are not included.

On 13 November 1959, an oral presentation was made to the Advisory Committee on Reactor Safeguards in order that specific questions could be answered relatively quickly. Mr Robbins of GE explained with the help of slides how the development work was applied to the prediction of peak dry well pressure, and a short movie was shown on the condensation of steam jets and pool action. The refueling building ventilation system was also described. During normal operation, ventilating air was exhausted from several locations in the building. If any one of several radiation monitoring devices detected an abnormal amount of radiation in the building, the building would be sealed off immediately from the outside atmosphere. The leakage rate tests for the pressure suppression system were discussed. Soap bubbles had been used to judge the amount of leakage from the suppression chamber under 125% of design pressure. Earthquake allowances were also described. It was the belief of PG & E that the structural concrete would not crack under seismic forces, and any minor cracking due to temperature changes or shrinkage would be remedied by the steel liners in the dry well and suppression pool. The results of this meeting were submitted as Amendment No. 4 on 16 November.

Amendment No. 5 was submitted on 30 November 1959. It contained answers to questions asked by members of the Advisory Committee on subjects such as the transient pressure analysis, the effect of air on the condensation of steam, the effect of using a scaled down model of the actual system, and the selection of 14 in diameter vent pipes.

Amendment No. 6, submitted 29 January 1960, contained the answers to several questions contained in a letter from the Commission dated 21 December 1959. The first question was about the basic hydrodynamic principles used in determining the behavior, size, and location of the steam nozzles in the suppression pool. The answer, in brief, was that the test results could be readily explained by established principles of heat transfer and hydrodynamics, as long as the actual facility bore a resemblance to the test facilities, the same principles could be applied with validity.

The second question was about the methods used to justify the design pressure of the principal containment components. The maximum dry well pressure could be computed by conventional thermodynamic consideration and steady-state flow calculations, calculated to be approximately 35 psig at the moment the water is expelled

EXHIBIT 17 **Application for AEC License**

BEFORE THE UNITED STATES ATOMIC ENERGY COMMISSION

Application of PACIFIC GAS AND ELECTRIC)
COMPANY for a Class 104b. License to)
Construct and Operate a Nuclear Reactor)
as a Part of Unit No. 3 of Its Humboldt)
Bay Power Plant)
_____)

Docket No. 50-133

Amendment No. 3

Now comes PACIFIC GAS AND ELECTRIC COMPANY (the
Company) and amends its above-numbered application by sub-
mitting herewith Amendment No. 3, which consists of
Addendum A and Addendum B to the Preliminary Hazards Summary
Report, dated April 15, 1959 (Exhibit B to said application).

The information contained in Addendum A supplements
and amends the Preliminary Hazards Summary Report and the
information submitted in Section II of Amendment No. 2 to
said application, dated July 22, 1959. Section II of
Amendment No. 2 presented a discussion of the pressure
suppression system for reactor containment. It also stated
that the containment section of the Preliminary Hazards
Summary Report would be submitted after the choice between
pressure suppression and conventional dry containment had
been made. Pressure suppression containment has been chosen
and a conceptual design has been selected for the nuclear
unit (the Unit) at the Humboldt Bay Power Plant (the Plant).
Addendum A describes the general features and requirements

1

EXHIBIT 17 *(contd.)*

for design of the pressure suppression system of containment. It also discusses major accidents and the protection afforded by pressure suppression and other safeguards. The Commission's concurrence with the technical specifications for the containment system (Section I.B) is sought.

Addendum B also amends the Preliminary Hazards Summary Report. It presents additional information on gaseous waste disposal and the revised method for reactor refueling.

In the event of a conflict the information in this amendment supersedes the information previously submitted.

Subscribed in San Francisco, California, this day of September, 1959.

Respectfully submitted,

PACIFIC GAS AND ELECTRIC COMPANY

By_____
                    President

RICHARD H. PETERSON
FREDERICK W. MIELKE, JR.
PHILIP A. CRANE, JR.
Attorneys for Applicant

By_____

Subscribed and sworn to before me
this   day of September, 1959.

_____(SEAL)
Notary Public in and for the
City and County of San Francisco,
State of California.

My Commission expires July 16, 1963.

2

from the submerged vent pipes, increasing until the steam–water flow out of the vents equals the flow into the dry well from the break.

The suppression chamber maximum pressure was determined primarily by the ratio of the initial volume taken by the air in the dry well and suppression chamber to the final air volume after all air had been transferred to and compressed in the suppression chamber.

The refueling building was structurally designed to resist loadings due to earthquake, wind, and live load. It was also designed to retain radioactive materials which might be released during an accident by maintaining it under a negative pressure of $\frac{1}{4}$ in of water under exposure to wind velocities up to 20 mph and barometric and temperature changes.

The third question of the Commission was about the condensation of superheated steam from 1 000 psi to 72 psi, and how it related to the 100 psi saturated steam used in the Moss Landing tests. Test results were that the steam jet temperatures which could occur at Humboldt would not differ greatly from those at Moss Landing.

The fourth question was whether it might not be possible for a portion of the steam to pass through the suppression pool without condensing – an impossibility.

The last question was about the size, location, and number of penetrations of the dry well, and how these were included in designing the dry well to withstand the initial blast of steam. The integrity of the dry well vessel is maintained by utilizing all-welded connections except for the top and bottom flanged access heads and the electrical and hydraulic control piping flanges.

GE, Bechtel, and PG & E met with the AEC again on 25 February. This time the subcommittee on reactor safeguards was present. There were questions on transient analysis, scaling, jet break-up, and tank vibration. These seemed to be answered to everyone's satisfaction. The AEC was not satisfied with information about dry well air flow through the vents. Bechtel was to work on this aspect for the next meeting. Further information was also requested on the effect of air on steam condensation. Also, additional calculations concerning peak pressures were asked for. There were some meteorological questions but, meteorological investigations were not yet complete. A hearing with the reactor safeguards committee was scheduled for 11 March. At this meeting information was to be presented on:

1. The literature search on the action of submerged jets and condensation
2. The investigation of effects of air on condensation; (some work had been done on this as early as Phase I by GE at the urgings of a consultant)
3. The literature search on heat transfer for steam–air bubbles
4. Testing of a $\frac{1}{48}$ full-size pool sector
5. Pool baffling
6. A testing program for the containment structure before and after initial operation
7. Meteorology.

Safeguard committee hearing

The hearing with the safeguards committee was held on 11 March as scheduled, and much of the information that had been requested was presented. Pacific Gas and Electric did not want to conduct the full-scale test of a $\frac{1}{48}$ segment of the system. They hoped to have enough convincing arguments to persuade the AEC that pressure suppression was acceptable without going to the effort and expense of doing a full-scale test of a segment.

The letter of 14 March 1960, from the Advisory Committee on Reactor Safeguards after a review of the first six amendments made the following comment:

'Presupposing continued generally favorable experience with boiling reactors of this type, it is the opinion of the Committee that the conceptual design of this boiling water reactor is adequate for this site with conventional pressure vessel type of containment. It is not clear how much of the total reactor system will be housed within the vapour suppression chamber. The information so far provided does not demonstrate the suitability of the steam condensing system. *Further tests are necessary*. The Advisory Committee on Reactor Safeguards believes that, while the concept has merit, it has not yet been demonstrated that the vapour suppression system proposed can be relied upon to protect the health and safety of the public at this site.'

The $\frac{1}{48}$-size reactor tests

It was at this point that PG & E decided to perform a $\frac{1}{48}$th full-scale test of the proposed pressure suppression system in an effort to convince the AEC of its satisfactory performance. To this end, they constructed at Moss Landing, a $\frac{1}{48}$-size reactor vessel, a $\frac{1}{48}$-size dry well, one full-size vent pipe, and a full-size segment of suppression chamber and pool. Water in the reactor vessel was heated to 1 250 psig saturated, which corresponded to the design pressure of the reactor vessel. A rupture orifice was then broken, allowing the steam and water to flow through an orifice $\frac{1}{48}$ the break size corresponding to the maximum credible operating accident. This flashing steam and water mixture passed to the dry well and, by way of the vent pipe, to the suppression pool. Maximum pressures during the transient were recorded for comparison with values used for design. In the suppression chamber, the highest test pressure obtained was 9·3 psig compared to the design value of 10 psig. For the dry well, the highest test pressure obtained was 36 psig and the design was 72 psig. This lack of agreement in the latter case apparently resulted from very conservative assumptions regarding the value of the orifice coefficient for a flashing water–steam mixture. The tests further confirmed that:

1. Condensation of a steam jet in water under the Humboldt design conditions was rapid and complete
2. Uncovering a vent pipe through agitation of the water in the pool would not affect condensation
3. Dry well air did not prevent prompt completion of steam condensation
4. The 14 in vent pipes are adequate in size and number
5. Ample water was available for condensation
6. At least twice the maximum credible break area could be handled by the Humboldt pressure suppression system without exceeding its capabilities; flows up to three times the maximum credible accident break area would be condensed by the suppression chamber, and
7. The initial pool water temperature could be at least 60°F (15°C) higher than the design value without affecting maximum dry well and suppression chamber pressures.

AEC approval

On 23 June 1960 Mr Whelchel described the tests and how the results were applicable to the Humboldt design. This finally convinced the AEC that the system was safe, and its approval was obtained, subject to some modifications. The modifications were primarily centered around the fact that the major portion of the primary piping was outside the dry well and was to be isolated in the case of an accident by single isolating valves. The committee suggested that double isolating valves should be used, and that as much of the primary piping as possible outside the dry well should be shrouded, with the shroud vented to the pressure suppression system PG & E agreed to these modifications and obtained AEC approval.

Before a construction permit could be granted, however, it was necessary to have a public hearing on the matter. This public hearing took place on 24 August 1960 and at it PG & E represented its case for the safe operation of the Humboldt plant. The results of this meeting were favorable, and on 9 November 1960, the AEC issued to PG & E a permit to construct a 50-megawatt direct cycle, internal circulation, boiling water nuclear reactor as a utilization facility.

Epilogue   **PART 7**

Our case officially ends with the AEC approval of the pressure suppression containment system proposed for Humboldt Bay Unit No. 3. A sketch of the completed facility is in Exhibit 18. It is interesting to note that through the pioneering work of PG & E and GE, pressure suppression containment is currently being widely used in single-cycle boiling water reactors, both in this country and overseas. These projects include:

1. The Tarapur station in India
2. The Nine Mile Point Station, Niagara Mohawk Power Corporation
3. The Oyster Creek Plant, for the Jersey Central Power and Light Company in New Jersey
4. The Dresden Unit No. 2, located at the existing Dresden Power Plant site near Chicago, for the Commonwealth Edison Company.

Pressure suppression was also proposed as a means of containment in PG & E's Bodega Bay unit, a unit which was never built. (See Exhibit 20)

It is interesting to note that the idea of using a toroidal suppression chamber was thought of by a draftsman.

References

1. 'Pressure Suppression Containment for Nuclear Power Plants', by C. C. Whelchel and C. H. Robbins, ASME Paper 59-A-215.
2. 'Predicting Maximum Pressures in Pressure Suppression Reactor Containment', by D. B. Barton, C. P. Ashworth, Earl Janssen, and C. H. Robbins, ASME Paper 61-WA-222.
3. 'Pressure Suppression Approved for Humboldt Bay', *Electrical World*, 21 November 1960.

EXHIBIT 18 Pressure suppression containment and refueling building for Humboldt Bay Unit No. 3

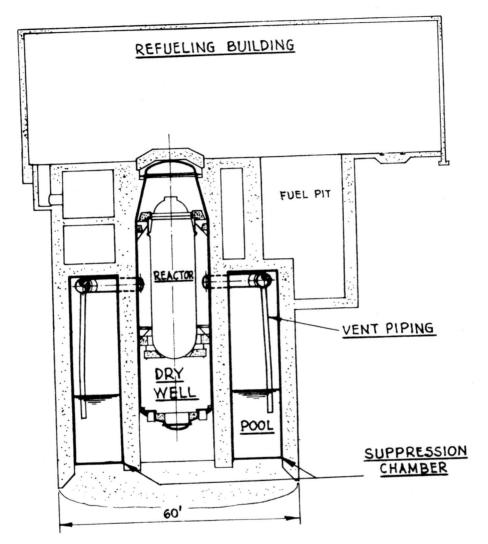

REFUELING BUILDING

FUEL PIT

REACTOR

VENT PIPING

DRY WELL

POOL

SUPPRESSION CHAMBER

60'

PRESSURE SUPPRESSION CONTAINMENT AND REFUELING BUILDING FOR HUMBOLDT BAY UNIT No. 3

EXHIBIT 19   Boiling water reactor

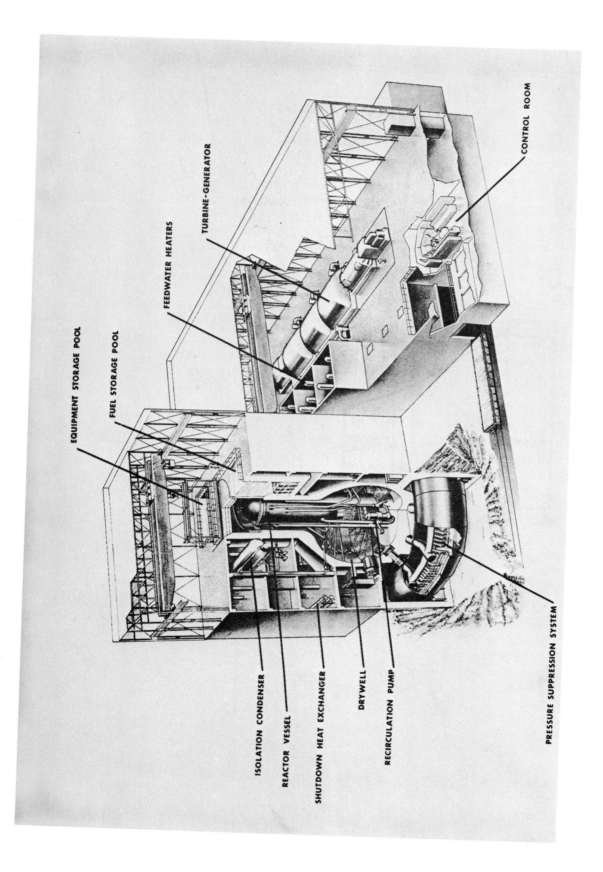

EQUIPMENT STORAGE POOL

FUEL STORAGE POOL

FEEDWATER HEATERS

TURBINE-GENERATOR

CONTROL ROOM

ISOLATION CONDENSER

REACTOR VESSEL

SHUTDOWN HEAT EXCHANGER

DRYWELL

RECIRCULATION PUMP

PRESSURE SUPPRESSION SYSTEM

EXHIBIT 20   Pressure suppression containment and refuelling building for the proposed Bodega Bay Unit

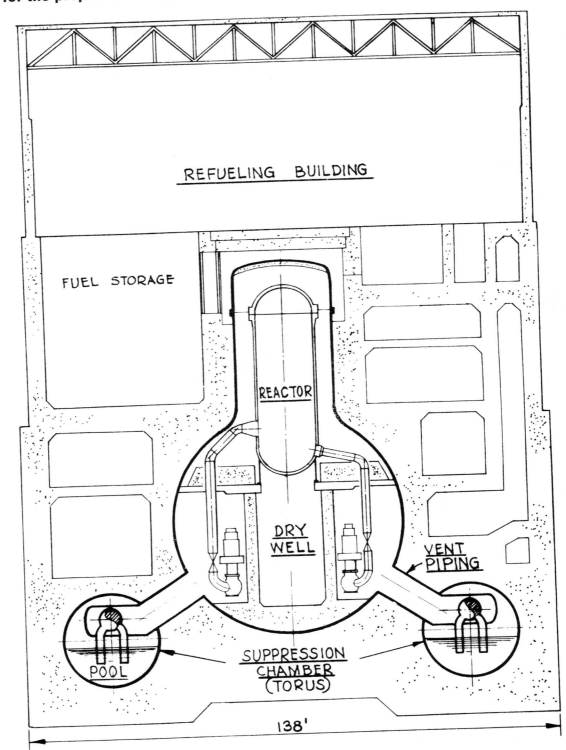

REFUELING BUILDING

FUEL STORAGE

REACTOR

DRY WELL

VENT PIPING

POOL

SUPPRESSION CHAMBER (TORUS)

138'

Appendix – Notes by C. P. Ashworth dated 3/5/59

This is a summary of various features of pressure suppression system arrangements. A number of pool, building, and dry well arrangements are considered.

Pools

Four types of pools have been proposed:

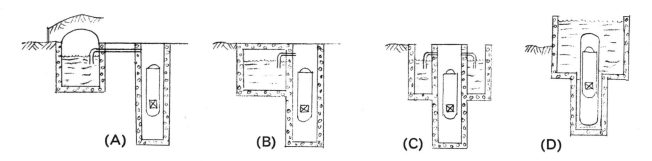

(A) (Bechtel) separate pool; (B) Pool to one side; (C) Concentric; (D) Submerged reactor

Arrangement A has been modified relative to the Bechtel scope in three ways:

1. The pool has been lowered to permit a horizontal run of large diameter pipe to minimize pressure drop between dry well and the pool.
2. The dry well is the steel-lined concrete biological shield rather than the separate steel container proposed by Bechtel. This is expected to give greater missile and shock protection.
3. The pool has been enlarged to about 12 000 ft³ air, 30 000 ft³ water compared to Bechtel's 7 000 air, 18 000 water. The new size will result in a pool temperature rise of 25°F and a pressure of 10 psig.

In order to provide for core flooding after a major accident, it is necessary to have at least 6 000 ft³ of water in the pool above the elevation which provides sufficient static head to make the water flow into the reactor. This is necessary to fill the dry well up to the top of the reactor core if the major accident is a rupture near the bottom of the reactor vessel. As a rule of thumb estimate, the pool water level should be at least 5 ft above the reactor steam outlet flange if core flooding is to be provided for.

Advantages and disadvantages of the four arrangements are as follows:

Arrangement	Advantages	Disadvantages
A	—	Differential settling of foundations.
B	—	Differential settling.
C	Easy to obtain required flow area. Can be used with a total enclosure.	May require moving reactor farther away from turbine.
D	Maximum missile and shock protection. Possibility of substructure rupture is eliminated. Least vertical height.	Possible activation of pool water. Must lower water level for refueling operation. Possible maintenance and operational problems.

Buildings:

Three types of reactor buildings have been proposed:

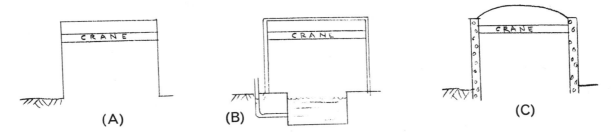

(A) Open (Scope); (B) Low pressure refueling enclosure; (C) Total

Arrangement C is cylindrical and could have either a steel or concrete dome and would need to be designed for about 2 psig.

Arrangement B might be rectangular and could be designed for about 0·1 psig. Excessive pressure in event of a refueling accident could be relieved by displacing water from the refueling pit through a standpipe to the outside as shown.

Advantages and disavantages are as follows:

Arrangement	Advantages	Disadvantages
A	Cheapest.	No protection in event of major refueling accident.
B	Adaptable to any pool arrangement. Same equipment layout as scope. Reactor is always enclosed but building is separate from pressure suppression enclosure.	Plant evacuation required in event of a major refueling accident.
C	Containment equal to existing dry type enclosures with shine shields.	Circular crane. Limited space. Provides more protection than considered necessary. Awkward fuel handling.

Dry well lids

In pool arrangement D the dry well lid is determined by the pool arrangement. In all the other pool arrangements the dry well lid is an important consideration.

The lid must provide a seal for the 50 psig expected maximum pressure in the dry well. Also it must be capable of withstanding missiles and shock waves. It must provide biological shielding. To meet all these requirements, the lid must be bolted down to withstand a 3 000 000-lb uplift and must be leak-tight and at least 5 ft thick, if concrete, or $1\frac{1}{2}$ ft thick, if steel.

Three types of lids have been proposed:

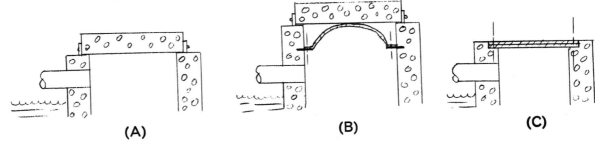

(A)   (B)   (C)

Arrangements A and B have a rolling concrete biological shield. The steel shield in C could be handled by the crane if it is more than one thickness. In B the rolling shield does not need to be bolted down because the shocks, leaks, and pressures are taken care of by a separate steel lid.

Arrangement	Advantages	Disadvantages
A	—	Possibly difficult to bolt down and make leak tight.
B	Heavy shield is not bolted down.	Must purge dry well before bolts can begin to be removed. Relative height of crane above pool water level is greater than for A or C.
C	No wheels or tracks. Thickness of steel required for shielding is probably ample to take the expected pressures with flat plates. Maximum missile protection.	

SUGGESTED STUDY

As you read:
1. Establish a chronology (time line) of events.
2. Design has been described as a decision-making process. After reading the case, list the major decisions which you consider to be important. Remember that a decision is a choice. It may or may not be an event.
3. The division of the development process into three phases is conventional. In your own words, describe what is involved in each phase.
4. List some of the design simplifications that were developed as phase II progressed.
5. Establish a morphology of the principal decisions.
6. What is the critical decision on which the success or failure of the design hinges? Who makes that decision?
7. Where does management enter the decision-making process?
8. In retrospect, could the $\frac{1}{48}$-scale test have been avoided? Should it have been planned as part of the original plan?
9. Of the engineers mentioned in the study, in your opinion, who performed best?

Case 8 Design and development of a bellows restraint unit

by P. J. Booker, A.M.I.E.D.

This case consists of Part 1 and Part 2 of the four-part article 'Principles and Precedents in Engineering Design' by P. J. Booker which appeared in The Engineering Designer, *September 1962. Part 3 of the original article provided further background, and part 4 gave a general analysis of the engineering design process.*

Permission to reproduce is gratefully acknowledged.

Introduction

Following the researches in design activity carried out by D. L. Marples [5], * the author undertook further studies on behalf of the Institution of Engineering Designers. At this stage no particular aspect of engineering was being investigated. Instead it seemed desirable to record and analyse as many varied design cases as possible with a view to establishing in a reasonably comprehensible form the essential nature of engineering design, whilst hoping that any findings would suggest further lines of enquiry.

A fair number of designs were investigated over quite a wide range of engineering products. For practical reasons only one design story is recorded here, one which is particularly suitable for demonstration purposes, in that the whole represents an almost self-contained microcosm of design. The type of design entity dealt with is of a specialized nature and, whilst the whole story tends to become complex, the principles upon which each version rests are conceptually simple, the differences generally being visually apparent. The history of the equipment dealt with extends only over a few years on the scale considered, during which time the number of versions produced is small enough to be able to keep track of all.

The story demonstrates different philosophies of design arising from the backgrounds and experiences of various individuals, design groups, and companies; it illustrates within one field of work the orthodox and creative approaches and the circumstances which led to each; it shows the use made of testing equipment, of development through prototypes, of buying out parts or complete units, the use made of technical papers, and many other aspects.

The approach is different from what it might have been, in that the story is not entirely in chronological order. Only one design is dealt with in any detail and the part proceeding this gives, effectively, the background as known to those engaged upon the recorded design. In fact, a number of other designs emerged in parallel through different companies, although no information was released until quite recently. These other designs are briefly discussed in a further part where the overall picture is completed.

The technique used is that of following through the development of a major project and to extract from this successively smaller design entities. This has been adopted both with respect to the story in Part 1 and in Part 2.

The material available would, of course, fill a book and a great deal of successive selection and condensation has been necessary to confine it to the limits of a paper.

Although only one special piece of equipment is dealt with, this paper should not be treated as one dealing with the technology of this equipment, but rather as one demonstrating the world of engineering design with this equipment as the medium.

A great many individuals and companies have given assistance and these are recorded under Acknowledgments at the end. Whilst every effort has been made to verify the factual content of this paper, the case is presented simply for demonstrative purposes and should not be regarded as a complete history. Nothing herein contained is necessarily endorsed by any of the individuals, companies, groups or other authorities mentioned.

PART 1 General background

The gas-cooled graphite-moderated reactor

The first atomic reactor was constructed in America during the war years. It consisted of a pile of graphite blocks built up into a rectangular prismoidal form with a number of horizontal channels for taking the uranium fuel.

* Figures in square brackets indicate references listed at the end of this case.

When Britain considered commencing an atomic energy programme in 1945, it seemed imperative that this should be based upon natural uranium as a fuel, and two moderators then seemed possible – heavy water and graphite. A moderator was necessary to slow down the products of fission to a velocity suitable for them to promote further fissions. Heavy water being expensive and slow in delivery, the British team chose graphite, especially as an air-cooled graphite-moderated pile had been successfully operated at Clinton, USA.

Sufficient uranium and graphite for two reactors was procured and construction of GLEEP, of about 100 kW output was commenced at Harwell in 1946, to be followed shortly by BEPO, of about 6 000 kW designed heat output. Atmospheric air cooling was adopted for the latter as it offered the least complication in producing such a reactor quickly.

In 1947 the virtues of finned fuel elements with gas cooling were recognized and it was decided that reactors cooled with atmospheric air could be made to yield plutonium for military purposes at the required rate and sooner than by adopting other possible systems. Two reactors were then built at Windscale, being effectively super-BEPOs.

When thoughts turned towards the production of useful power from atomic energy, one obvious line to follow was that of developing the atomic pile, of which a considerable amount of knowledge had been accumulated. The basic idea was to recirculate the air or other cooling fluid round a closed circuit, so that the heated gases passed from the reactor to a heat exchanger, where heat would be given up to turn water into steam, and then returned to the reactor for further heating and recycling. Providing that the steam was hot enough and gave sufficient pressure, it could then be used in much the same way as steam produced by more conventional means. Recirculation was necessary, of course, otherwise the heat held by the gases emerging from the heat exchanger would be lost.

Simple as this fundamental concept was, it posed a vast number of sub-problems. Because of the radio-active nature of power by fission, the heat-exchanger and much other equipment had to be completely separated from the reactor, the latter being virtually embedded in thick concrete known as a biological shield. The reactor and heat exchanger, in consequence, had to be connected by pipes or ducts so that the coolant could flow in circuit through both. The fluid flowing in these ducts had to be as hot as general design conditions would allow in order to get a reasonable thermal efficiency. Apart from the temperature, there was also the matter of the quantity of heat to be transported. Some liquids were considered but whilst these are good conveyors of heat they are, in general, strong neutron absorbers compared with gases and could lead to a reactor never reaching criticality unless enriched fuel were used. In the end, carbon di-oxide gas was selected because it was the best which could be obtained sufficiently pure in bulk at a reasonable cost. The choice of a gas coolant meant that it had to be used under fairly high pressure in order to cope with the quantity of heat. Indeed, since at any time there are limits to the size of heat exchangers and ducts which can be built, and because beyond certain speeds of gas flow the pumping power requirements become excessive, to handle the quantity of heat it became necessary to have a number of heat exchangers and duct systems connected to the reactor vessel.

Using a gas under pressure made every part of the circuit effectively a pressure vessel. The maximum pressure which could be used depended mainly on:

1. How small the reactor, and so the enveloping pressure vessel, could be made
2. The thickness of steel plate which could be safely and confidently welded to form the pressure vessel.

In the original studies it proved easier to support a reactor and pressure vessel in the upright position than in the horizontal, so that the horizontal fuel channels were

EXHIBIT 1

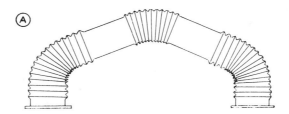

FIG. A   Inherently flexible gas ducts with corrugations at bends

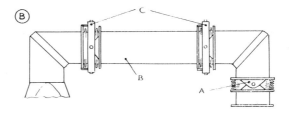

FIG. B   Full 'kinematic' gas ducting as used on industrial gas turbines

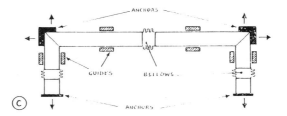

FIG. C   Bellows used to take up axial movement due to thermal expansion, with end loads taken on anchors

EXHIBIT 2   Layout of ducting

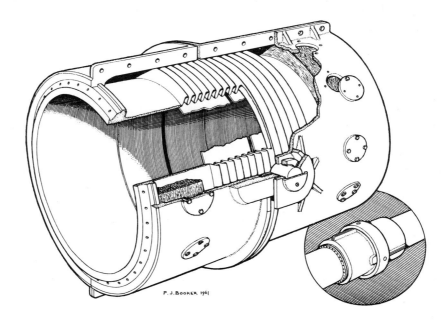

P. J. BOOKER 1961

replaced by vertical ones. This led also to the present layout of vertical heat exchangers and ducting in vertical planes.

Problems of large gas ducts

The problems which arise in designing large gas ducts for nuclear power stations stem from a number of different and often conflicting requirements which fall broadly into the following categories:

1. Taking up the thermal expansion of the ducts and vessels
2. Ensuring leak-tightness and the strength of the ducts as pressure vessels
3. Minimizing the power required to move the gas round the circuit.

The maximum stress in the pressure vessel shell is made up roughly of two components; stress due to internal gas pressure, and stresses at branch flanges or nozzles due to bending moments or other forces transmitted by the ducts through expansion or other causes. If the ducts raise very high stresses at the nozzles then the gas pressure of the system has to be reduced until a safe overall stress level is obtained.

For an efficient system, therefore, the ducts should produce minimum loads at the points where they enter the pressure vessel and heat exchanger and this is accomplished generally by making the ducts flexible. In, say, 1950, there were two main ways of making ducts flexible. Both had been developed for heavy industrial gas turbines and had grown out of previous steam turbine and plant practice. The first method was to utilize the bends in the ducting to give inherent flexibility; corrugations were then often introduced into the duct material in the vicinity of the bends to reduce stiffness and so increase flexibility (Exhibit 1, Fig. A). The other method was to introduce virtually 'pin joints' at certain points so that the ducting could move on 'mechanism' principles.

About the latter, Feilden wrote in 1958 [1]:

'In industrial gas turbines, a very important objective is to obtain what, for want of another word, has become known as "kinematic" construction. When this term is applied to gas turbine ducting components, it indicates that a design is used in which temperature variations are accommodated without the development of the very large thermal stresses which might occur if a rigid attachment were used. A simple example of ducting of this type is shown in Exhibit 1, Fig. B, which may be regarded as a completely flexible system, within the limits of the maximum permissible deflections of the three metallic bellows used in it. Though bellows are sometimes used to accommodate axial movements in pressure pipework, it was decided to restrict their movement in this gas turbine application to one of flexture, which would keep the stress in the bellows at a low value. The problem was then to carry the substantial end-loading due to the internal pressure in the duct, which in this design amounted to about 4 tons. In one of the bellows (A) in Fig. B (Exhibit 1) a simple trunnion mounting was used, which only allowed the bellows to flex along the axis of its pivots, and took up the longitudinal expansions in the horizontal pipe (B), or in the parts to which it was connected. To ensure a fully kinematic construction, gimbal rings were used to carry the end loads developed in the two remaining bellows (C), and these gave a characteristic analogous to a ball joint – i.e., freedom of flexture in all directions, but complete restraint of axial movement.'

The alternative method mentioned by Feilden, in which expansion is taken up by axial movement in the bellows is illustrated at Exhibit 1, Fig. C. Here the pipework is supported in guides which allow axial movement while the bellows compress to accommodate the expansion in the pipe sections. This is an appropriate method for small-scale work or when pressures are low, but as each section of pipe is free except for the bellows connection the end loads due to internal fluid pressure have to be taken on anchors – and these loads grow rapidly with increases in duct diameter and pressure.

In the Calder Hall nuclear power station, which might be considered as the prototype for subsequent stations, the size of duct settled upon was 4 ft 6 in diameter constructed of $\frac{3}{8}$ in plate to hold a gas pressure of 100 psi. The layout of this ducting is shown in Exhibit 2. At the time no information was available on the manufacture or behaviour of large corrugated type ducts, especially how they might behave over long periods; in consequence, the only reasonable solution to the problem of giving the ducts flexibility for thermal expansion was the kinematic one using tied bellows.

As with the pressure vessel and duct-work, convoluted bellows have stresses arising from two causes; the internal pressure of the gas giving hoop stress and, added to this, stresses due to bellows deflection. To keep the overall stresses in the bellows as low as possible, the deflection from the neutral position was halved by pre-stressing the bellows in the cold condition in the opposite direction. In practice this meant artificially contracting the duct-work – or giving it a 'cold draw' – and this was effected by closing each section of the circuit with a piece of ducting specially made to template. The early stages of heating then lowered the bellows stresses by moving the bellows to the neutral position; further heating then took them beyond this point and stressed them in the opposite sense. This has, naturally, become standard practice.

Calder Hall gas ducts and bellows restraint units

From a special issue of the *Journal of British Nuclear Energy Conference* [2] published in 1957 and devoted to Calder Hall, was the following:

'The principal problems associated with the large 4 ft 6 in dia gas ducts ... were those concerned with maintaining a leak-tight system and ensuring sufficient flexibility so that despite the very considerable expansion differentials involved, the loads and moments on the reactor vessel, heat exchangers and circulators to which they were connected, were kept to very low limits. At the same time the pressure drop in the system had to be kept to a minimum.

'The general design was based on roughly comparable conditions experienced in large gas-turbine installations where most expansion problems between two points could be covered by the introduction of a bend with three sets of bellows units tied against end pressure and hinged to permit flexing. The ducting is considered as rigid links with the joints at the bellows units, the angular rotation at the joints being a function of terminal displacement of the vessels (due to expansion and differential settlement) and expansion of the duct lengths themselves. Where flexibility is required in more than one plane in a circuit, a gimbal type of bellows construction is used. All circuits were constructed to Lloyd's Class 1 requirements.

'... The problem of carrying the concentrated end load of approximately 150 tons for the design pressure of 125 lb/in^2 gauge from the pins to the flanges in such a way as to prevent distorting stress concentrations, demanded much thought. Model tests were made on several possible solutions, and in the design which was finally evolved an axially split and bolted casing completely containing the bellows proper was spigotted to the flanges, thus serving to distribute the highly concentrated pin loads more or less uniformly round the flanges. Note that the casings were stiffened circumferentially at the joints to avoid undue deflection of the hinge pin and so remove the possibility of its seizing. The bellows pins are of nitrided steel in lead-bronze bushes.'

The pins were lubricated with molybdenum disulphide grease.
This restraint unit is illustrated in Exhibit 2.

EXHIBIT 3   Family tree

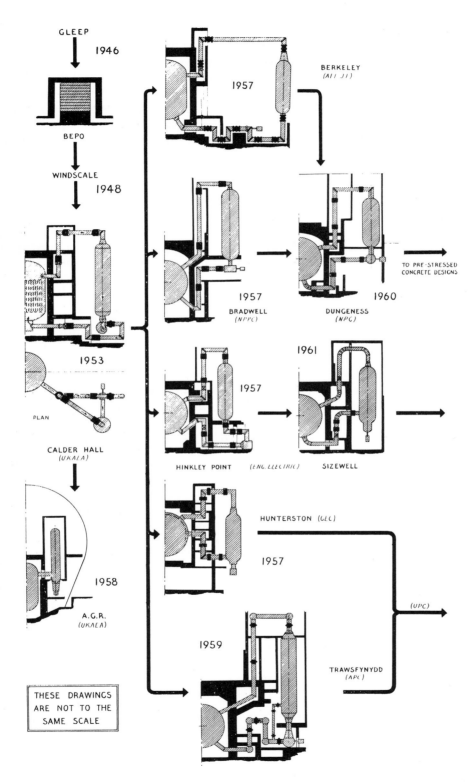

Showing the development of gas-cooled, graphite-moderated reactors for power generation

Commercial nuclear power stations

Calder Hall having served as a successful prototype, British industry began to organize itself for the purpose of designing and constructing commercial nuclear power stations. A number of consortia were formed and these eventually tendered for the first orders from the Central Electricity Generating Board.

The AEI–John Thompson Group began work on the Berkeley station and the Nuclear Power Plant Co. Ltd. started on the Bradwell station. Shortly afterwards the GEC–Simon-Carves Atomic Energy Group won the contract for the Hunterston station (Scottish Electricity Generating Board), while the English Electric–Babcock & Wilcox–Taylor Woodrow Atomic Power Group undertook the design and construction of the station at Hinkley Point. The GEC Group also won the contract for a station at Tokai Mura in Japan, and the NPPC similarly undertook the building of a nuclear power station in Italy at Latina. Somewhat later Atomic Power Constructions Ltd. were formed and they, in due course, were awarded the contract for the Trawsfynydd station in Wales.

Since there are a very large number of variable parameters to settle and much information becomes available with the passing of time, naturally the detailed layouts of these power stations differ considerably, though all have basic similarities founded upon the Calder Hall prototype. The general arrangements of pressure vessels, ducts and heat exchangers for one circuit, somewhat simplified, are shown in Exhibit 3 for some of these stations.

In 1960 the AEI–John Thompson Group and the NPPC went into partnership to form The Nuclear Power Group, and they started work on the construction of a second generation station at Dungeness. The English Electric Group also started work on a second generation plant at Sizewell and, to complete the picture, APC and the GEC Group went into partnership as United Power Company Ltd.

Whilst there will be changes in future stations due to the use of prestressed concrete pressure vessels, all the stations depicted in Exhibit 3 are very similar, although there are successive improvements over time. All the first and most of the second generation stations use 'kinematically' designed ducting incorporating tied-bellows units developed within each Group by different design teams. It is not possible to make any direct comparisons between these different restraint units, partly because insufficient information had been released, and also because each design had to be made to suit the requirements and properties of a particular duct layout. Many of the designs are separated in time and in the intervening periods there were continual increases in gas pressures from 125 to about 300 psi and in gas temperatures from 350° to 410°C (662° to 802°F). Duct diameters also increase from 4 ft 6 in to 6 ft 6 in. The situation was dynamic rather than static and the various units produced are not interchangeable.

Published information on large bellows restraint units

The Nuclear Power Plant Co. Ltd. gave the job of developing suitable bellows restraint units for the Bradwell station to C. A. Parsons & Co. Ltd., who had previously worked on the Calder Hall units and which was a member Company of the Group. Subsequently a Paper was read and published in which the following is recorded [3]:

'(The Calder Hall hinge-pin type) of construction was not adopted for the Bradwell bellows because it was considered to have limited development prospects owing to its high cost and the tendency of this cost to increase rapidly with size. The design chosen incorporates a simple flexible tongue the ends of which are welded into the adjacent ducts. A central cross-brace is provided at each end of the tongue to ensure that the tongue remains plane when flexed. In order to provide protection against the catastrophic consequences of a tongue failure under pressure, each bellows unit has two tie rods attached to brackets at each end of

EXHIBIT 4   Flexible tongue bellows restraint unit

Developed by Parsons for the Bradwell nuclear power station gas ducts. Diameter 5 ft; end load about 220 tons from gas pressure of 140 psi. The design of individual units varies somewhat according to their positions in the duct circuits

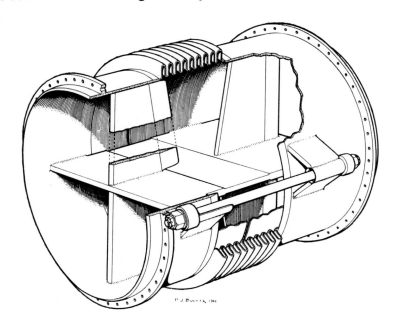

the unit. These bars are normally unstressed but they are designed to take full gas force in the event of a sudden brittle failure of the tongue. Their proportions have been so chosen that they dissipate by slight yielding the work done by the gas force. This device substantially reduces the stresses in the stiff brackets to which they are attached. (Exhibit 4).

'A disadvantage associated with the use of a tongue for tying a bellows is that the joint has no fixed centre of rotation. This does not lead to any difficulties where the bellows joint is deflected purely by a couple with no shear-type forces at right angles to the plane of the tongue. Shear forces can impose large extra deflection stresses on a bellows. These stresses are highest for the convolutions at each end of the bellows and are additive algebraically to any existing stresses in these convolutions due to rotation of the joint. This difficulty with shear forces can easily be overcome by the addition of a light pair of hinges on the centre line of the bellows. These hinges do not carry any axial load but are subject to a small lateral load which at Bradwell would not exceed five tons. The effect of shear forces is very small in the majority of bellows at Bradwell and hinges are unnecessary. In the three joints where they could be used with advantage it was decided that on balance it was cheaper to use a somewhat larger number of convolutions than to equip a small number of bellows with hinges.'

In the authors' reply to the discussion, Dr A. T. Bowden and Mr J. C. Drumm said that they were fully appreciative of the fact that an internal tongue restraint was not a universal solution to the general problem.

'... They were very conscious of the limits of the field of applicability of the design, particularly those associated with the use of that type of restraint for temperatures significantly higher than the 400 to 420°C (752° to 820°F) figure which was common to the current generation of civil reactors in Britain. They were aware that it was possible to design a hinged restraint which would give perfectly satisfactory service in a reactor gas circuit. The same could also be said of restraints which relied on the use of two or four external tie bars, externally mounted flexible tongues, or rolling action by a pair of interlocking hinge pieces mounted externally on the axis of rotation of the bellows. During the preparation of the design of the Bradwell station those various alternatives had been examined and the conclusion had been reached that for the particular conditions prevailing in the gas circuits of that station a flexible tongue restraint would represent a solution to the bellows problem which was at once simple, adequate for the duty and cheaper than competing designs.'

A fair amount of development work was necessary and the Paper recorded some of this together with information on the various tests carried out. Quoting from the Bowden and Drumm Paper:

'The final design adopted for the Bradwell bellows tongues was chosen after a number of tensile and bending tests had been carried out on tenth-scale models. These model tests showed that longitudinal tension in the tongue gave rise to an area of severe stress concentration in the region of the inboard ends of the welds which connect the tongue to the duct wall. It was decided to reduce the magnitude of the local peak stress in this region by using $2\frac{1}{2}$ in thick plate with a central flexible portion which is machined down to $1\frac{1}{2}$ in thickness. Peak bending stresses occur at the ends of the flexible length. It will be noticed ... (Exhibit 4) ... that these ends have been kept well inboard of the areas of stress concentration close to the tongue welds ...

'Strain gauge readings taken during a pressure test on a prototype 60 in bellows to four times design pressure indicated that the magnitude of the stress concentration factor at the weld ends was 7. ... At one time it was hoped that by suitably profiling the edge of the tongue it would be possible to reduce this high stress concentration factor. However, a series of ten tests on models with different fillet radii showed that a substantial reduction in the stress concentration factor could not be achieved along these lines.

'In the early stages of the design of the gas circuit from Bradwell it was decided to build a full-size bellows joint and to subject it to an extensive series of tests with a view to proving the suitability of this type of bellows for service in the coolant circuit of the reactor. It was also hoped that the tests would provide useful data on which to base future designs.'

Very little information concerning comparable hinge-pin types of bellows restraint units had been published at this time, but some interesting views were aired in the Discussion and Communications attached to the Bowden and Drumm Paper. For instance, from Mr J. J. Haftke (Babock & Wilcox Ltd.):

'... the authors had stated the basic requirements which the tying device should fulfil as being absolute reliability, a small resisting moment and the absence of need for maintenance or lubrication over the life-time of the plant. It could well be argued that in a design for which long-term operating experience was not available, it could not be certain that maintenance or repair would not be required at some time. It was perhaps truer to say that the tying device incorporated in the Bradwell bellows was incapable of being maintained, if by maintenance was meant a procedure aiming at slowing down the rate of normal wear and tear deterioration. The more conventional design of restraint, incorporating a hinge pin carrying the full load, so far as could be predicted – and there was a considerable volume of experience available – should not require any maintenance during the life-time of the plant. Should, however, that prediction prove optimistic, such maintenance would be quite a simple operation by comparison with that which would be involved if the Bradwell type of restraint were to show signs of deterioration earlier than expected.

'In addition the ability provided by the hinge pin design for the periodic examination, during shut-down, of the pin surfaces and of testing their freedom of rotation seemed to be a very significant advantage. It would be quite unjustified to place similar faith on the value of periodic inspection in the case of the tongue-type restraint where the signs of trouble which would be looked for would be cracking of a nature which could well pass unnoticed in any non-destructive examination. . . .'

'The limitations of the hinge-pin type of restraint were presumably not as imminent as the authors believed at the time of the design of the Bradwell bellows, since they had been successfully developed, for instance, for the Hinkley Point power station where the duct diameter was 6 ft 6 in and the operating pressure was in excess of 200 lb/in². It was also quite apparent that these conditions by no means represented the limit of capability of that type of simple restraint.'

Also from Mr V. Westermann (Engineering Appliances Ltd.):

'Perhaps the most serious objection to the flexing plate . . . appeared to have been dismissed far too lightly in his opinion. That was the lack of any fixed centre of rotation. The authors had commented that it did not lead to any difficulties where the bellows joint was deflected purely by a couple with no shear type forces. In a complete ducting system he thought that there would be very great difficulty in eliminating all forces of that nature, and with that type of unit, light restraining hinges would appear to be essential. With the existing arrangement there was no certainty about which axis the bellows would bend, or even that the bellows would tend to bend truly rather than take up a combined offset and bending position. There was also no resistance to any turning moment which might tend to twist the bellows radially about the axis of the unit. . . . The machining and placing of the tongue became a precision operation incompatible with the general "plumbing" attitude inevitable in any large duct-work where working accuracies of $\frac{1}{8}$ in were extremely good. The use of the bending plate also led to a great overall length; an equivalent hinged unit would only be approximately 5 ft long. The authors' unit would appear to be 8 or 9 ft long at least.

'The only advantage of the bending plate would appear to be the constant moment obtained throughout its life, although that might be affected by corrosion or radiation on the bending plate. His practical test figures using hinges were nevertheless very similar to those of the authors, a 5 ft 6 in unit pressurized to 230 lb/in² and deflected through one degree required a maximum moment of about 500 ton/in to deflect. . . . After 5 000 movements that figure was reduced to less than half that value.'

To round off this series of pros and cons, the following is taken from the authors' reply:

'The mode of flexure of such bellows (in the particular case) could be predicted with sufficient accuracy. . . . It was quite true that the torsional stiffness of the tongue was low compared with that of the bellows. In carefully erected uniplanar duct-work . . . there were no torsional forces present which were likely to give rise to significant stresses in any but the most fragile bellows. They would point out that hinged restraints were by no means immune to the failings mentioned by Mr Westermann. Because of the clearance at the hinge pins and the transverse flexibility of the hinge arms, a considerable portion of the transverse forces present in a duct line would be taken by the convolutions of a bellows. . . . They would also suggest that the usual designs of bellows hinge gear made but a small contribution to the overall torsional stiffness of a hinged bellows.'

PART 2

Design and development of the frusto-conical bellows restraint unit

Initial ideas

The tender for the Trawsfynydd nuclear power station, which had been submitted to the Central Electricity Generating Board, was accepted in August 1959. Work began on site almost immediately and from thereon design and procurement had to be fitted into a time table, the first reactor and associated plant being planned to go on load in October 1963.

In the preliminary layout of the ducts, hinge-pin type bellows restraint units had been specified, the size and properties of these being assumed from units known to exist. When the time came to investigate the possibilities of procuring bellows units, one of Atomic Power Constructors' associated Companies, Richardsons, Westgarth & Co. Ltd. (RWG), still had some available capacity and they suggested that they might submit a tender. RWG had never tackled restraint units of this size before, but like so many other firms dealing with such units, there was a considerable experience of designing and manufacturing flexible ducting on the gas- and steam turbine scale. Since the criteria were simply that a suitable design of unit should be available at the right time and price, it was agreed in January 1960, that RWG should be given the opportunity to submit a design of restraint unit, and a first technical meeting was called at which, in effect, a preliminary specification was given.

One of the important points to emerge was that any restraint designed by RWG should be comparable in length to those postulated in the duct layout. This was about 5 ft, a figure based upon the shortest hinge-pin unit currently believed to be available. The shortest unit was chosen in the first instance because this led to a more compact ducting arrangement and thus a more compact building. A great many decisions affecting the ducting, the building and the arrangement of other equipment had already been taken, and an increase in bellows unit length would have necessitated quite uneconomic changes. A preliminary figure for tolerable bending or restoring moment for full deflection was also given, as well as ducting size, gas temperature, pressure and other necessary data.

Work was started at RWG immediately. At this time four consortia were well advanced with first generation nuclear power stations. Whilst these had been written up extensively in the technical press, the emphasis was mainly on the nuclear aspects rather than on the incidental engineering. Whilst a bellows restraint unit is important – as is everything in a design – it is still a 'detail' with respect to the whole station. The information was meagre and approximated to that given in the quotations of Part I.

Whilst it seemed to be technically possible to produce a suitable hinge-pin unit, it was almost a foregone conclusion that the development work necessary would make such a unit costly and not competitive with models already available from companies with some years' experience (Exhibit 6 – Solution 1). It seemed better, if possible, to design a unit which was not in direct competition with nearly identical units. In other words, if development was necessary, as seemed likely, the time and money thus spent should be on some new design which had a further development potential.

Apart from the hinge-pin type, a second precedent was also known of, namely, the Parsons' flexible tongue restraint designed for the Bradwell station. This unit had been developed for a 5-ft diameter duct with a nominal gas pressure of 140 psi. The Trawsfynydd units would be for 5 ft 6 in ducts with a nominal gas pressure of 240 psi. Allowing for fault and other conditions, the end loads for the two cases would be about 200 and 450 tons respectively. Calculations were put in hand for a preliminary assessment of a flexible tongue design. To take the higher gas pressure a somewhat thicker tongue would be required for the higher tensile load. To keep the additional

EXHIBIT 5   Section of Trawsfynydd ducting

Showing position of bellows in the cool duct. The spherical elbow bends were designed parallel to the restraint units by the same team and were not shown on early duct layouts

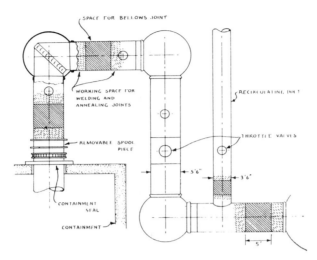

bending stresses low enough, a larger radius of curvature would be needed, leading to a longer tongue for a given deflection angle – say, about 3 ft. No calculations were carried out on the difficult stressing problem of anchoring the tongue, but using the Parsons' unit as a guide it seemed likely that 2 to 3 ft would be necessary either side of the tongue. The idea was, therefore, considered unfeasible for a unit less than 8 or 9 ft in length (Exhibit 6–Solution 2).

In the meantime, thrusting out for new ideas, the Chief Designer had put forward the idea of a unit working on torsion instead of bending. Preliminary sketches were made and calculations showed that the idea was probably feasible for the project in hand. The torsion bar would be under complex bending and shear loading as well as torsional loading, and this was not considered to be an ideal solution, particularly as it would probably have little development potential (Exhibit 6–Solution 3).

After digesting the results of these preliminary investigations, two further ideas were put forward. One was a modification to the flexible tongue solution, aimed at reducing the length of tongue required; this was put forward, naturally enough by the engineer who had made the flexible tongue calculations. In this proposal, the tongue would be replaced by two tongues one above the other each half the thickness of the single tongue (Exhibit 6–Solution 4). The tongues would be pre-stressed by bending them slightly in the opposite sense. During deflection, one tongue would bend more, building up bending stress but being partly relieved of tensile stress, while the other tongue straightened out, taking more of the tensile load while being relieved of bending stress. This would allow the tongue length, and so that of the whole unit, to be shortened. Preliminary calculations showed that the idea was probably feasible; on the other hand, the anchoring and correct pre-stressing of the tongues might present difficulties.

The Chief Engineer suggested as another idea the possibility of using a conical lattice of rods. The precedent for this was rather remote and the similarity was only apparent when viewed at a high level of abstraction. Some years previously, when the Chief Engineer and Chief Designer were working on a gas turbine design, during testing a peculiar unpredicted mode of vibration occurred on a shaft as in Exhibit 6(A). Investigations led them to the conclusion that the frusto-conical shape of the casting holding the shaft was the cause. In simplifying calculations, this had been replaced by an 'equivalent' cylindrical piece. However, the behaviour of the frusto-cone turned out to be rather different from that of a cylinder. Whereas a cylinder

EXHIBIT 6　The solutions

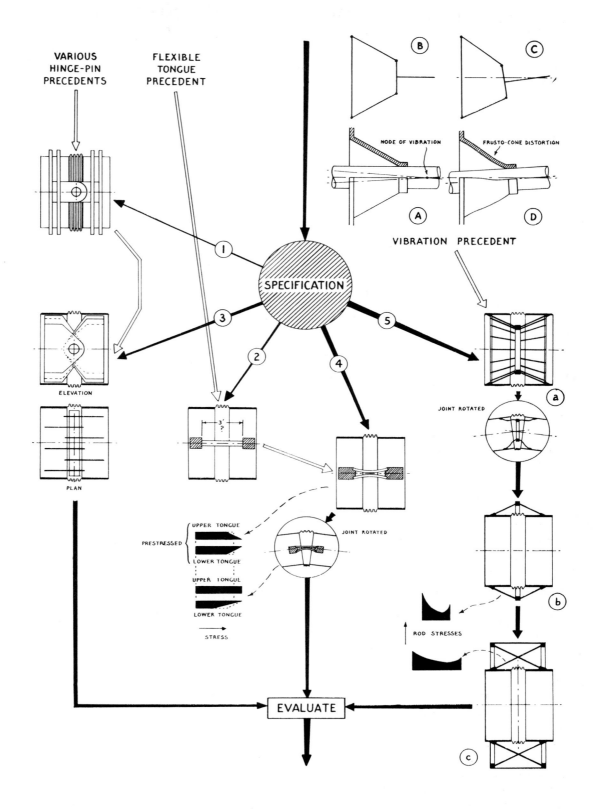

Illustrating the solutions considered; their derivations, relationships and development

behaved similar to a beam, the frusto-cone tended to 'side-slip', so that up and down movement of the bearing also caused an angular movement. This is demonstrated in (B) and (C) by pin-jointed links. The actual distortion of the frusto-cone, exaggerated, is shown in (D). In other words, whilst in the axial direction the frusto-cone had the same properties and rigidity as an equivalent cylinder, this shape had another property of being able to side-slip, causing an angular movement. The small angular movement of the shaft was recognized to be similar to that required on the duct joint.

The abstract frusto-conical surface was replaced by a number of suitably arranged rods and these were initially considered inside the ducting, effectively replacing the tongue or double-tongues. The novelty of the arrangement, however, posed a number of problems, mainly on account of their being so many variables – the length and angles of the rods could be varied, as well as the number and diameter of the rods (Exhibit 6–Solution 5(a)).

Like the tongue case, each rod would need to take a tensile load plus a bending stress. Initially, to keep the tensile stress down, the small angle of 25° was assumed. Within the overall length limitation this appeared to give rods of a reasonable length, which seemed desirable to keep bending stresses low. Taking 1-in diameter rods and conventional stress values, etc., preliminary calculations showed that this arrangement called for about 60 rods in each conical frustum. The bending moment of these rods in rough calculations appeared to be very low, being only a fraction of the maximum allowable bending moment for the complete unit. The rods and central anchorage would offer an obstruction to gas flow and would be inaccessible. However, the solution in principle appeared to be eminently feasible with a wide margin for development.

The gas flow objection could be overcome in the obvious way by placing the conical lattices on the outside of the duct. The circular anchorage where the cone bases met would be considerably larger and heavier, but this seemed to be a small price to pay for a completely free bore and for ease of accessibility (Exhibit 6–Solution 5(b)).

Prototype design

The possibilities of this new idea were such that preliminary drawings for a prototype were embarked upon. Stress calculations (dealt with below) based now upon a more definite arrangement, showed that a 25° cone angle was reasonable but that rather longer rods were desirable to keep rod stresses low. It was soon seen that this could be done by making the two conical lattices interpenetrate one another. Indeed, by this device it seemed possible to shorten the unit length below 5 ft if necessary. (See Exhibit 6–Solution 5(c).)

At this stage, far more attention had to be given to the design of the rods and this fell into a number of categories – such as stressing, fixing, manufacturing methods, and so on, each affecting the others to a greater or lesser extent.

When a rod is used as a beam, generally the load is known and the deflection varies according to the dimensions given to the beam. In this case a rod is used in such a way that the deflection is known and the force necessary to give this deflection – which shows up in the whole unit as the bending or restoring moment – is dependent upon rod shape and size. Another criterion is the maximum stress level in the rods. Both restoring moment and stress were calculated very approximately when judging feasibility of this idea; however, at that time nominal stress levels assuming static deflection were assumed, and the arrangement was too loosely defined to work out any reliable figures.

What should be the maximum allowable stress in the rods? This was a difficult question. These restraints would be virtually stationary for long periods, but Bowden had estimated that they would move through at least 240 full deflection cycles in a 20-year life. Each unit would also have to be tested with a number of full deflection cycles under pressure. The conditions were thus ones in which fatigue was of importance, although the variations in stress through movement might be widely spaced in time. Fatigue conditions mean lowering the maximum stress level from what would be permissible in static conditions, but whilst fairly reliable figures were available for cases of millions of reversals, no data was available for only a few thousand reversals – besides which generalized data of this kind tend to become unreliable in this range. Some reasonable assumptions had to be made taking into account the method of manufacture, surface finish, and so on, specimen rods later being put to the test.

The stressing of these rods was not as straightforward as might at first sight appear. The rods would be in tension to take the gas end load, giving them a basic stress. The rods would also be bent, effectively as two cantilevers joined together, giving an additional bending stress. There was yet a third effect; the tensile load acting through the bent rod would naturally tend to straighten it, an effect which increased the bending moment at the anchors – increasing both the bending moment of the whole unit and increasing the maximum stress which occurred at the anchorages. It soon became apparent that 1-in diameter rods would give excessive stresses and bending moments. Instead, three times as many rods, each with one-third cross-sectional area, were postulated – 180 rods of 0·6 in diameter. The maximum stress at the rod ends could be further reduced by increasing the diameter at the ends; every postulated change of this kind, however, altered the rod deflection shape, the bending moment and the stress levels.

In the meantime, the overall form of these rods could not be decided until the method of fixing was known and the method of manufacture.

Some of the detailing stages are demonstrated in principle in Exhibit 7. If the rods were looked upon statically the arrangement in (A) seemed fair enough in principle. However, examination of the whole arrangement showed that only the first one or two could be fitted. In general the rods would have to pass through the annular ring to be fitted. So (A) was modified to (B), providing a nut at both sides, the remainder of the rod being such that the inboard nut could be fitted. This arrangement also had to be modified, of course, to provide means of turning the rod so that thread at the other end could be screwed home, to provide means of locking the nuts, to give the necessary clearances and so on, as in (C).

Because of its shape and because a uniform tensile strength was required throughout, machining from bar seemed to be called for. On the other hand, the fatigue conditions, as yet unassessed, suggested that a surface free from machining marks would be advantageous. RWG engineers were, however, familiar with 'Newallastic' studs, used in other fields of design, which were cold rolled and particularly fatigue resistant. Although a tie rod was very much longer than any known stud of this kind, this method of production seemed worth investigating and an enquiry was made to Messrs A. P. Newall & Co. Ltd.

The process of producing 'Newallastic' studs and bolts is essentially one of cold hammering or swaging heat-treated and ground alloy steel in rotary machines, which reduce the stud or bolt shank to a size slightly below the root diameter of the thread. The reduction in the shank diameter, of course, increases the length of the stud, so that fine control of the process is necessary. The surface produced by swaging is extremely smooth and entirely free from tool marks, while a certain amount of compressive stress is retained, due to cold working, which is beneficial in providing fatigue resistance. The threads are rolled on circular die machines and this again produces threads of great strength.

The properties of cold swaged studs were just what were wanted in the tie rods, although the length of the latter, some 42 in, and the amount of reduction were somewhat greater than anything previously dealt with. As mentioned earlier, there was a considerable overlap of the various aspects affecting the design of these tie rods and, in fact, at the time Newall's engineers started working with those of RWG, the actual number of steps and their diameters had not been finalized, since this was as

EXHIBIT 7   Simplified stages in the design of the frusto-conical restraint tie rods

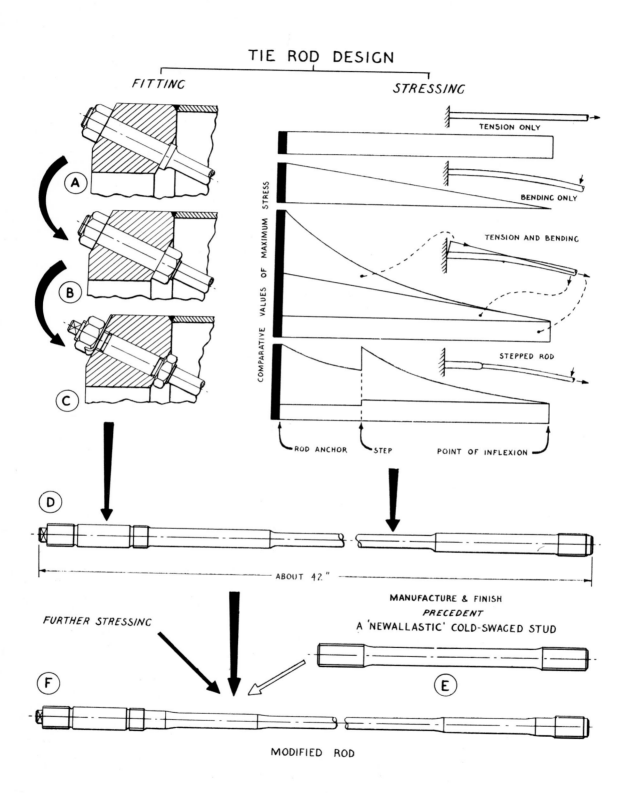

TIE ROD DESIGN

FITTING

STRESSING

TENSION ONLY

BENDING ONLY

TENSION AND BENDING

STEPPED ROD

COMPARATIVE VALUES OF MAXIMUM STRESS

ROD ANCHOR   STEP   POINT OF INFLEXION

ABOUT 42"

MANUFACTURE & FINISH
PRECEDENT
A 'NEWALLASTIC' COLD-SWAGED STUD

FURTHER STRESSING

MODIFIED ROD

much a matter of what could be manufactured as what was theoretically desirable. It was finally decided that the rods should be made in a 75-ton steel and should be reduced at each end in three steps, one to below thread root diameter, followed by two further reductions.

Even so, the actual production of these rods presented a number of problems to Newall's engineers. Because of their length, an extension of about 14 in took place during the processing from the original blank to the finished product, and to maintain the accurate swaged lengths on each operation, special limiting devices had to be incorporated in the machine feeds and the whole technique almost completely reversed from that normally employed. Furthermore, where a very long swaged portion was required – as was the case with these rods – extreme care had to be exercised to maintain an accurate and constant diameter, as a deviation of as little as 0·005 in would result in quite a considerable difference in the length of the swaged portion and so of the whole rod length.

EXHIBIT 8   Prototype bellows restraint unit

Diameter 5 ft 6 in; end load 420 tons from gas pressure of 240 psi. At this stage only 'a bellows' is shown, no particular make or kind having been chosen

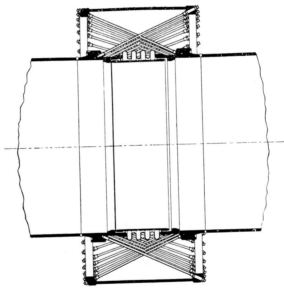

After a number of experiments had been made, suitable rods were produced and, in the words of Newall's Technical Director, '. . . quite a considerable amount of credit is due to the machine operators themselves, who showed tremendous enthusiasm and skill in producing what proved to be a first-class job.'

Naturally, all the other parts of the prototype arrangement had to be progressed through similar stages to finalize the form of each, and each of these parts provided further problems to production staff and so on.

In the meantime, as soon in this process as the general form became clear enough, RWG submitted preliminary drawings (approx. Exhibit 8) to APC with an estimated price and a meeting was called to assess the position. Since the tender had been drawn up, some particulars of two comparable bellows restraint units had been published – the flexible tongue type in the Bowden and Drumm paper and a tie-bar type in some technical journals. The very existence of these as well as the arguments put forward by their designers raised doubts about the suitability of hinge-pin type units for such critical application. Very few units had, in fact, been tested at that time for conditions comparable to those which would exist at Trawsfynydd. The

Parsons and AEI units were far too long to accommodate. Furthermore, commercially there was something to be said for a unit produced within the Group's own Companies. At this stage the design appeared to be promising and RWG were asked to go ahead with development on condition that restraint units could be supplied eventually at a cost comparable to that of alternative units, and it was agreed to produce a prototype for testing by the end of May 1960.

When the first rods arrived from Newall's, tests showed that these had rather better strength than might have been anticipated and, as the holes had not then been

FIG. 1

drilled in the prototype under construction, the number of rods was reduced by a third to 120 in each lattice. A special test rig had also been designed and built for fatigue testing these rods, and the fatigue properties of the specimen rods turned out to be so good that in assembling the prototype the number of rods was again reduced by a quarter to 90, by simply omitting every fourth rod (see Fig. 1).

A suitable convoluted bellows had been procured from Teddington Aircraft Controls (Bellows Division) and, as far as possible, calculations were made to predict the bending moment of the complete restraint unit. The bending moment of the bellows was known and the moment of the conical lattice of rods could be predicted fairly closely as a result of the calculations already made and from experiments on single rods. A comparatively unknown quantity, however, was that concerned with the 'pumping action' of the unit, a function of the change of enclosed volume when deflected. Whilst some estimations had been made, so far this had been given little attention as any change in volume, especially for a very small deflection, would undoubtedly be small. However, even shortenings of the order of thousandths of an inch when working against a gas load of 450 tons could give a not insignificant component to the overall bending moment.

If one imagined the rods pin-jointed at their anchors, one could predict a small shortening as a property of the geometry. However, the rods were not pin-jointed and it was difficult to tell whether such calculations were significant or not. Again, the bending of a rod would cause its two ends to approach one another by a very small amount and this would probably be reflected in a shortening of the unit, although by how much was difficult to tell. Some simplifying assumptions had to be made, though in making these the very smallness of the changes in lengths led to difficulties in assessing the order of importance of various effects. However, upon the assumptions made a preliminary figure for bending moment due to 'pumping action' was arrived at. Whilst this was higher than might have been anticipated, there was a fair chance that the properties of the prototype would fall within the limits of the specification.

Prototype test

The prototype having been completed to time, tests were put in hand. In the main these were tests for strength and safety undertaken to prove feasibility to the insurance authorities, APC and the ultimate customer, the CEGB. Apart from pressure tests, etc., the prototype was moved through 1 250 full deflections under pressure and temperature, and a further 1 250 full deflections under pressure. A number of other tests were also carried out in the interests of gaining a fuller understanding of the new design's properties.

One of the first tests was to measure the overall restoring or bending moment with the unit under full pressure and this turned out to be rather higher than expected – nearly twice the calculated value. This was of some temporary concern as the provisional figure for allowable bending moment in the specification had been reduced by about 20%.

It seemed that the trouble probably was to do with the 'pumping action'. The method by which this action's effect on restoring moment had been derived was, therefore, re-examined. It became apparent that some of the effects previously considered to be of a secondary nature were probably as important as other effects, and a new method of calculating this component of bending moment was worked out. The values then arrived at for the prototype were very close to the measured value of restoring moment. A difficulty then arose; the new calculations were based upon the predicted shortening of the unit. It was possible that some of the high bending moment recorded was due to structural effects. These, however, could not be isolated since, unpressurized, the measured bending moment due to the rods would be smaller than when in tension, whilst, pressurized, this component of bending moment would be added to the 'pumping action' component. However, the actual change of volume of the unit could be measured. The restraint was accordingly filled with water and the change of volume upon deflection ascertained by measuring the overflow. These measured values tallied within a few per cent of those calculated and the revised restoring moment theory was thus considered to be sufficiently accurate for prediction purposes.

Revised design

While the total restoring moment of this prototype was excessive for the project in hand feasibility had been proved and the final theory of restraint shortening was capable of analysis and of showing the effects of the variable parameters. At the start when stress levels were considered of paramount importance, a small cone angle and long rods seemed desirable. An analysis of the new theory, however, showed that shortening of the unit, now an important criterion, was a function of rod length, of the ratio of rod length to cone apex distance and the cone angle – all inter-related. So, for overall bending moment considerations, the rod length was decreased and the cone angle increased to 50°. To counteract what would otherwise lead to higher stress levels, the number of rods was again increased to 120. These alterations changed the overall properties of the unit, reducing the total restoring moment well within the specified value, whilst leaving the rod stresses very little different from those in the prototype.

These changes in geometry made interpenetration of the frusto-cones unnecessary and new drawings (Exhibit 9) incorporating the revised layout were produced and submitted to APC with a quotation.

It was appreciated also that some minor shear and/or torsion loads might fall on the unit and that the arrangement of rods in itself was inadequate to transmit these across the restraint (i.e., prevent their falling on the bellows. It was, therefore, proposed to replace the one inner sleeve with two sleeves (similar to those shown in Exhibits 2 and 4) and to arrange interlocking dog-teeth where they met.

Reports of the tests having been examined by the CEGB and the insurance authorities, and the design principles having been approved, APC engineers got down to the job of examining this new restraint layout in terms of its suitability for incorporation in their duct system. The design as a whole had a number of useful properties. The bore was clear for uninterrupted gas flow; there was no mechanical movement; the flexing elements were external; and the joint was universal, equivalent to a gimbal ring hinge-pin unit. The latter property, whilst not essential in this case, was a very useful one, as slight errors in the positioning of ducts, etc., would have a minimum effect. By this time the duct design had advanced to the stage where it was possible to estimate fairly closely the shear loads likely to fall on to these units. The worst case was occasioned by the thrust of the recirculation ducts and the figure for shear load appeared to be higher than that for which the dog-teeth arrangement would be suitable.

This was pointed out to RWG; the figure given for anticipated maximum shear load was, to them, surprisingly high and it soon became apparent that the dog-teeth arrangement would be inadequate in this particular case. The inner sleeves were only to be fabricated from light plate, easily distortable under heavy load. A light pair of hinges could be added which would give a centre of rotation but without carrying any of the end load. This seemed, however, to be inelegant in comparison with the main design and there was not a lot of space available. A simple, general solution was called for which, amongst other things, would retain the universal property of the joint.

Having had success so far with flexing rods, it was suggested that this problem might be similarly solved. The frusto-conical lattice, viewed at a high degree of abstraction, was simply the use of rods, which were inherently stiff along their axes but flexible in bending, to resist some forces but not others. The idea, then, was to synthetize another arrangement of rods to give the desired properties. Shear and torsion had to be resisted in a plane – so that rods were considered lying in this plane (Exhibit 10). They had to be suitably anchored to the duct ends and allow of normal joint rotation. Various arrangements became possible, and the most appropriate one to emerge was to use four rods lying in the centre plane perpendicular to the unit's axis, one end of each rod being attached to one duct end and the other to the other duct end, as shown in Exhibit 10 (and Exhibit 11). This idea had to be developed, of course, through stages very similar to those detailed for the lattice rods, etc. (These anti-shear units are a design entity in themselves and insufficient information is available to reconstruct their design.)

EXHIBIT 9   Revised restraint design

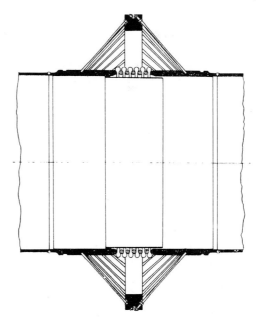

Omitting proposed internal dog-teeth arrangement to transmit shear loads

EXHIBIT 10   Effects of torsion and shear

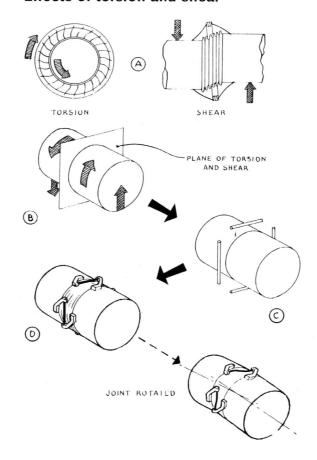

Effects of torsion and shear on frusto-conical lattices and arrangement of rods for anti-shear units. (*See also* **Exhibits 11, 12**)

Production design

At a meeting with APC engineers the final design incorporating the additional anti-shear modification was examined and approved, and in July 1960, RWG were given the go-ahead for design and production. It is recorded that 'detailed design and testing proceeded'; however, this covered a great deal of work. For instance, the anti-shear modification had to be translated into a suitable hardware form and specimen

EXHIBIT 11   Frusto-conical restraint units

Comparison of frusto-conical restraint units for main and recirculating gas ducts. Notice different cone angles, number, length and diameter of rods for different end loads and angles of deflection

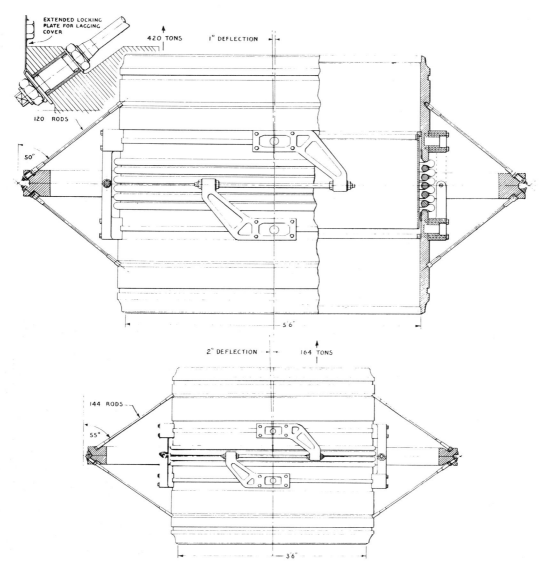

rods and brackets had to be designed, made and tested. More attention had to be paid to every piece of detail, bearing in mind that though the prototype was available as a precedent it had had to serve only for a few weeks whilst the production units would have to last for years. Corrosion and many other factors had to be considered in the choice of materials in juxtaposition and provision had to be made for fixing lagging to the units. The flanges in the production version and the annular ring were considerably different from those of the prototype and called for different production methods, and so on.

Parallel with the designing of the main production units of 66 in bore, of which 72 were required at Trawsfynydd, another design for a similar unit of 42 in bore was put in hand for the recirculation ducts, 36 such units being required. Whilst this followed closely the main design, many of the parameters, apart from size, had to be altered to give different amounts of deflection and different restoring moments. The two units are shown for comparison in Exhibit 11.

Whilst the two designs continued to develop, everyone concerned had necessarily to keep reviewing the designs as a whole and in detail as time went by and more information came to light, to ensure that so far as was humanly possible nothing had been overlooked or forgotten. Modifications were made from time to time, sometimes suggested by test results, sometimes as the result of reappraisals of previous work. For instance, the bellows chosen was of the type using equalizing rings; tucked away amongst the published communications with the Bowden and Drumm paper were the following remarks:

> '. . . the need for equalizing rings between the convolutions . . . Very careful dirt exclusion would also be necessary as the presence of a hard particle within the equalizing rings would tend to damage the bellows material. . . .'

Whatever the source may have been, eventually someone queried whether there was a possibility of dirt getting between the equalizing rings and the bellows convolutions, and if so whether it could lead to premature failure. No positive answer could be given, but it was considered to be a risk not worth taking, so a modification was made to the design in which the convoluted bellows were covered within an aluminium foil envelope.

The final design was approved both by the CEGB and the insurance authorities. Tests were carried out on the pre-production unit in June 1961, when the design was satisfactorily proved, and the first production unit was completed on 3 July 1961.

EXHIBIT 12   First production unit

Acknowledgments

In collecting the material for this paper, the author has received very willing and helpful co-operation from many quarters, although it is to be noted that the views expressed are entirely those of the author and are not necessarily endorsed by the Companies concerned.

Very great thanks are due to the Directors and Staff of Messrs Richardsons, Westgarth and Company Ltd., Wallsend, Northumberland, and of Atomic Power Constructions Ltd., and to the Central Electricity Generating Board for permission to publish this paper.

Thanks are also due to the authors of papers quoted, and to Messrs. C. A. Parsons & Co. Ltd., Associated Electrical Industries (Heavy Plant Division) Ltd., Messrs Babcock & Wilcox Ltd., A. P. Newall & Co. Ltd., Teddington Aircraft Control (Bellows Division) Ltd., and Engineering Appliances Ltd. (for Industrie-Werke Karlsruhe) for providing copies of photographs and much background information.

The author would also like to thank Mr D. L. Marples for making available research material not used in his paper (Ref. 5), and in particular Dr A. Frankel, Mr J. A. Lorett, Mr G. N. Doyle, Mr J. J. Haftke and Mr V. Westerman for their personal help and interest.

References

1. Feilden, G. B. R.: 'A Critical Approach to Design in Mechanical Engineering', Bulleid Memorial Lectures, 1959: Univ. of Nottingham, Lecture No. 1, pp. A23–4.
2. Bowden, A. T. and Martin, G. H.: 'Design of Important Plant Items', *J. Brit. Nucl. Energy Conf.*, Vol. 2, No. 2, 1957.
3. Bowden, A. T. and Drumm, J. G.: 'Design and Testing of Large Gas Ducts', *Proc. I. Mech. E.*, Vol. 174, No. 3, 1960, pp. 119–57 (Read and discussed 4 November, 1959).
4. 'Expansion Joints', *Nuclear Engg.*, Jan., 1959, pp. 26–31.
5. Marples, D. L.: 'The Decisions of Engineering Design', *Engg. Designer*, I.E.D., Dec., 1960.
6. Booker, P. J. and Prince, J.: 'A Special Purpose Mechanical Design', *Engg. Designer*, I.E.D., March, 1961.
7. British Patent Spec. No. 812,652.
8. German Patent Spec. No. 1,069,965.
9. 'A Tie-Bar Duct Expansion Restraint Design', *Engg. Designer*, Feb., 1962.

SUGGESTED STUDY

1. Design has been described as a decision-making process. List the major decisions.
2. Establish a morphology of this project. Break it down into component activities and show the interrelation of these activities (refer to Exhibit 6).
3. Suggest alternate concepts for accommodating the expansion and contraction of the ducts.
4. Estimate and compare the costs of various concepts.
5. Estimate the flow restriction caused by bellows and restraints.
6. How much motion is caused by a temperature change at 752°F (400°C) in the given configurations?
7. How does the compressive stress produced in swaging improve fatigue resistance?
8. Specify a scale model test of the frustro-conical restraint unit. What geometric size factor will be suitable? Shall the same number of tie rods be used? How do these choices affect stresses, forces, moments?

Case 9

Development of a circular strike plate

by Richard C. Bourne

Prepared with support from the National Science Foundation for the Stanford Engineering Case Program. Assistance from the Schlage Lock Company and Mr Ernest L. Schlage is gratefully acknowledged.

This case is available in pamphlet form as ECL 114 from the Engineering Case Library, Room 500, Stanford University, Stanford, California 94305.

PART 1

Background

Introduction

In March 1963, Mr Ernest L. Schlage, Director of Research at the Schlage Lock Company, San Francisco, California, and son of the inventor of the modern door lock, pondered the design of a new strike plate.* The new 'pre-hung' door concept had opened the market for a strike more easily installed than the traditional rectangular model; the construction industry wanted a strike for which they could prepare the jamb with a single boring operation, rather than the present routing† or hand chiselling. Mr Schlage later commented, 'Our investigations showed that several of our competitors already held patents on circular strikes, but these designs did not provide sufficient allowance for door sag. We hoped to develop a superior strike which would not infringe these patents, and which might even be patentable itself.'

Mr Marron Kendrick, President of the Schlage Lock Company, added, 'For years we have concentrated on development of the lock assembly, neglecting the strike as an insignificant part of the system. Our marketing group had recently reported, however, that the introduction of a bore-in strike could now spur a substantial growth in sales of Schlage residential locks. When I first reviewed some of our preliminary research work on this project, I was amazed at what improvements the engineers were making on an item so common and apparently simple that it had resisted basic change for at least a century.'

The Company

The Schlage‡ Lock Company, founded in 1920 with a working capital of $30, was, by 1925, turning out 20 000 locks a month; it now produces that many in a day and is the world's largest producer of door locks. Originating in a San Francisco loft with a total employment of four, the company today employs 1 700 people in its modern manufacturing plant. Schlage limits its product line to door locks for residences and commercial buildings, related hardware such as strike and escutcheon plates, and, in limited quantities, tools for the installation of these products.

The initial growth of the company resulted from the excellence of its first product, a remarkable invention patented by Walter Schlage. Ernest Schlage, presently serving the company as Vice President and Director of Research states, 'My father revolutionized the lock industry with the introduction of the cylindrical door lock. Today, we and all our competitors manufacture locks which are, in their basic features, substantially identical to my father's designs of the 1920s.' Walter Schlage, a San Francisco engineer of German birth, served as chief engineer until his death in 1946; during his life he was awarded 140 patents. His son Ernest, with engineering degrees from MIT and Stanford, has obtained more than 100 patents.

While patents protecting the basic cylindrical lock have long since expired, the Schlage Lock Company has gained and maintained a position of leadership in the industry by producing locks of very high quality, and, through its Research Division, by accurately forecasting and meeting future requirements for locks.

Of the three engineering groups at Schlage, Research, with 13 employees, includes approximately one-fourth of the total number of engineers. The largest group, Production Engineering, is engaged in the design of tools, dies, and jigs for the manufacture of Schlage products; the third group, Product Design, modifies and improves existing products, and is roughly equal in size to the Research Division.

* The strike plate or 'strike' is located in the face of the door jamb, and receives the extended lock latchbolt to hold the door in a closed position.
† Routing is a power woodcutting operation in which a rotary bit cuts a relatively shallow bore and then moves laterally to produce a depressant plane surface of variable shape.
‡ The pronunciation of 'Schlage' has been Americanized to rhyme with 'vague'.

Securing the door

Most doors are held in a closed (or 'latched') position by a system comprised of a retractable latchbolt located in the door edge and a strike plate fastened to the door frame, the plate being cut with a hole to receive the latchbolt. The latchbolt is spring-loaded (hence sometimes called a 'springlatch') to remain in the extended position, and is wedge-shaped at its end, to be cammed back into the door upon contacting a lip on the striker plate. The latchbolt is also retracted by turning the unlocked door knob. When security is desired, doors are equipped to *lock* as well as *latch*. In older devices, locking was accomplished by turning either a key or thumbturn in the chassis of the lock, thus moving a heavy bolt into a second hold in the strike plate. Unlike the springlatch, this bolt cannot be forced into the retracted position by wedging a blade or strong card between the door edge and frame; for this reason it is called a 'deadbolt'. For conditions of maximum security, additional deadbolt units may be installed auxiliary to the springlatch assembly.

Before introduction of the Schlage cylindrical lock, most door locks were of the mortised type (see Exhibit 1), so called because a cavity for the device had to be chiselled, or 'mortised', in the door. These locks were handfitted and expensive to produce. The hammer and chisel installation process produced alignment problems, took too long, and sometimes seriously weakened the door.

The basic breakthrough made by Walter Schlage was the introduction of a door lock composed of interlocking chassis and latch units, both cylindrical in shape. This 'cylindrical door lock'* is installed simply by boring two intersecting holes in the door, one from the face (usually $2\frac{1}{8}$ in in diameter) and one from the edge ($\frac{7}{8}$ in diameter). The latch unit is inserted and secured from the edge; the lock chassis and spindle portion is then inserted from the face, and interlocks with the latch assembly. Fastening the mounting plate, the decorative escutcheon and then the second knob on its spindle secures the door lock in place.

A further innovation by Walter Schlage was the provision for door locking with a single latchbolt. The Schlage 'deadlatch' (Exhibit 2) combines springlatch and deadbolt characteristics in one compact assembly, which fits in the same bored hole size as the plain springlatch unit. A small spring-loaded plunger is held back by the strike plate, whose opening is shaped to receive the bolt but not the plunger. An internal mechanism locks the bolt in the extended position when the plunger is thus restrained, so that, as with a deadbolt, the door cannot be unlatched by sliding a blade between door and frame.

An added impetus to the great success of the basic Schlage lock has been its 'button-locking' feature, now found on almost all door locks. This clever design ended the need for an interior key, and enables a room's occupant to tell at a glance whether the door is locked. The interior knob was always free to open the door even when the exterior knob was locked. This panic-proof feature allowed immediate egress at all times.

The installation of the cylindrical lock in just two holes left no place for locating the key cylinder except inside the exterior knob. This location proved to be very convenient for the user and greatly simplified the design of the mechanism.

Despite the continued use of the basic Schlage lock, the industry retains a dynamic outlook. There is a constant effort by lock companies and independent inventors not only to improve the design of the cylindrical lock but to invent a new, lower cost, superior lock so as to revolutionize the lock industry as did Walter Schlage half a century ago.

---

* Not to be confused with the 'pin-tumbler cylinder', into which the key is inserted, which is also used on almost all of today's door locks, and was invented by Linus Yale, Jr. in 1865.

EXHIBIT 1 Old catalogue page showing mortise locks

MORTISE CYLINDER LOCKS

HEAVY LINE

MR—MADISON DESIGN LOCK SETS

COLD FORGED BRONZE (.048 GA.)

OFFICE OR CLASSROOM LOCKS

Nos. 7705½ MR and 7737 MR-MKD

One pair knobs, inside and outside 1653. Cylinder ring 98. Escutcheons, inside 7875½ MR, outside 7876 MR. Spindle 1635—4½ inch. Lock 7705½ N or 7737 N. Master keyed to Master Key No. LB 17400 xM.

No.	Finish	Per Set
OP 7705½ MR—Office Doorlock	Dull Bronze	**$28.00**
OP 7737 MR—Classroom Lock	Dull Bronze	**31.50**

SCHOOL HOUSE LOCK

No. 7737 N

For classroom doors. Always operative from inside. Flat front. Fronts can be beveled at any angle from flat to ⅛ inch in 2 inches. Thickness of door should be specified. One bronze cylinder. Bronze hubs for ⅜ inch swivel spindle. Bronze bolt, ½-inch throw. **Size** — Lock, depth, 4 in.; height, 5⅝ inches; thickness, ¾ inch; 2¾-inch Backset. Faceplate, 1¼ x 8 inches with 1¼-inch strike.

Operation: Inside at all times by knob. Key on outside cylinder permits operation of knob outside or makes outside knob stationary as desired.

One in box—weight 3¼ lbs.

No.	Finish	Each
OP 7737 N	Dull Bronze	**$22.00**

Furnished with 2 Keys, No. 267.

SECTIONAL LOCK SET FOR ENTRANCE DOORS

Outside with Cylinder, Inside with Turn Knob

Set No. 1-6745

One-piece wrought bronze and brass knobs. Roses, key escutcheons and turn knobs wrought bronze and brass. With metal knobs. Lock 6745N, Inside and Outside Knobs 1773, Inside and Outside Roses 554, Wrought Bronze and Brass Cylinder Ring 98, Turn Knob 125, Spindle 1635—4½ in.

One set in box, weight 3¾ lbs.; 6 sets in case, weight 25 lbs.

No.	Finish	Per Set
B 1-6745	Polished Brass	**$19.15**
OP 1-6745	Dull Bronze	**19.15**

OFFICE LOCK

BRONZE FRONT — HEAVY BOLTS

No. 7705½ N

Flat front with guarded bolt and dead locking stop. Fronts can be beveled at any angle from flat to ⅛ inch in 2 inches. Thickness of door should be specified.

One bronze cylinder. Bronze hubs for ⅜-inch swivel spindle. Bronze bolts—½-inch throw. **Size** — Lock, depth, 4 inches; height, 5 inches; thickness, ¾ inch; 2¾-inch Backset. Faceplate, 1¼ x 8 inches, with 1¼-inch strike.

Operation: From outside by key, both sides by knobs. Outside knob is set by lower stop in face of lock and can be released only by pushing in upper stop.

One in box, weight 3 lbs.

No.	Finish	Each
OP 7705½ N	Dull Bronze	**$20.00**

Furnished with 2 Keys, No. 267.

FRENCH DOOR LOCK SET

Wrought Bronze and Brass

Set No. 1-5134, Bit Key 1½-Inch Backset

With No. 5134 Lock, No. 1772 Knob, **No. 1119 Lever Handle** and No. 710 Key Escutcheon. No. 906 steel key. Packed one set in box, with screws.

Set No.	Finish	Weight per Doz. Sets	Per Doz. Sets
B I-5134	Polished Brass	23 lbs.	**$82.00**
OB I-5134	Dull Brass	23 lbs.	**82.00**
OP I-5134	Dull Bronze	23 lbs.	**82.00**

MORTISE FRENCH DOOR LOCKS

Bronze Front and Strike

No. 5134

Spring on the hub is made extra heavy to hold up lever handle. Latch and dead bolt have [illegible] inch throw. Reverse, flat front, by removing cap. Bronze hub for [illegible] inch straight spindle, heavy bronze bolt. **Size** — Lock: depth, 2 inches; height, 3⅝ inches; thickness, ½ inch, 1½-inch Backset. Faceplate, 5 x 13/16 inches with ⅞-inch strike. No. 906 steel keys. 12 key changes. Wrought steel tumblers. Packed 6 in a box. Weight per box, 6½ lbs.

No.	Finish	Each
B-5134	Polished Brass	**$5.90**
OB-5134	Dull Brass	**5.90**
OP-5134	Dull Bronze	**5.90**

EXHIBIT 2   **Typical cylindrical lock assembly**

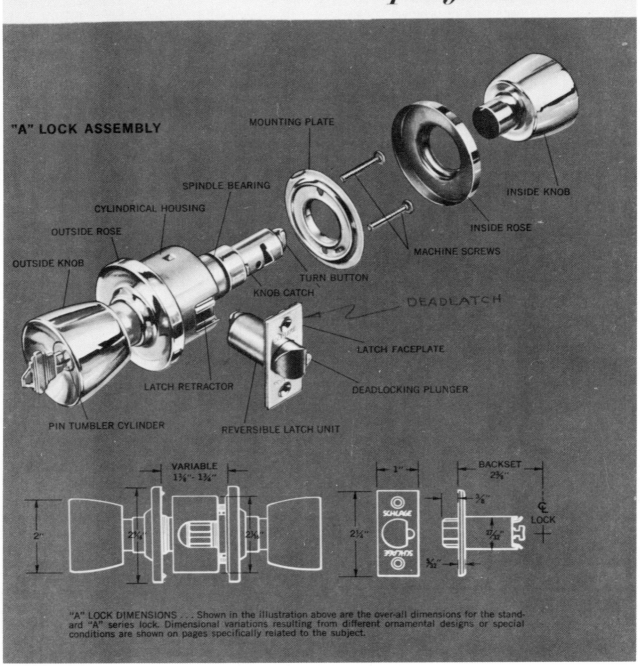

△LOCKS
specifications

"A" LOCK ASSEMBLY

MOUNTING PLATE

SPINDLE BEARING

CYLINDRICAL HOUSING

OUTSIDE ROSE

OUTSIDE KNOB

TURN BUTTON

KNOB CATCH

INSIDE KNOB

INSIDE ROSE

MACHINE SCREWS

DEADLATCH

LATCH FACEPLATE

LATCH RETRACTOR

DEADLOCKING PLUNGER

PIN TUMBLER CYLINDER

REVERSIBLE LATCH UNIT

VARIABLE
1⅜" - 1¾"

2"

2¼"

2¼"

1"

2¼"

BACKSET
2⅜"

⅜"

¹⁷⁄₃₂"

³⁄₃₂"

CL
LOCK

SCHLAGE

"A" LOCK DIMENSIONS . . . Shown in the illustration above are the over-all dimensions for the standard "A" series lock. Dimensional variations resulting from different ornamental designs or special conditions are shown on pages specifically related to the subject.

EXHIBIT 3　**Strike plates**

STRIKES

990
Standard Strike
SIZE: 2¼″ high
LIPS: Full Lip

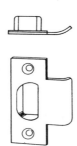

951
Standard Box Strike
SIZE: 2¾″ x 1⅛″ x ³⁄₃₂″ thick
LIPS: ⅞″ through 4″ in ⅛″
　　　graduations

951R
**Standard ⅜″ or ½″ Rabbeted
Box Strike**
SIZE: 2¾″ x 1⅛″ x ³⁄₃₂″ thick
LIPS: ⅞″ through 4″ in ⅛″
　　　graduations
(Strike shields may be ordered separately.
Specify finish and depth of rabbet.)

952
Standard Box Strike
SIZE: 2¾″ x 1⅛″ x ³⁄₃₂″ thick
　　　(For use with deadlocks)

952R
**Standard ⅜″ or ½″ Rabbeted
Box Strike**
SIZE: 2¾″ x 1⅛″ x ³⁄₃₂″ thick
　　　(For use with deadlocks)
(Strike shields may be ordered separately.
Specify finish and depth of rabbet.)

953
Protected Back Box Strike
(For 1⅜″ and 1¾″ doors only)
LIP: ¹⁵⁄₁₆″ for 1⅜″ doors
LIP: 1¼″ for 1¾″ doors

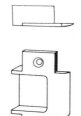

954
Cast Open Back Strike
SIZE: 2¾″ x 1⅛″ x ⅛″ thick
LIP: 1″ (For 1¾″ thick doors)

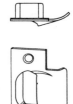

955
Raised Lip Box Strike
SIZE: 2¾″ x 1⅛″ x ³⁄₃₂″ thick
LIPS: 1¼″ through 4″ in ⅛″
　　　graduations

All Schlage strikes are reversible and are furnished complete with screws. Lip strikes, with the exception of the 953 and 954, are furnished with 1¼″ lip as standard. Rabbeted door frames and rabbeted latch bolts require rabbeted strikes. Rabbeted strikes for hollow metal doors and other special strikes can be made to order.

To accurately determine the length of lip required for a strike, measure from the center line of strike to the end of door jamb as shown in the illustration. This dimension, plus ¼″ for curved lip jamb clearance, is the required lip length. When ordering, specify strike number, finish and lip size.

ORDERING EXAMPLE:

10	960	3	1½″ LIP
Quantity	Catalog No.	Finish	Detail

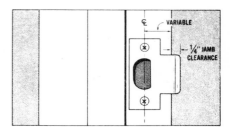

EXHIBIT 4   Typical latching arrangement

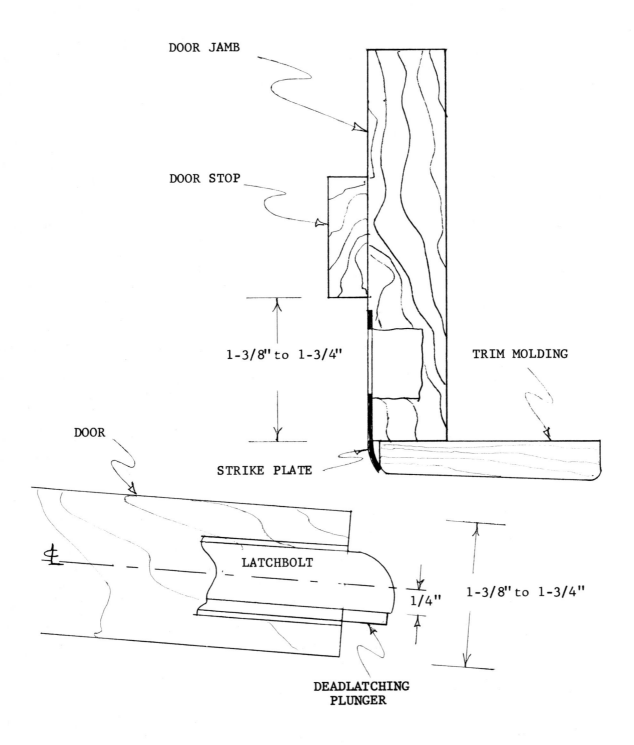

DOOR JAMB

DOOR STOP

1-3/8" to 1-3/4"

TRIM MOLDING

DOOR

STRIKE PLATE

LATCHBOLT

1/4"

1-3/8" to 1-3/4"

DEADLATCHING
PLUNGER

Strike plates

The mortise-lock, with its separate springlatch and deadbolt, required a rather large strike plate with two receiving holes. The introduction of Schlage's deadlatch, of course, meant that only one hole was required, but most strikes are still rectangular in shape (although smaller) and are usually mortised-in by the carpenter to fit flush with the surface of the door jamb. Thus, the installer is required to chisel out one deep recess for the latchbolt, and a second very shallow and accurately cut recess for the strike plate itself. For better appearance, many strikes are equipped with a 'strike box' so that raw wood cannot be seen through the latch receiving hole. Exhibit 3 shows various Schlage strikes.

The primary functions of the strike plate are to cam the latchbolt into the retracted position, and to provide a hole to receive the latchbolt. The strike must be strong and well-anchored to withstand the high force which may develop when the door slams shut and rebounds off the doorstop; this force is the largest to which the latchbolt and strike plate are normally subjected, and is received, on the strike plate, entirely by the straight front edge of the latch receiving hole.

Secondary functions of the strike plate are to provide compensation for misalignments between latchbolt and receiving hole; such misalignments may result either from inaccurate installation or from shifts of door and frame with time.

A vertical margin is provided in the receiving hole of most strike plates to allow for both door sag, which results from the entire weight of the door being carried at the hinged edge, and frame sag, which often occurs as a building settles. Standard rectangular strikes are designed with up to one-half inch ($\pm\frac{1}{4}$) sag allowance.

Lateral adjustability is also desirable, for if the strike is located too near the doorstop strip, the latchbolt cannot enter the receiving hole; and if the strike is located too far from the stop, the door will rattle when latched. With the conventional Tee-shaped strike, however, lateral adjustment can be accomplished only by removing the strike, extending the mortise in the direction of the desired strike movement, re-fastening the strike, and using putty or plastic wood to fill the newly exposed gap.

Exhibit 4 shows a door and jamb assembly in a section view through the latchbolt and strike plate. Door thicknesses vary from $1\frac{3}{8}$ in to $1\frac{3}{4}$ in for standard residential doors; the latchbolt is always centred (within normal installation tolerances) in the door. The flat face of the latchbolt is about $\frac{1}{4}$ in from the bolt centerline; the strike plate is located such that when closed, the door face is flush with the jamb edge.

The need for a new strike

In early 1963, Schlage, engineers began investigating the possibility of developing a radically new, 'bore-in' strike. Salesmen and distributors noted that more and more housing developments were using 'pre-hung' doors, already hinged in the frame when delivered to the construction site. With this system, boring for the latch hardware is done at the mill; usually the latch unit and strike plate are installed there as well, to hold the door closed during transport. The knob-chassis unit is installed on the job, since prior assembly would make stacking of the pre-hung doors more difficult, and could cause marring of the door surfaces. For pre-hung doors, the latch receiving hole in the jamb is usually machine bored with a bit, but a router must be used to mortise a shallow cavity for the rectangular strike plate. In terms of both time and equipment required, the routing process is more expensive than a boring operation.

One simple way to provide less expensive strike installation would be to surface-mount the strike; that is, manufacture the strike of thin metal to be fastened to the surface of the jamb (not mortised-in). The latchbolt receiving hole could then be circular and no chiselling or routing would be required for the strike plate itself. Mr Schlage noted that, 'Not long ago, a new strike aimed at the pre-hung door industry was brought out by a competitor of ours. It required the boring of a single hole, and was not mortised in flush, but rested on the surface. We experimented with this idea too, but decided that it was not really a satisfactory solution. Surface-mounted strikes have been tried by us before, and they have never been well-accepted.' Drawbacks to surface mounting are poor appearance, possible interference in cases where very little clearance exists between frame and door, and, perhaps most important, low strength. The metal must necessarily be thin, and the forces exerted on the strike have to be carried to the frame entirely by the fastening screws.

In January 1963, Mr Schlage had begun sketching possible designs for a new bore-in strike. Among these were the surface-mount explorations previously referred to. Mr Schlage had first considered a trapezoidal strike with integral strike box, as shown in Exhibit 5(a). With the strike lip merging into the body of the strike, this design would have a very clean appearance, and the strike box would strengthen the system; the front edge of the strike box would contact the wood and thus carry some of the load to the frame. Mr Schlage considered fastening the strike through the bottom of the box; this would permit a strike of reduced dimension, since the rim would no longer need to be wide enough to include fastening screws. However, most residential jambs are made from 1 in stock (which, after planing, is about $\frac{3}{4}$ in thick). Since strikes were usually installed at the mill, it was not permissible for any part of the installed strike to extend beyond the back of the jamb (this would interfere with jamb installation at the job site). This problem necessitated either very short screws (through the strike box bottom) or a new fastening scheme. Shorter screws would have very few threads engaged in the wood, so Mr Schlage began sketching other fastening possibilities. Some of these are depicted in the pages from his daily notes included as Exhibit 6.

Possible patent conflicts

After rejecting the surface-mount concept for their new bore-in strike, Schlage engineers began the design of a flush-mounted circular strike. A prototype of one of the initial circular designs is shown in Exhibit 5(b). For installation, this strike would require the boring of two concentric holes. Concurrently with these initial designs, Mr Schlage initiated a patent search to see what inventions had previously been claimed in the area of bore-in strikes. 'Maintaining an awareness of competitive patents is part of the game of inventing,' notes Mr Schlage. 'When designing new products, one always hopes that his good ideas are original and were not previously patented by others working in the same field. Avoidance of patent infringement becomes a very important consideration by the designer if he is to spare his company expensive litigation.'

Basically, a patent is made up of three parts: the illustrations, the description, and the claims. The first two are included as explanatory background material for the claims; it is the claims which legally describe and bound the invention. Each claim is a single sentence made up of a number of distinct phrases called 'elements'. A patent claim is said to be infringed by a device which embodies features described by *every* element of that claim. If the device lacks even one feature described by one of the elements, the device does not infringe that claim. Stated legally, 'Avoidance of an element of a claim avoids infringement.' Mr Schlage explained, 'You try to write your claims as short as possible. Long claims with many elements usually leave loopholes, but if you see a claim written in six or seven lines, you say "This is a basic claim".'

The patent search turned up two competitive patents of significance to the new project; the first of these (Exhibit 7) appears to contain basic claims to a flush-mounted circular strike; the circular Schlage prototype would clearly infringe on this patent. The strike described in the second patent was an improvement on Russell's; although he obtained a patent, the inventor could not produce his strike without first obtaining rights to the basic strike patented by Russell.

EXHIBIT 5(a)   Photograph 'trapezoidal prototype'

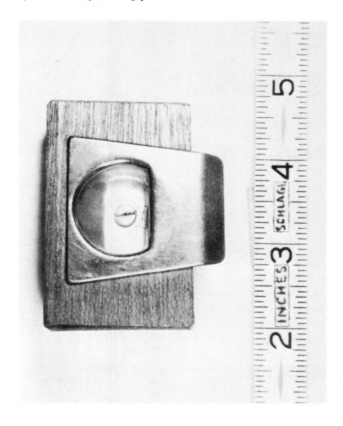

EXHIBIT 5(b)   Photograph 'first circular prototype'

EXHIBIT 6   Mr Schlage's sketches of fastening schemes dated 11 and 14 January 1963

Friday, January 11, 1963

354 days
follow

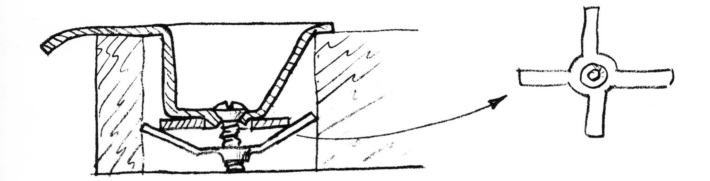

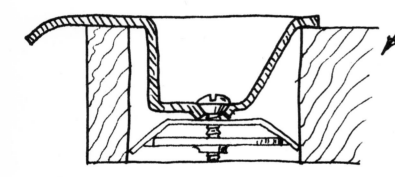

This design, although more difficult to insert into the hole, has the advantage that it draws or pulls the strike down into the hole as the screw is tightened.

E L Schlage Jan 11, 63

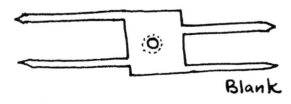

Blank

Folded
up.

EXHIBIT 6 *(contd.)*

Monday, January 14, 1963

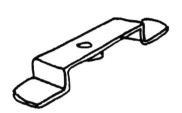

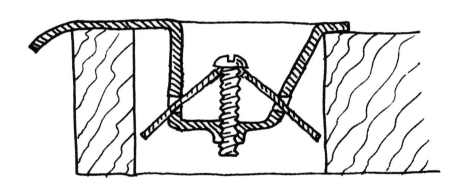

E L Schlage
Jan 14, 1963

Tightening screw causes legs to bite into wood around hole and to draw strike down flush.

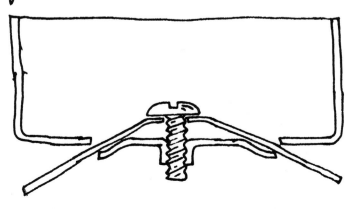

Model made.

EXHIBIT 7   Russell Patent 2,937,898

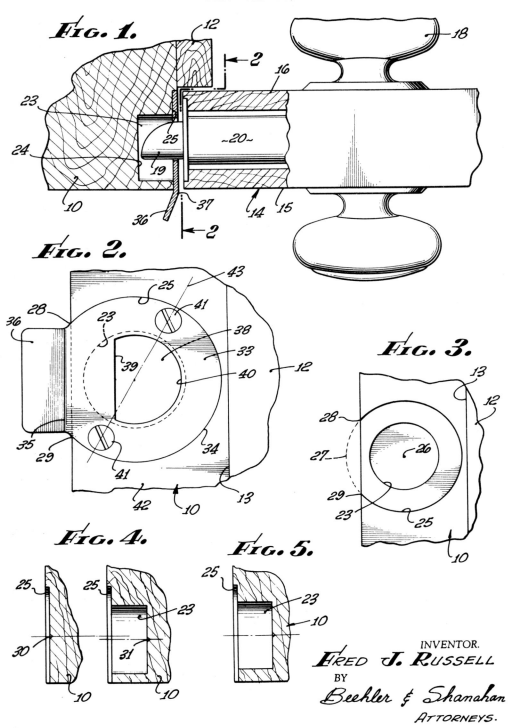

May 24, 1960   F. J. RUSSELL   2,937,898

CIRCULAR STRIKE INSTALLATION

Filed Jan. 28, 1958

FIG. 1.

FIG. 2.

FIG. 3.

FIG. 4.

FIG. 5.

INVENTOR.

FRED J. RUSSELL

BY

Beehler & Shanahan

ATTORNEYS.

EXHIBIT 7 *(contd.)*

United States Patent Office

2,937,898
Patented May 24, 1960

1

2,937,898

CIRCULAR STRIKE INSTALLATION

Fred J. Russell, 3800 Don Felipe Drive,
Los Angeles, Calif.

Filed Jan. 28, 1958, Ser. No. 711,615

3 Claims.   (Cl. 292—340)

The invention relates to latches, locks and lock hardware and has particular reference to a strike plate for use with such hardware against which the customary latch bolt strikes and engages to hold a door in latched position.

One of the elements which has been a factor of some consequence in the installation of door hardware has been the shape of strike plates which has necessitated careful chiseling to provide square and rectangular apertures so that the strike plate of corresponding configuration can be made to fit snugly and neatly in the aperture. In fine residential and commercial building construction, the fitting of strike plates of the character mentioned takes on the character of fine cabinet work which is often in practice somewhat beyond the ability of an average workman called upon to perform such tasks.

Moreover, the accelerated employment of power tools has taught that where hardware is such that it can be mounted in openings prepared by drilling tools, appreciable time can be saved.

It is therefore among the objects of the invention to provide a new and improved circular strike installation which involves both a method of mounting a strike plate on a door frame by effective, rapid means and a properly designed strike plate of inexpensive construction which is especially well adapted to employment with the method.

Another object of the invention is to provide a new and improved circular strike mounting whereby the frame of a door can be prepared for reception of a strike plate neatly and accurately by relatively simple means and tools which will result in a very workman-like installation even though performed by a person not especially well skilled in cabinet work.

Still another object of the invention is to provide a new and improved circular strike installation which incorporates a specially formed strike plate so constructed that it can be accurately, neatly and effectively mounted in and over recesses which have been formed by circularly revolving tools, the strike plate being moreover such that once installed, the installation is sturdy and rugged to a material degree and thereby capable of withstanding frequent and prolonged jarring such as would normally occur with the opening and closing of a door.

Still another object of the invention is to provide a new and improved means and accompanying device which by virtue of the construction and mode of installation is capable of minimizing to a material degree the cost of installation as well as the attendant cost of the appropriate strike plate hardware item.

With these and other objects in view, the invention consists in the construction, arrangement and combination of the various parts of the device together with the method for achieving the same whereby the objects contemplated are attained, as hereinafter set forth, pointed out in the appended claims and illustrated in the accompanying drawings.

In the drawings:

Figure 1 is a horizontal view partially broken away showing a

2

door and door frame in closed position and illustrating the mounting and operation of the circular strike installation.

Figure 2 is a fragmentary elevational view of the circular strike plate installation taken on the line 2—2 of Figure 1.

Figure 3 is a fragmentary elevational view showing the recess in the frame ready for reception of a circular strike plate.

Figure 4 is a fragmentary sectional view showing two successive steps in forming the recess for reception of the circular strike plate.

Figure 5 is a fragmentary sectional view showing a modified method for the preparation of a recess for reception of a circular strike plate.

In an embodiment of the invention illustrated in the drawings there is shown a door frame 10 upon which is mounted a door stop 11 spaced from an edge 12 of the door frame so as to leave an area therebetween bounded at one vertical side by the edge 12 and on an opposite vertical side by a face 13 of the door stop.

A door 14 is shown having an outer face 15 and an inner face 16 which is mounted in a conventional fashion to swing within the frame 10. Outer and inner knobs 17 and 18, respectively, are mounted in a conventional fashion for manipulation of a latch bolt 19 which is contained in a conventional latch bolt casing 20 in the door 14.

In the frame, as shown to good advantage in Figures 1 and 3, there is provided a clearance hole 23 having a diameter substantially less than the distance between the edge 12 of the frame and the face 13 of the stop. The depth of the clearance hole 23 is sufficiently great so that a bottom 24 of the hole permits full extension of the latch bolt 19.

A second hole 25 of relatively larger diameter overlies the location of the clearance hole 23. This hole 25 which is of relatively greater diameter may be concentric with respect to the hole 23, in the form of invention illustrated in Figures 1, 2, 3 and 5. It will be noted particularly that the diameter of the hole 25 is greater than the distance between the edge 12 of the frame and the face 13 of the stop 12. Moreover, by locating a center point 26 more nearly the edge 12 than the face 13, the wall of the hole 25 will break through the edge 12 as shown to good advantage in Figure 3 where the outermost portion of the path of a tool (not shown) for making the hole 25 is shown by a broken arcuate line 27. Breaking through the edge 12 in this fashion provides a shallow opening identified by an upper edge 28 and a lower edge 29, as shown in Figures 2 and 3.

In boring holes 23 and 25 in the event separate bits or drills may be employed, it is advantageous to first bore the hole 25 which is of relatively large diameter and thereafter bore the hole 23 which is of relatively smaller diameter. In the event that the holes are to be concentric, the bit or drill in each instance will be centered at the same center point 26. Where the centers are concentric, it may on occasions be advantageous to make use of a step drill of conventional form (not shown) thereby to bore both of the holes in a single operation and by use of a single tool. The method employing a step drill may be more particularly effective where material comprising the frame 10 is something other than wood.

On some occasions it may be desirable not to have the relatively larger hole 25 concentric with the relatively smaller hole 23. Under such circumstances it is most advantageous to first bore the hole 25 about its center 30. Thereafter the hole 23 can be bored about its center 31 which may be offset a slight distance

EXHIBIT 7 *(contd.)*

2,937,898

3

from the center **30**. By drilling the hole **25** first, a portion of the material of the frame **10** will be left remaining in order to effectively center the tool for boring the relatively smaller hole **23**. In this last described method as in the order of procedure first indicated, the location of the center **30** of the hole **25** will be more nearly the edge **12** of the frame than the face **13** of the stop.

Regardless of the manner or order of boring of the holes **25** and **23**, the mounting of a circular strike plate **32** is substantially the same.

The strike plate itself comprises a frame-engaging portion **33** which is flat and which is provided with a perimeter **34** which is a partial circle. At one side of the frame-engaging portion, namely, a side **35**, is a strike lip **36** which is composed of the same sheet material and is bent over as indicated in Figure 1 in a substantially conventional fashion. The frame-engaging portion, however, is made to extend toward the strike lip a distance such that the line **37** where the bend takes place extends outwardly beyond the edge **12** of the door frame **10**. In conformance with the circular patterning there is provided in the frame-engaging portion a recess **38** for reception of the latch bolt **19** which has a straight edge **39** and a semi-circular edge **40**.

For fastening the strike plate to the door frame there are provided screws **41** which extend through appropriate countersunk screw holes in the strike plate, details of which have been omitted. It is significant, however, that the screws are located on opposite sides of a diametric line **42** through the center of the frame-engaging portion **33** in a direction parallel to the edge **12**. The screws may be considered as lying upon an axis **43** which lies oblique with respect to the diametric line **42**. By mounting the screws in this relationship, the tendency of the strike plate to tilt as a result of having the lip struck repeatedly by the latch bolt is substantially minimized.

The installation herein described will, when employed with a strike plate having a substantially circular frame-engaging portion, make possible the mounting of such a strike plate on a door frame ready for installation with extreme rapidity and result in an appropriate saving in time and money to the ultimate consumer. As is apparent, by making use of perfectly circular mounting holes, modern power tools can be used to the greatest possible advantage.

While I have herein shown and described my invention in what I have conceived to be the most practical and preferred embodiment, it is recognized that departures may be made therefrom within the scope of my invention, which is not to be limited to the details disclosed herein but is to be accorded the full scope of the claims so as to embrace any and all equivalent devices.

Having described my invention, what I claim as new and desire to secure by Letters Patent is:

1. A circular strike mounting comprising a door frame having a straight edge on one side, a door stop on the frame and an area of predetermined width between the stop and the opposite edge of the frame, a relatively shallow hole of diameter between opposite curved portions greater than the width of said area bored into the frame at a location intersecting said opposite edge and forming a recess in said opposite edge of length less than the diameter of said shallow hole, said opposite curved portion having points of intersection with said edge at oblique angles, a relatively deep clearance hole concentric with said shallow hole, said clearance hole having a diameter smaller than said width and being located entirely within said area, and an apertured strike plate of sheet metal of uniform thickness and

4

having a frame-engaging portion of substantially the same diameter as said shallow hole, said frame-engaging portion being positioned in said shallow hole, a recess in the strike plate *concentric* with both said holes and having a straight side parallel with said straight edge, and a portion of said strike plate partially overlying said clearance hole and lying partially in said recess, the said rim portion having an extension disposed in an oblique direction protruding beyond said opposite edge and forming in strike lip.

2. A circular strike mounting comprising a door frame, a door stop on the frame and an area of predetermined width between the stop and the opposite edge of the frame, a relatively shallow hole of diameter greater than the width of said area bored into the frame at a location intersecting said opposite edge, and forming a recess in said opposite edge having a length less than the diameter of said shallow hole, said hole having a continuously curved wall of uniform radius with ends terminating at opposite ends of said recess, a relatively deep clearance hole concentric with said shallow hole, said clearance hole having a diameter smaller than said width and being located entirely within said area, and an apertured strike plate having a frame-engaging portion of substantially the same curvature and diameter as said shallow hole, a straight edge on said frame-engaging portion coincident with said recess, said frame-engaging portion being positioned in said shallow hole and a rim portion joined to the frame-engaging portion at said straight edge, said rim portion having a location protruding beyond said opposite edge, a strike lip on said rim portion and screw holes for reception of screws lying on a centering line oblique relative to a diametric line drawn through said strike plate in a direction parallel to said opposite edge.

3. In a door frame having a stop thereon and an edge on the opposite side of said frame spaced from said stop, a method of outfitting said door frame for operation with a latch bolt comprising simultaneously boring concentric holes in the frame between said stop and said edge, making one of said holes relatively shallow and of relatively larger diameter and making the other of said holes relatively deep and of smaller diameter, locating the centers of said holes near enough to said edge so that said one hole when bored cuts through said edge forming an opening and avoids cutting said top and so that said other hole is spaced inwardly from said edge leaving a portion of the frame between the hole and said edge, forming a strike plate of sheet material with a curved perimeter of the same curvature and depth as said one hole and forming on one side of the plate a strike plate having a junction with the plate of the same length as said recess and tilting said lip from the plane of said plate, forming a latch bolt aperture in said strike plate concentric with the curved perimeter and of diameter smaller than said perimeter, forming one straight side on said latch bolt aperture parallel to the junction of said lip with said plate, placing said plate in said relatively shallow hole so that the lip extends through the recess and a portion of the plate adjacent said one straight side overlies a part of said other hole, and anchoring said plate to the frame.

References Cited in the file of this patent
UNITED STATES PATENTS

1,553,531	Hoffman	Sept. 15, 1925
2,112,909	Jackson	Apr. 5, 1938
2,272,241	Fendring	Feb. 10, 1942
2,448,293	Berini	Aug. 31, 1948
2,656,205	Fletcher	Oct. 20, 1953

With the Russell patent appearing to discourage further development of the first circular prototype, Mr Schlage began compiling an up-to-date set of design criteria for the new strike. He felt that the jamb preparation should be possible by boring along a single axis, and should provide for flush mounting of the strike. The strike should be reasonably low in cost, yet attractive and strong enough to withstand high rebound forces. It should preferably provide sufficient allowance for door sag (now lacking on both the Schlage prototype and the patented Russell strike). The newest criterion, of course, was that the new strike must not infringe the Russell patent.

Note : Part 2 describes the steps taken which resulted in the design and introduction of a new circular strike meeting all of the above requirements.

Part 2   The Design of the Strike Plate

'The revolutionary aspect of our circular strike is the fact that the main bored hole, the hole that accepts the strike box, breaks out. We obtained a really basic patent on this idea and we have been able to stop infringers,' states Mr Schlage.

After carefully examining the Russell patent, Mr Schlage had listed three ways to produce a similar strike without infringement:

1. *Curve* the edge which contacts the latchbolt.
2. Align the fastening screws *vertically* rather than inclined.
3. Make the strike lip part of the rim itself, rather than an 'extension' of the rim.

None of these changes, however, would permit a significant improvement over previous bore-in strike concepts in terms of sag allowance, which appeared to be limited by the available width on the door jamb. To be compatible with the latchbolt location in the door, the bolt receiving hole should be offset toward the frame edge (as seen in the Russell patent, Fig. 3, Exhibit 7), rather than centred between doorstop and frame edge. This meant that for narrow door installations, even a one inch diameter bolt receiving hole would come very close to the frame edge.

Mr Schlage suddenly realized that it might make good sense for the bolt receiving hole to be large enough actually to break through the edge of the frame. While this meant the hole would show on the front edge of the raw door frame, most frames are finished with a 'molding strip' to cover the joint between wall and frame. This strip could be butted against the back of the strike lip to hide the bolt receiving hole. Mr Schlage re-examined the Russell patent and found it clearly specified that the smaller, deeper hole was to be located 'entirely within' the area between the doorstop and frame edge.

Mr Schlage emphasized that this 'breakthrough' had more value than just its patentability, 'It made our circular strike superior, because our latchbolt receiving opening could be larger, particularly in the vertical direction. Previous circular strikes had very little provision for sag. That is very important. We have as much allowance for sag as on our standard strikes.'

After Mr Schlage suggested the basic concept, other members of the Research staff evolved specific design proposals. A number of these are summarized on the Project Review sheet included as Exhibit 8. During the five weeks from 15 April to 20 May, a final detail design was developed based on the results of prototype tests. Schlage has mechanical equipment to carry out over 45 tests on door locking parts; strikes are usually subjected to 'door slam' and 'kick' tests, which indicate a strike's strength and holding power in the jamb, and to wear tests, which determine how well

EXHIBIT 8   Circular strike project review sheet

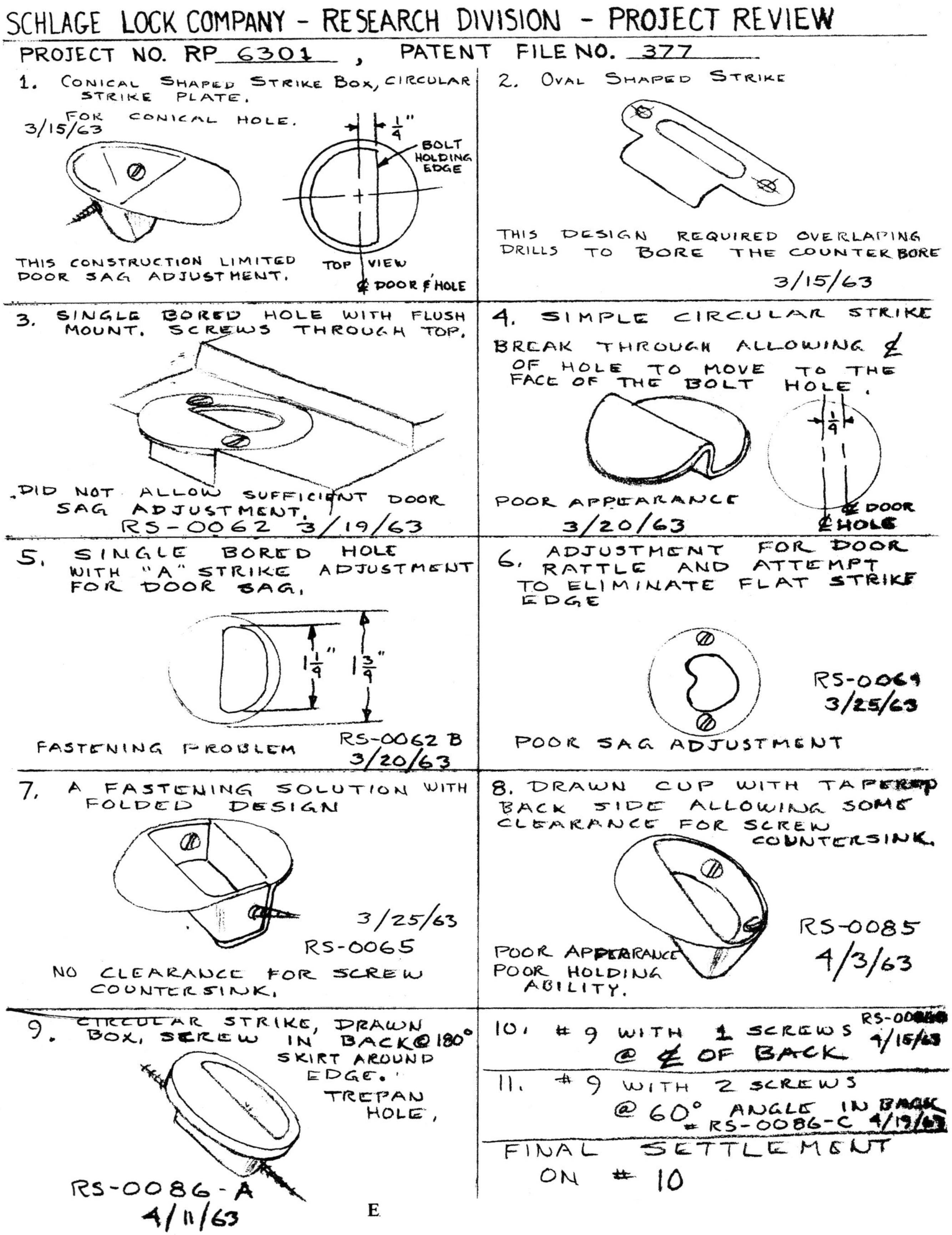

SCHLAGE LOCK COMPANY - RESEARCH DIVISION - PROJECT REVIEW

PROJECT NO. RP 6301 , PATENT FILE NO. 377

1. CONICAL SHAPED STRIKE BOX, CIRCULAR STRIKE PLATE.

FOR CONICAL HOLE.
3/15/63

BOLT HOLDING EDGE
$\frac{1}{4}$"

TOP VIEW
¢ DOOR ¢ HOLE

THIS CONSTRUCTION LIMITED DOOR SAG ADJUSTMENT.

2. OVAL SHAPED STRIKE

THIS DESIGN REQUIRED OVERLAPING DRILLS TO BORE THE COUNTERBORE
3/15/63

3. SINGLE BORED HOLE WITH FLUSH MOUNT. SCREWS THROUGH TOP.

DID NOT ALLOW SUFFICIENT DOOR SAG ADJUSTMENT.
RS-0062 3/19/63

4. SIMPLE CIRCULAR STRIKE

BREAK THROUGH ALLOWING ¢ OF HOLE TO MOVE TO THE FACE OF THE BOLT HOLE.

$\frac{1}{4}$"

POOR APPEARANCE
3/20/63
¢ DOOR ¢ HOLE

5. SINGLE BORED HOLE WITH "A" STRIKE ADJUSTMENT FOR DOOR SAG.

$1\frac{1}{4}$" $1\frac{3}{9}$"

FASTENING PROBLEM
RS-0062 B
3/20/63

6. ADJUSTMENT FOR DOOR RATTLE AND ATTEMPT TO ELIMINATE FLAT STRIKE EDGE

RS-0069
3/25/63

POOR SAG ADJUSTMENT

7. A FASTENING SOLUTION WITH FOLDED DESIGN

3/25/63
RS-0065

NO CLEARANCE FOR SCREW COUNTERSINK.

8. DRAWN CUP WITH TAPERED BACK SIDE ALLOWING SOME CLEARANCE FOR SCREW COUNTERSINK.

RS-0085
4/3/63

POOR APPEARANCE POOR HOLDING ABILITY.

9. CIRCULAR STRIKE, DRAWN BOX, SCREW IN BACK @ 180° SKIRT AROUND EDGE.
TREPAN HOLE.

RS-0086-A
4/11/63

10. # 9 WITH 1 SCREWS @ ¢ OF BACK
RS-0086
4/15/63

11. # 9 WITH 2 SCREWS @ 60° ANGLE IN BACK
RS-0086-C 4/19/63

FINAL SETTLEMENT ON # 10

E

EXHIBIT 9   Door kick test

STANDARD TEST METHOD

2.21 DOOR KICK TEST

Purpose:

1. To determine the effect on a lock of a sudden impact applied to a closed door simulating kicking the door or pushing the door with the shoulder.

2. To compare lock strength with door and door frame strength.

Procedure:

1. Install lock in test door.

2. Close door.

3. Raise 25 lb. hammer 2", then drop onto impacter.

4. Repeat step 3 with hammer raised in 1" increments, i.e., 3", 4", 5", etc. Maximum drop height to be determined at each test.

5. After each hammer drop observe for any damage, loosening, or shifting of lock unit, latch, strike, screws, door, or door frame.

Record:

1. Weight of hammer: 25 lbs.

2. Height from which hammer was dropped, inches. (See sketch).

3. Appearance and condition of lock unit, latch, strike, screws, door and door frame after each hammer drop.

4. Thickness and type of door.

Test Equipment:

1. Test door.

2. Kick fixture.

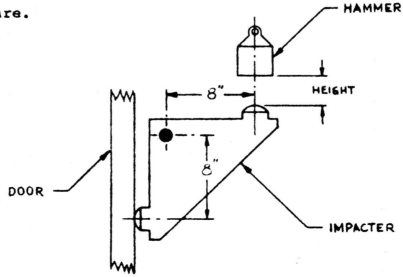

the strike's finish stands up under the abrasive action of the latchbolt. A description of the standard kick test is included as Exhibit 9.

As anticipated, fastening proved to be a problem. Like the earlier trapezoidal prototype strike, the circular strike would have an integral strike box. However, since the receiving hole broke out, schemes such as those considered earlier (Part 1) could not be used. To achieve adequate sag allowance, the strike had to be fastened through the strike box, for if fastened through a rim, the rim would necessarily become too large for narrow door installations – that is, the doorstop limits the maximum diameter of the rim which can be used. Screws through the strike box walls had to be placed at an angle which would permit access with drill and screwdriver. This meant the screw or screws had to be installed through a wall which slanted inward toward the bottom of the strike box. The conical strike box suggested in concept 1 of Exhibit 8 would permit an acceptable screw angle; however, the conical wall, if carried around to the bolt-receiving wall, would severely limit sag allowance.

Schlage engineers solved this problem by forming the strike box wall of intersecting sections of a cone and cylinder. One anchoring screw was located at the rear (conical) wall, which met cylindrical sections extending from the front (flat) bolt-contacting wall. The single fastening screw would not be required to carry the entire load to the jamb; an outer skirt on the strike would fit into the outer groove of a 'trepan' hole to transmit much of the force to the frame. This trepanned hole required the design of a tool to bore three holes of different depths and diameters on the same axis, as shown on the installation sheet (Exhibit 10).

Several prototypes of the one-screw, trepan-hole model were made (concept 10 in Exhibit 8); the Research group members were enthusiastic about both its appearance and solidity when fastened. On 20 May 1963 the drawing included as Exhibit 11 was approved for pilot run production of the new strike.

Mr Schlage was later awarded extensive patent rights in connection with the new circular strike; excerpts from his patent are shown in Exhibit 12. In discussing the design phase of the strike, he emphasized that several innovative features in addition to the larger sag allowance should contribute to its commercial success. The integral strike box scheme, while practically required from a security standpoint (without a strike box to establish the depth of the bolt receiving hole in the jamb, it might be bored too shallow), carried several additional benefits, the most marketable of which was probably its fine appearance. The new strike also promised to be cheaper to produce than earlier strike box assemblies, since only one piece of metal would be involved (previously the strike box had been a separate piece which had to be fastened to the strike), and since the strike box would add considerable strength to the bolt receiving edge, thus permitting the strike to be made of thinner metal with no loss of strength.

Another novel feature of the Schlage circular strike was the skirt designed to fit into a trepan hole. Coupled with the integral strike box, the presence of the skirt gives the strike an appearance of solidity compatible with Schlage's reputation for quality hardware. When located (as recommended) in a trepan hole, the strike withstood an extra 50 lb force in the kick test as compared to the alternate (non-trepan) installation (see Exhibit 10).

Production problems

The initial batch, or 'pilot run' of circular strikes was produced by the Larkin Specialty Manufacturing Company of South San Francisco. Mr Schlage explained that frequently the pilot run of a new product is produced outside the company; 'The Production Engineering group is understandably not always enthusiastic about the new products we develop. Their staff and equipment are kept busy enough by the items already in production, so we usually hear some grumbling from them when we release a new product, often to the effect that it's impossible to produce the part

EXHIBIT 10   Installation sheet

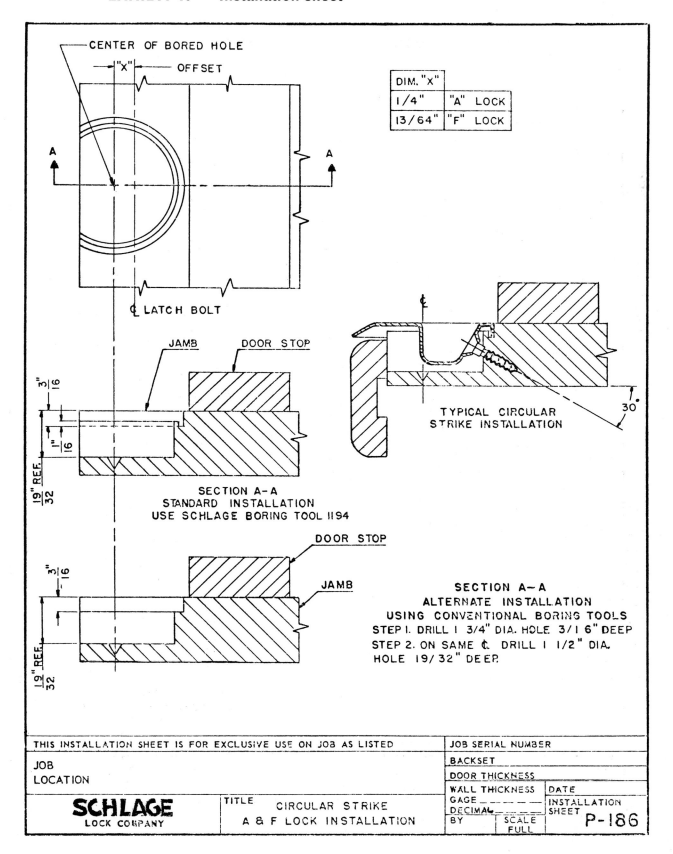

CENTER OF BORED HOLE

"X" — OFFSET

DIM. "X"	
1/4"	"A" LOCK
13/64"	"F" LOCK

A — A

¢ LATCH BOLT

JAMB   DOOR STOP

$\frac{3"}{16}$

$\frac{1"}{16}$

$\frac{19"}{32}$ REF.

SECTION A—A
STANDARD INSTALLATION
USE SCHLAGE BORING TOOL 1194

TYPICAL CIRCULAR
STRIKE INSTALLATION

30°

DOOR STOP

JAMB

$\frac{3"}{16}$

$\frac{19"}{32}$ REF.

SECTION A—A
ALTERNATE INSTALLATION
USING CONVENTIONAL BORING TOOLS
STEP 1. DRILL 1 3/4" DIA. HOLE 3/16" DEEP
STEP 2. ON SAME ¢ DRILL 1 1/2" DIA.
HOLE 19/32" DEEP.

THIS INSTALLATION SHEET IS FOR EXCLUSIVE USE ON JOB AS LISTED

JOB SERIAL NUMBER

JOB
LOCATION

BACKSET

DOOR THICKNESS

WALL THICKNESS

GAGE _ _ _ _ _

DECIMAL _ _ _ _

BY | SCALE FULL

DATE

INSTALLATION SHEET

P-186

SCHLAGE
LOCK COMPANY

TITLE   CIRCULAR STRIKE
A & F LOCK INSTALLATION

EXHIBIT 11   Circular strike production drawing

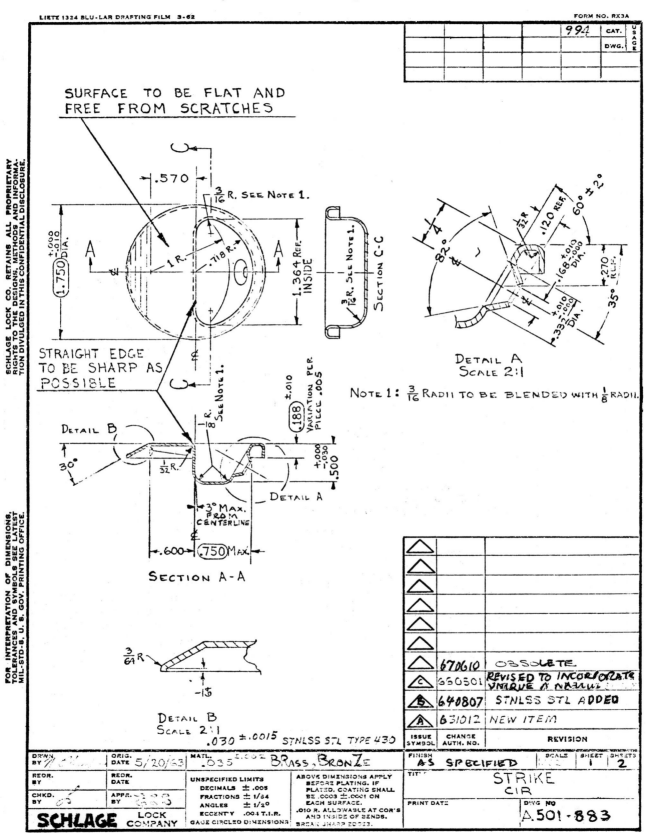

EXHIBIT 12   Excerpts from Schlage Patent 3,159,416

Dec. 1, 1964

E. L. SCHLAGE

CIRCULAR STRIKE

3,159,416

Filed Sept. 26, 1963

3 Sheets—Sheet 1

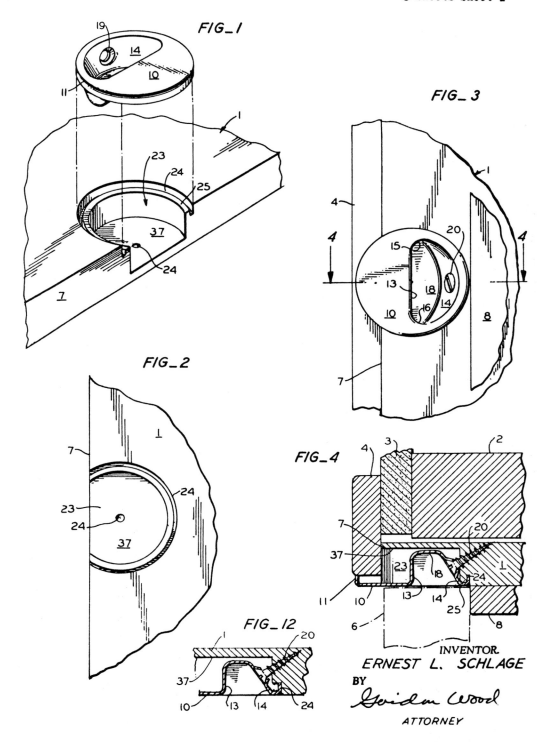

FIG_1

FIG_3

FIG_2

FIG_4

FIG_12

INVENTOR.

ERNEST L. SCHLAGE

BY

ATTORNEY

EXHIBIT 12 *(contd.)*   'Excerpts from Schlage Patent'

United States Patent Office

3,159,416
Patented Dec. 1, 1964

1

3,159,416
CIRCULAR STRIKE
Ernest L. Schlage, Burlingame, Calif., assignor to
Schlage Lock Company, a corporation
Filed Sept. 26, 1963, Ser. No. 311,815
17 Claims.   (Cl. 292—340)

This invention relates to strike plates for door locks and more particularly to a strike having a substantially circular form.

The main object of the present invention is the provision of a strike which lends itself to installation into a bored circular opening formed on the door jamb so as to obviate much of the expense and time required in chiseling rectangular mortises and openings which heretofore have been necessary in the installation of the conventional strikes.

Another object of the invention is the provision of a strike that is substantially circular in shape and which is provided with means for rigidly securing it in a bored opening in the jamb, said means providing optimum resistance to its being pulled or torn out of the door jamb.

Another object of the invention is the provision of an improved circular strike of the type installed flush with the surface of the jamb.

Still another object of the invention is the provision of a strike and a method of installing the same which method lends itself to the use of power driving boring equipment suitable for prefabrication of door frames on which the strikes may be installed prior to shipping to a building site.

Yet another object of the invention is the provision of a strike having a latch bolt receiving opening which provides a maximum clearance in a vertical direction for the associated latch bolt thereby permitting larger installation tolerances and also allowing for vertical misalignment of the door and for door sag.

Yet another object of the invention is the provision of a circular strike which lends itself to rotation about the axis of the strike to permit the adjustment required to prevent door rattle in the event shrinkage results after the door has been installed.

Yet another object of the invention is the provision of a circular strike in which a strike box may be formed integrally therewith if desired.

Other objects and advantages will be apparent from the following specification and drawings wherein:

FIG. 1 is an exploded perspective showing one form of the invention in association with the recess formed in the jamb and adapted to receive the strike therein. The jamb is shown horizontally disposed as it may be during the door frame building operation.

FIG. 2 is a side elevation of a portion of a door jamb showing the recess adapted to receive the strike therein with the frame, casing molding and doorstop omitted.

FIG. 3 is a side elevation of a jamb including casing molding and doorstop showing the strike installed.

FIG. 4 is a horizontal cross section of the structure of FIG. 3 taken in a plane indicated by lines 4—4.

FIG. 5 is a front elevation of a modified strike adapted to receive two securing screws.

FIG. 6 is a side elevation of the strike of FIG. 5.

FIG. 7 is a side elevation of a boring tool adapted to form a recess that receives the flange formed on the strike.

FIG. 8 is a side elevation of the tool of FIG. 7.

2

FIG. 9 is a perspective of a modified form of strike provided with an inclined lip for engaging the latch bolt.

FIG. 10 is a perspective of a strike formed without a strike box.

FIG. 11 is a perspective of another modified form of strike.

FIG. 12 is a view similar to FIG. 4 but showing a modification of the recess that receives the strike.

FIG. 13 is a perspective of still another modified form of strike.

FIG. 14 is a cross section of the strike of FIG. 13 showing the same installed on a jamb.

FIG. 15 is a perspective of another modified form of strike.

In detail, and first with reference to FIGS. 3 and 4, the invention is adapted to be employed on the vertically extending jamb 1 which is generally secured to a rough framing member 2 adjoining a wall panel 3. Usually a length of casing molding 4 is secured by nails passing through the wall panel 3 to the frame 2.

In FIG. 4 the associated door 6 is indicated in dot-dash lines and the outer face of such door is generally coplanar with the outer edge 7 (FIG. 1) of the jamb. The inner face of door 6 is adapted to abut a stop 8 secured to the jamb 1.

It will be noted that the width of the jamb between the outer edge 7 and the stop 8 is available for installation of the strike and as will subsequently be understood the present invention makes optimum use of the relatively narrow space available, and which space is roughly equal to the width of the door.

One form of the strike of the present invention comprises a circular plate having a main planar portion 10 provided with an integral peripherally extending flange 11 which not only stiffens the plate but provides a means for enchancing the securement of the strike to the jamb in a manner that will subsequently be described. Integrally formed with the strike plate is a strike box having a relatively long, flat side 13 which extends diametrally of the plate and an opposite slantingly disposed side 14 which is connected to flat side 13 by curved junctures 15, 16 (FIG. 3). Said junctures and opposed sides are integrally connected to the bottom 18 of the strike box.

Centrally of the slanting side 14 there is provided a countersunk hole 19 for receiving therethrough a wood screw 20 for securing the strike to the jamb 1.

To obtain optimum resistance to the strike being pulled out of the jamb 1 the latter is formed in the manner shown in FIG. 1. A relatively deep clearance hole 23 is bored about an axis 24 (FIG. 2) for receiving the strike box therein. Radially outwardly of the central clearance hole 23 there is provided a circular groove 24 which is of a width to receive the flange 11 of the strike. Between the clearance hole 23 and the annular groove 24 an annular ridge 25 is provided the outer surface of which is spaced inwardly from the outer face of the jamb 1 a distance about equal to the thickness of the material forming the strike plate so that when the strike is installed the outer face of the planar portion 10 is coplanar with the outer face of the jamb 1.

Since the axis about which the hole 23 and groove 24 are formed is closer to the edge 7 than the radius of the clearance hole 23, it will be noted that both the hole 23 and the groove 24 break out of the edge 7 as best seen in FIG. 1. Referring to FIG. 4 it will be seen that when the strike is installed the molding 4 may be placed against the free edge of the overhanging flange 11 so as to close up the opening in the edge 7 caused by boring the above mentioned hole and groove.

EXHIBIT 12 *(contd.)*

The above operation is conveniently performed by means of a boring tool shown in FIGS. 7 and 8. To the arbor **30** of said tool there is secured by means of screws **31** a relatively flat blade **32** which is shaped and formed with cutting edges to provide a pair

The above specifically described preferred forms of the invention should not be taken as restrictive since it will be apparent that various modifications in design may be resorted to without departing from the following claims.

35 I claim:
1. A strike for a door comprising:
 a circular strike plate provided with a peripherally extending flange at right angles thereto,
 a strike box formed integrally with said plate and having an
40 open side for receiving a latch bolt therethrough,
 said box being provided with one side adapted to be engaged by the corresponding face of a latch bolt positioned alongside said one side when such door is closed,
45 the side of said box opposite said one side being apertured to receive a securing screw therethrough.
2. A strike according to claim 1 wherein the portion of the annular juncture between said plate and said flange that is remote from said opposite side is formed with a slanting surface for engagement by the associated latch bolt when the door is
50 closing.
3. A strike according to claim 1 wherein said one side extends diametrally of said plate to provide maximum clearance for such latch bolt.
4. A strike assembly for a door jamb comprising:
55 a circular strike plate including a strike box integral therewith received by said jamb,
 said jamb being provided with a pair of coaxial bored holes,
 the larger of said holes being relatively shallow and adapted to receive said plate therein and the smaller of said holes
60 being relatively deep to receive said box therein,
 said plate being provided with a peripherally extending flange substantially equal in width to the depth of said larger hole and adapted to be received in said larger hole.

of cutting teeth **33** for forming the annular groove **24** and a lower cutting edge **35** for forming the clearance hole **23**. Between the cutting teeth **33** and the central portion of the blade **32** cutting edges **34** are provided for forming the outer face

 . . . etc.

5. In combination with a door jamb
 a circular strike plate including a strike box integral therewith received on said jamb,
 said plate being provided with a peripherally extending flange substantially at right angles to the plane of said plate,   5
 said jamb being formed with an annular groove for receiving said flange therein and being formed with a central clearance hole concentric with said groove and inwardly thereof for receiving said box therein.
6. A strike assembly for a door comprising: 10
 a door jamb,
 a circular strike plate including a strike box integral therewith received on said jamb,
 said plate being provided with a peripherally extending flange substantially at right angles to the plane of said plate,   15
 said jamb being formed with a counterbore for receiving said flange therein and being formed with a central clearance hole concentric with said counterbore and inwardly thereof for receiving said box therein.
7. A strike assembly comprising: 20
 a strike including an integral strike box,
 a door jamb having a straight edge adjacent the outer face of an associated door in closed position,
 a door stop on said jamb and spaced from said edge a distance substantially equal to the width of said door,   25
 the face of said jamb between said edge and said stop being provided with a bored hole having a radius greater than the spacing between said edge and the center of said hole whereby said hole breaks out through said edge of said jamb,
 said hole being of sufficient depth to receive said strike and 30 box therein with the outer face of said strike substantially coplanar with said face of said jamb,
 said strike being provided with a peripherally extending sidewall bottoming in said hole.

Claims 8 thru 16 have been deleted from this exhibit.

3,159,416

7

whereby said portions engage the side walls of said recess and resist outward movement of said plate in response to rebound forces impressed on said plate by said latch bolt.
17. A strike assembly adapted to cooperate with the latch bolt of an associated door comprising: 5
 a door jamb having a vertically extending straight edge adjacent the outer face of such door in closed position,
 a door stop on said jamb and spaced from said edge a distance substantially equal to the width of such door,
 the face of said jamb between said edge and said stop being 10 provided with a hole of sufficient depth to receive such latch bolt therein and having a radius greater than the spacing between said edge and the center of said hole whereby said hole breaks out through said edge of said jamb,
 a substantially planar strike plate secured to said jamb,   15
 said strike plate being formed with an arcuate peripheral portion having a center of curvature coincident with the center of said hole when said plate is so secured to said jamb,
 said plate being provided with an aperture in registration with said hole for receiving such latch bolt therethrough,

8

the periphery of said aperture including a straight edge extending parallel to said edge of said jamb and substantially through said center of curvature to permit optimum vertical clearance between said hole and the associated latch bolt, and
 means integral with said plate at said aperture extending into said hole for receiving a fastener for securing said plate to said jamb.

References Cited by the Examiner
UNITED STATES PATENTS

490,440	1/93	Jacobus	70—99
1,001,082	8/11	Samuelson	292—340
2,272,241	2/42	Fendring	292—340
2,401,854	6/46	Berry	292—341.9
2,861,660	11/58	Ensign	70—99 X
2,937,898	5/60	Russell	292—340
2,964,346	12/60	Check	292—340

20 ALBERT H. KAMPE, *Primary Examiner.*

economically. We knew this circular strike would be no exception, as it requires a deep draw of rather complicated shape, and a secondary operation is required to countersink the strike box for the fastening screw. By subcontracting the pilot run of the circular strike, the company would be able to test the marketability of the new product without having to commit expensive production facilities to the project at once. This plan would also eliminate some of the headaches involved in the development of production tooling.'

Bids were solicited from a number of local manufacturing companies, on lots of both 5 000 and 50 000 units. Of the 12 companies given the opportunity to bid, 10 actually did. There was wide variation in the bids received; the highest bid was $4\frac{1}{2}$ times the lowest. Larkin submitted the second lowest bid, with the lowest unit labour cost, meaning that their tooling estimate was higher than that of the lowest bidder. On 31 May, Larkin was awarded a contract to produce 5 000 of the circular strikes; they won the contract because it was felt that their more expensive tooling would mean less difficulty in producing the strikes within the required tolerances. It was agreed under the terms of the bidding that Schlage would supply the material from which the pilot run of strikes would be produced. During the three weeks that Larkin had indicated they would need for tooling, Schlage shipped them two large rolls of the proper size material (3 in wide, 0·035 in thick). One roll was a 'cartridge brass' (70% copper, 30% zinc), the other 'commercial bronze' (90% copper, 10% tin).

Larkin's low bid had been based on the conviction that they could produce the integral strike box with a single 'draw'. Drawing is a forming operation performed on sheet metal by a punch and die system located in a heavy 'punch press' which develops the forces necessary to deform the metal. A strip of sheet metal is fed between the punch and die by a mechanical feed system which is coordinated with the punch action. As the strip stops under the punch, a holding ring lowers against the sheet metal and clamps it. The punch is then forced into the sheet metal, which is deformed into the die. The holding ring must clamp just tightly enough that some of the metal is allowed to flow from under it without forming wrinkles as it moves toward the edge of the die cavity.

More than one draw may be required to achieve a finished piece, depending on the type of metal and its desired final shape. When Larkin began a test run with their single draw system, they were dismayed to find that the punch consistently broke through the bottom of the strike box. Larkin reported this problem to Schlage, suggesting that perhaps annealing the material would improve the results. The testing group at Schlage's Research Division carried out annealing tests on several Schlage production materials. They found from these tests that their 111-2350 cartridge brass could be formed in one draw when annealed for 15 min at 750°F (399°C); they also found that their 111-2500 brass could be formed in one draw without annealing.

About 600 lb of the 111-2500 brass was delivered to Larkin during the last week of June. Meanwhile, Larkin had proceeded to anneal the rest of the original roll of brass, after which the strikes could be formed in one draw; but an orange-peel effect was noticed on the material after drawing and the entire roll had to be scrapped. Five thousand strikes were successfully produced from the second shipment of material and delivered just in time to satisfy several urgent customer demands.

The pilot run of circular strikes was generally received with enthusiasm by Schlage customers, and the company decided to set up their own production facilities for the new item. Production Engineering designed a seven station progressive die to form the basic strike; a second operation formed the screw countersink. The new die was constructed by an eastern die shop which specializes in carbide dies and was put in service during July 1964 without further incident.

Mr Schlage later summarized the first circular strike project. 'The new strike proved to be very popular with several large pre-hung door mills, particularly because of its appearance, allowance for sag, and ease of installation. There were even indications that door mills which did not normally furnish Schlage locks would purchase this circular strike for use with other brands of locks, a practice unheard of in the industry.

'Some problems which later developed resulted from the inability of some of our customers to install the strikes as accurately as necessary; these customers were having problems with door rattle and with strikes tilting up due to improper fastening screw angle. Also, in a few instances where strikes were installed for use with thick doors, the latchbolt, during closing of the door, would contact the trim moulding before the strike lip, thus scoring the moulding. We suspected that we would eventually have to provide a strike which was not susceptible to these problems.'

Note: Part 3 describes efforts to modify the circular strike in the attempt to correct the above-mentioned problem.

Part 3   Modifying the New Strike Plate

In August 1963 as the new Schlage strike was first being delivered to the waiting distributors, Schlage initiated two new research projects concerned with further development of the circular strike. The purpose of the first project was to develop an adjustable circular strike; the second would investigate the possibility of modifying the original circular strike to be moulded of plastic. Both projects were envisioned as long-range ones in comparison to the fairly rapid development of the first circular strike, which had responded to a more urgent customer demand.

Later field results showed that adjustability for door rattle would be desirable due to the problems experienced by some mills in correctly locating the bored hole, but Mr Schlage had recognized this potential difficulty even before the marketing of the strike. 'It would no longer be possible to adjust the strike by removing, relocating the hole, and refastening. True, the circular strike could be rotated slightly to provide some adjustment, but this made it look haphazardly installed. Also, only one edge of the latchbolt then contacts the strike, causing a twisting movement of the bolt which might shorten the life of the lock.' While the title of this research project was 'Circular Strike, Adjustable' it was hoped that the resulting design might be applied to the standard Schlage rectangular strikes as well. In recognition of this fact, the name of the project was changed in November to 'Adjustable Strike'.

Neil Clumpner, a research engineer who had done some of the detail work on the first circular strike (in fact, he had prepared the final production drawing), began sketching possible designs for both projects. He concentrated his design efforts on trying to devise a method of adjustability which could be produced economically. Many adjustable strikes had been designed in the past, by Schlage and others; in fact, a search located over twenty patents concerned with lateral adjustment of strikes (illustrations from several of these patents are shown in Exhibit 13). However, few of these designs had ever been produced because of their high cost of manufacture; the cost criterion became even more important for the circular strike, because it was aimed at the pre-hung door manufacturers whose product was sold primarily to builders of inexpensive tract housing.

Mr Clumpner worked about half-time on the two projects through mid-October, sketching over 40 possible strike designs. As he sketched the various possibilities, he presented them to Mr Ralph Neary, a senior engineer under whom he was working. Mr Neary then selected the best of these for Mr Clumpner to prepare in more detail. When this had been done, Mr Neary would present them at a weekly Research meeting which would then make its recommendations as to whether or not the design

EXHIBIT 13   Patented adjustable strikes

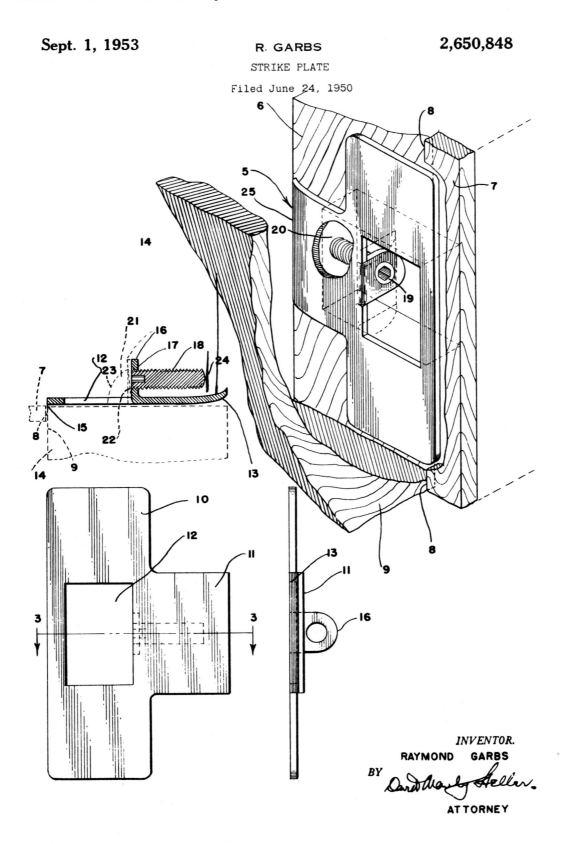

Sept. 1, 1953   R. GARBS   2,650,848

STRIKE PLATE

Filed June 24, 1950

INVENTOR.

RAYMOND   GARBS

BY

ATTORNEY

EXHIBIT 13 *(contd.)*

April 30, 1957   A. SCHOEPE   2,790,667

ADJUSTABLE STRIKE

Filed Feb. 8, 1954   2 Sheets—Sheet 2

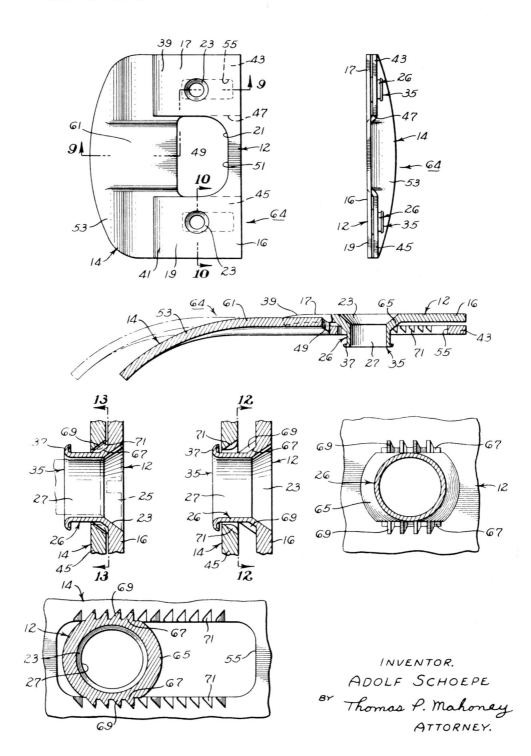

INVENTOR.

ADOLF SCHOEPE

BY *Thomas P. Mahoney*

ATTORNEY.

EXHIBIT 13 *(contd.)*

Oct. 31, 1961   J. ROYALTY   3,006,677

ADJUSTABLE STRIKE PLATES

Filed Nov. 27, 1959

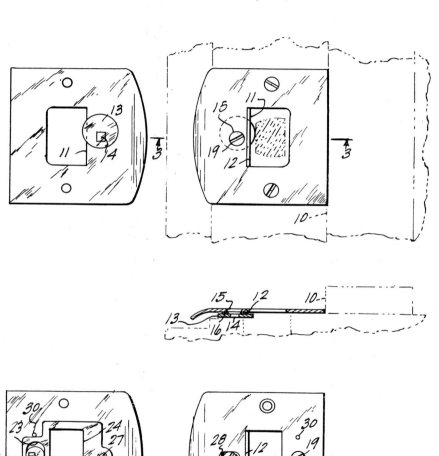

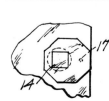

INVENTOR.

John Royalty

BY *Philip A. Smidell*

Attorney

EXHIBIT 14   Five concepts for adjustable strike

SCHLAGE LOCK COMPANY RESEARCH DIVISION	RESEARCH PROJECT: Cylindrical Strike, Adjustable		RP 6308 DATE: 8-1-63	SH. 1 OF

DESIGN NO.	DATE	DESCRIPTION (SKETCH)	RS. NO.	NO. OF MODELS	RT. NO.	PAT. FILE	REMARKS (CONCLUSIONS)	INIT.
		Note: Early sketches or copies of sketches of adjustable strikes are included in this file. These sketches are dated 1950, 1953, 1959 and 1960.						
1.	8/13/63				1.	145	Two pieces. Strike had no strike box, box was second piece. strike had adjustable lip.	NC
2.	8/16/63		0362 0361 0360 0363			146	Two pieces. Strike lip was movable within strike body	NC
3.	10/7/63		0383 0385 0384		1	163	Two pieces Strike lip. Metal with bendable adjustment tab. Strike body. Plastic. with and without sarrated perimeter.	NC
✳ 4	7/26/63	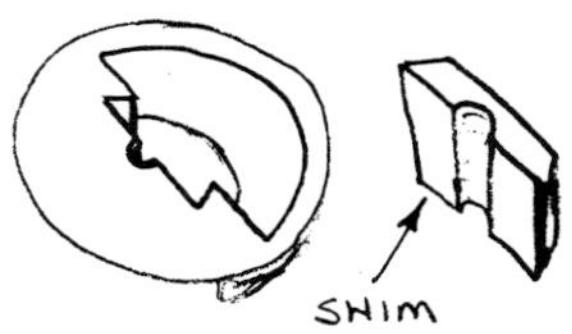 SHIM	SKETCH N.C.				PLASTIC. OR CAST METAL	✳
✳ 5	7/29/63	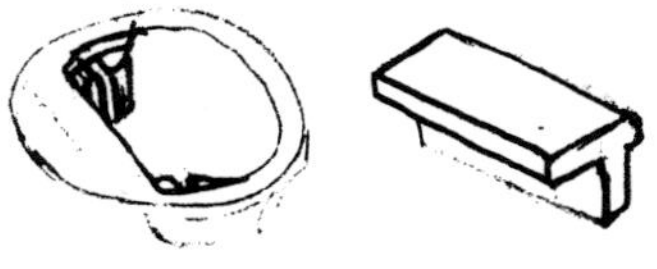	SKETCH N.C.				PLASTIC OR CAST METAL.	✳

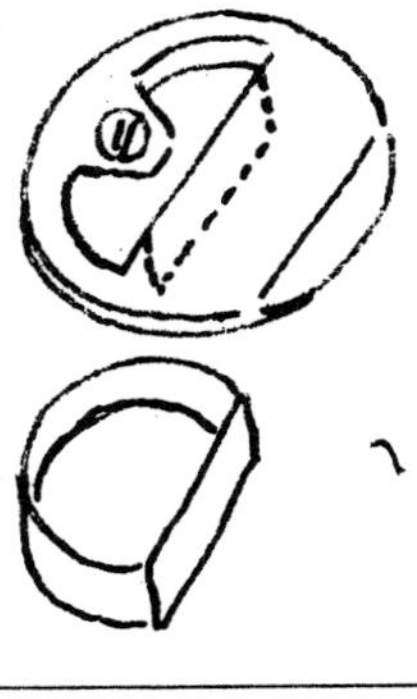

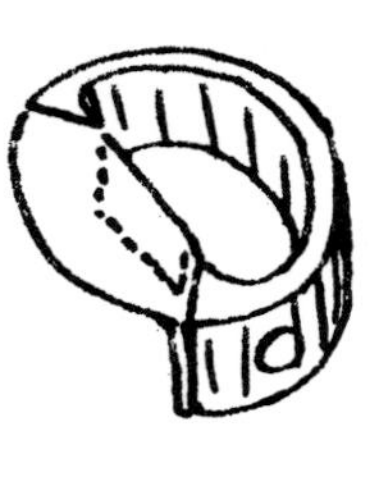

✳ INDICATES PARTS OF FILE TRANSFERRED TO RP 6501

should proceed to the modelling stage. Exhibit 14 shows the first five designs in the adjustable strike project which were carried to the detail drawing stage.

While several of Mr Clumpner's designs were modeled, none of them proved to be acceptable. An engineer in the testing group also submitted a few designs after carrying out tests on several earlier models. Mr Vern Bartels, present Manager of the Research Division, later commented, 'As a project proceeds through the stages of modeling and testing, engineers in these departments often come up with design ideas which they develop on their own.' However, none of these designs were modeled.

After October, both projects began to stagnate. The plastic strike project was soon closed, even though several strikes were successfully modelled and had performed well in tests. Indications at that time were that plastics still did not have a quality image before the public; also, fire conditions might soften the strikes and allow doors to open, permitting a fire to spread more easily. Schlage may again pursue the development of an all plastic strike in the future, particularly when injection moulding techniques for the heat-resisting thermosetting plastics are more fully developed.

The adjustable strike project slowed down due to personnel problems. Mr Clumpner left the Research Division in December; the project was reassigned in January 1964 to Dave LaField, a young engineer just hired by the company.

Mr LaField decided very quickly that the only way to make an inexpensive adjustable strike would be by using bendable tabs. He began by sketching possible 'tab' configurations for the standard rectangular strikes, then carried the same basic idea over to the circular strike. Several pages from his initial drawings are included as Exhibit 15. After developing this proposal, Mr LaField summarized his design considerations, in which he emphasized the importance of an adjustment which the homeowner could recognize (several years after installation) and perform with simple tools (hammer and screwdriver) without removing the strike. He also felt that an adjustable strike should not have separate parts which could be lost.

Due to a personality conflict with the then Research Manager, Mr LaField left the Research Division in late January and his proposal was not modeled. Several of the other engineers felt that his design had definite shortcomings. Since the tab was of heavier material than the strike body, it was felt that the body of the strike would be deformed during attempts at adjustment. It also appeared unlikely that the homeowner could achieve all of the configurations suggested, and the cost would certainly be quite high, as a new die would be required for the strike body, due to the enlarged strike box.

Nothing more was done on the project until April, when field reports began to realize Mr Schlage's premonitions concerning adjustability. Mr Schlage began to think about the adjustability problem himself and soon came up with the idea of simply punching two slots in the bolt receiving face of the present circular strike, as shown in Exhibit 16, to permit some adjustment for door rattle. Also shown is a sketch of a possible punching tool. It took only 5 hours to model this design; during May, extensive tests were carried out to measure its strength and adjustability. Discussion and results of the tests are included as Exhibit 17. Mr Schlage later commented on this design. 'It appeared that this modification would solve some adjustability problems, particularly door rattle. While the production people first said they could not cut the slots, Jim Maher in our model shop built a tool which cut them very nicely, so that we foresaw a very small additional production expense for adjustability. One drawback brought out by the tests was that door slams could change the adjustment, but this would be a problem with any scheme involving bending metal. Any piece adjusted by prying with a screwdriver will deform under a slam.'

After results of the tests were considered, the Research Department decided to release the modified strike for production. However, the then General Manager of the company, Mr Levinger, vetoed the modification on the grounds that field demand

EXHIBIT 15   Lafield's bent-tab adjustable strike

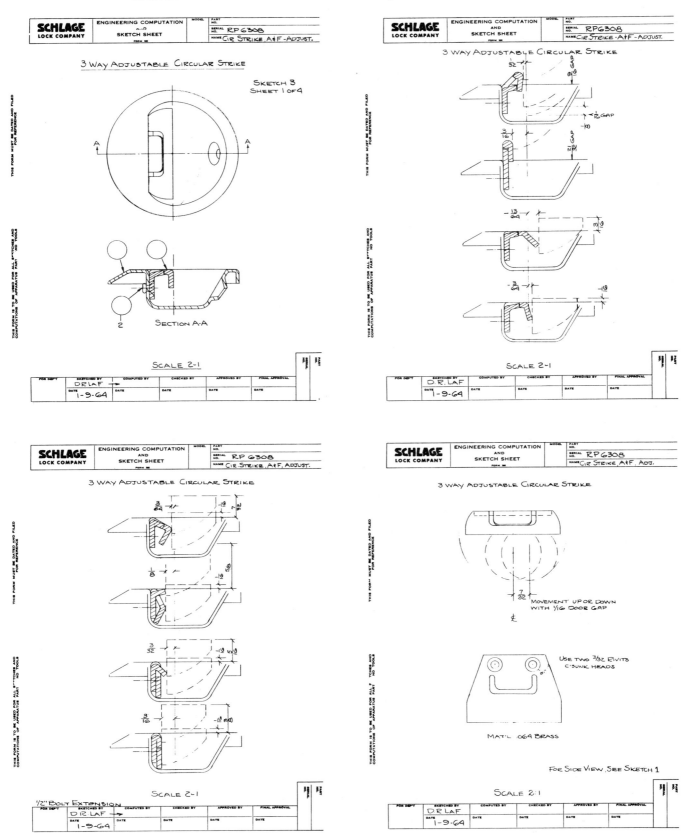

EXHIBIT 16   Slotted strike concept

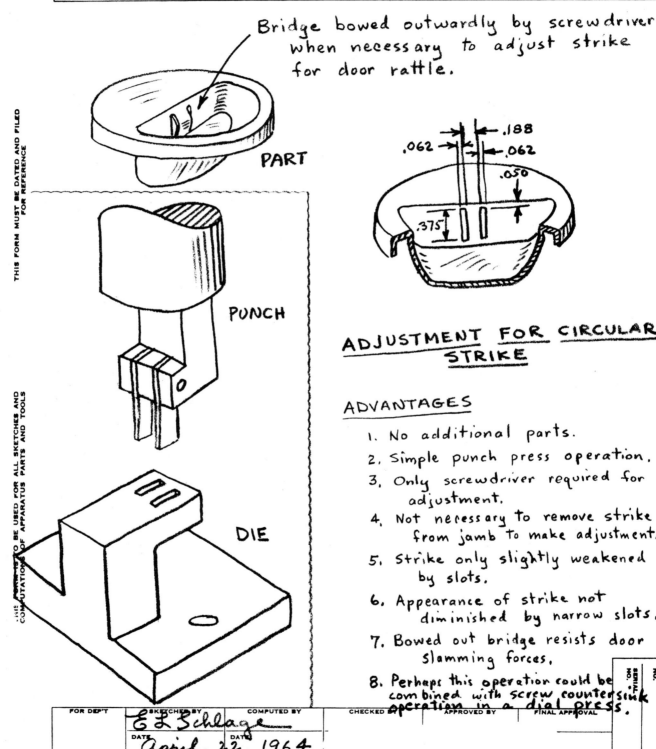

SCHLAGE LOCK COMPANY	ENGINEERING COMPUTATION AND SKETCH SHEET FORM 356	MODEL	ART).
		SERIAL NO.	
		NAME	PATENT FILE 391

Bridge bowed outwardly by screwdriver when necessary to adjust strike for door rattle.

PART

PUNCH

DIE

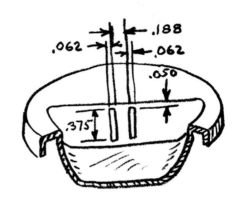

.188

.062   .062

.050

.375

ADJUSTMENT FOR CIRCULAR STRIKE

ADVANTAGES

1. No additional parts.
2. Simple punch press operation.
3. Only screwdriver required for adjustment.
4. Not necessary to remove strike from jamb to make adjustment.
5. Strike only slightly weakened by slots.
6. Appearance of strike not diminished by narrow slots.
7. Bowed out bridge resists door slamming forces.
8. Perhaps this operation could be combined with screw countersink operation in a dial press.

THIS FORM MUST BE DATED AND FILED FOR REFERENCE

THIS FORM TO BE USED FOR ALL SKETCHES AND COMPUTATIONS OF APPARATUS PARTS AND TOOLS

| FOR DEP'T | SKETCHED BY E L Schlage DATE April 22, 1964 | COMPUTED BY DATE | CHECKED BY | APPROVED BY | FINAL APPROVAL | SERIAL NO. | PART NO. |

EXHIBIT 17   Excerpts from test report on slotted strike

SCHLAGE LOCK COMPANY RESEARCH DIVISION	TEST REPORT	RT 196
		6-4-64 Page 3

Discussion:

Although it was possible to obtain .110" adjustment, in another case the adjustment bridge cracked and failed at .085". To prevent this cracking the maximum allowable adjustment should be 5/64". There was no difficulty in inserting a flat screwdriver blade and rotating it to obtain the desired adjustment. Of course, reasonable care must be exercised or the strike can be pried out of the door jamb or the bridge can be ripped by the screwdriver blade.

If the adjustment is made at the strike installation, a Phillips screwdriver is used for installation and a flat blade screwdriver for adjustment. This should not be objectionable.

Both light and medium slams depressed the adjustment bridge but the test door was 1¾" x 32" solid core. As a measure of slam severity, on F 26 a medium slam of this size and weight door cracked the door jamb. Of course, the condition of the jamb, such as wood grain, is also a factor.

The peak of the bridge is approximately ¼" from the strike surface and rotating the screwdriver gives this shape to the bridge:

This means that with a ⅛" or larger door gap and using a ⅜" throw latch unit the tip of the bolt will rest short of or at the bridge peak. In the warped door test the tension against the flat of the latch bolt caused the bolt to retract and the door to open. This same condition would exist on any strike with bendable tab except the tab surface would not be concave so the bolt would be contacting the tab over a larger surface.

Conclusion:

The adjustable circular strike will adjust to a maximum 5/64" and will probably prove most satisfactory on hollow core inside doors.

Reasonable care must be exercised when making the adjustment to prevent prying the strike out of the jamb or ripping the adjustment bridge with the screwdriver blade.

A badly warped door might not stay latched if the maximum adjustment was made in the strike. Tension against the flat of the latchbolt resting on an oblique surface might cause the bolt to retract itself. The maximum adjustment and warping would not exist together initially but if a maximum adjustment has been made for rattle, the door can warp later. Then the two conditions would exist together.

The adjustable bridge can be reset several times without metal failure.

Howard Scheffel
Howard Scheffel

wasn't yet strong enough and the cost was too great; he felt that adjustment by rotation was adequate. The project was then closed.

In February 1965 the Product Committee, composed of the General Manager and leaders from sales and all three engineering groups, recommended that Mr Schlage's 'bulge-type' (slotted) adjustment feature be improved for marketing; a new research project was set up, entitled 'Circular Strike, Adjustable'. Several variations of the slotted arrangement were tried, but the pilot run of 5 000, produced in early May, was essentially identical to the first design. The project was placed on inactive status for six months to allow time for evaluation in the field.

In the November Product Committee Meeting, the Sales Manager presented a summary of customer reaction to the modified strike; the general conclusion was that, while it was an improvement over the original circular strike, the customer would prefer adjustment in two directions (for both door rattle and failure to latch, rather than just the former). The sales manager also noted that the slotted strike did nothing to prevent the strike lip from tilting up as a result of improper fastening screw angle. As a result of these considerations, the slotted strike did not go into production, and the adjustable strike project suffered another slow-down. Mr Schlage later commented, 'At that point we appeared to be at an impasse. The general manager of the company had already expressed his view that even the slotted adjustability feature would be too expensive; yet now he could not understand why we had not developed a "better" solution. He now requested the Research and the Product Design engineering divisions to develop completely new design strikes from an entirely new aspect which would remedy the problems encountered with the original circular strike.' The requirements of this new strike can be summarized as follows:

1. Retain the single bored hole installation, large sag allowance, pleasing appearance, low cost and adequate strength.
2. Provide a bolt holding surface in the strike which can be adjusted to permit latching without rattle.
3. The bolt holding surface must be easily adjusted in the field and must retain its position even after moderate door slams.
4. Overcome the occasional problem of the latchbolt contacting the wood molding before the metal strike plate during closing of the door.
5. Overcome the tendency of the strike plate to lift out of its recess in the jamb upon tightening of an improperly directed fastening screw.

Note: Part 4 describes the work culminating in the development and adoption of the adjustable circular strike which satisfied all the above-mentioned requirements.

The Adjustable Circular Strike   PART 4

In the first quarter of 1966, both research and product design worked on separate versions of an adjustable circular strike. As the weeks went by, the general manager grew more and more impatient for a solution to the problem. During this time the Research Department came up with a scheme for adjustability involving a movable slide to contact the latchbolt. Earlier versions of this idea had apparently been conceived by engineers in the model shop; the U-shaped metal slide arrangement sketched in Exhibit 18 was designed by the Research Department Manager. A model of a movable slide was made which was presented to the Product Committee in March. This model was installed with two screws to overcome the tilt-up problem.

Management response to the model was mixed. In general they thought it was

EXHIBIT 18   Movable slide concept

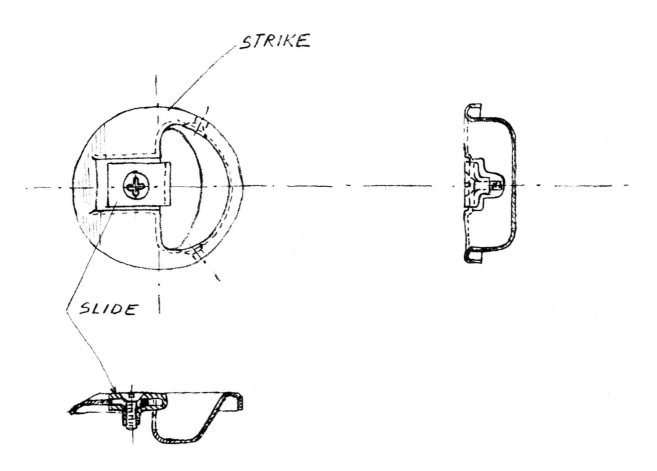

STRIKE

SLIDE

2-21-66

RP 6501

1:1
CIRCULAR STRIKE
ADJUSTABLE

RS- 1262

EXHIBIT 19   First serrated slide concept

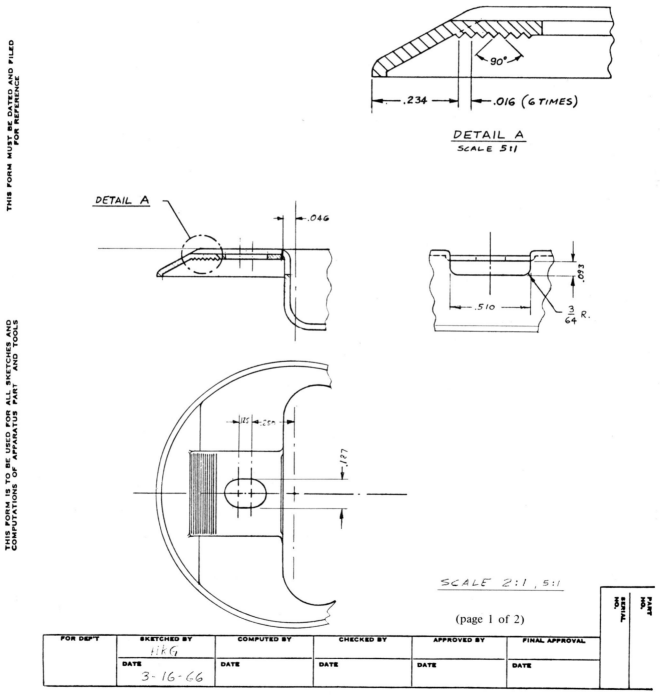

(page 1 of 2)

EXHIBIT 19 *(contd.)*

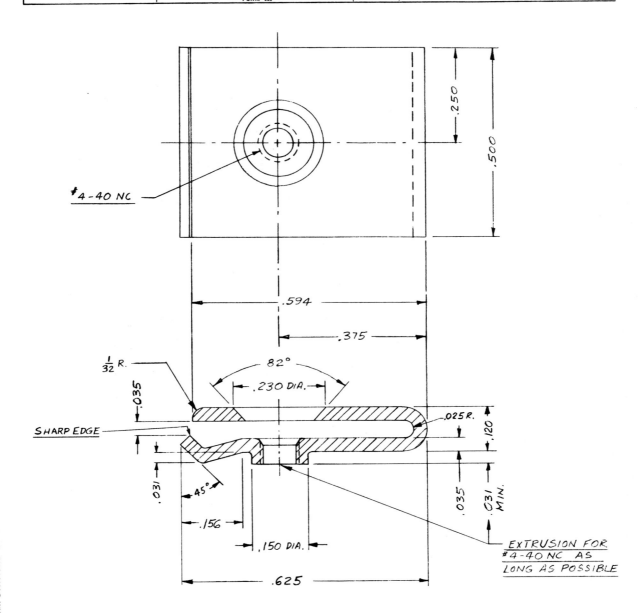

SCHLAGE LOCK COMPANY	ENGINEERING COMPUTATION AND SKETCH SHEET FORM 356	MODEL	PART NO.	RS-1278
			SERIAL NO.	RP 6501
			NAME	SLIDE

‡4-40 NC

.250

.500

.594

.375

1/32 R.

82°

.230 DIA.

.025 R.

.035

.120

SHARP EDGE

.031

45°

.035

.031 MIN.

.156

.150 DIA.

EXTRUSION FOR #4-40 NC AS LONG AS POSSIBLE

.625

SCALE 5:1   MATERIAL: .035 BRASS

(page 2 of 2)

THIS FORM MUST BE DATED AND FILED FOR REFERENCE

THIS FORM IS TO BE USED FOR ALL SKETCHES AND COMPUTATIONS OF APPARATUS PARTS AND TOOLS

FOR DEP'T	SKETCHED BY HKG	COMPUTED BY	CHECKED BY	APPROVED BY	FINAL APPROVAL	PART NO. RS-1278
	DATE 3-16-66	DATE	DATE	DATE	DATE	SERIAL NO. RP6501

the best solution yet presented, but they were worried about the cost. Research was told not to exhibit the strike to sales until production estimates were available. However, on the basis of this model, the project was tentatively reassigned to Research and the strike was further developed. On the basis of door slam tests, serrations were added between the slide and strike body to prevent slipping of the slide (Exhibit 19).

One of the Schlage's largest customers was having particular difficulty with the circular strikes; they had used several thousand slotted strikes but still were not completely satisfied and threatened to change to a competitive brand of hardware. In June Mr Schlage made a trip East to attempt to pacify this customer; on seeing the serrated-slide adjustable strike, the customer's people were most impressed and stated they would continue to handle the Schlage line if this strike were made available to them as soon as possible. At Product Committee meetings during Mr Schlage's absence, however, the general manager summarized his views on the adjustable strike, stating that he felt no 'simple and economical' solution had been presented to management; he advocated purchasing rights to a competitive strike very similar to that shown in the Russell patent (Exhibit 7) but with a bendable tab to engage the latchbolt for adjustment. Since their strike had no integral strike box, it could be made of heavier metal; thus the tab was less apt to bend (changing the adjustment) under a heavy slam.

When Mr Schlage returned and learned of the impending negotiations over the competitive strike, he communicated with the general manager and convinced him that such negotiations were unnecessary, as Schlage should retain the large allowance for door sag; if this were done, they need not fear infringement. Before a decision was made between a simple bent-tab scheme and the serrated slide, the product design group submitted their version of the adjustable circular strike. This model, shown at the left in Exhibit 20(a), was made of thicker material (0·050 in) and had no strike box. It fit into a deeper hole bored to a single diameter, and had a bendable tab for adjustment. Indications were that this strike could be produced for only slightly more than the original circular strike; the Product Committee decided to see how this strike would be received by the same customer recently visited by Mr Schlage. The customer found this new version acceptable, and Product Design was authorized to complete development of the strike and order tooling for a pilot run as soon as the drawings were finalized. The project was scheduled for completion in February 1967. Meanwhile, the Research strike was put on a six months' inactive status. A month later, however, management had a change of heart, and the serrated-slide strike was given first priority.

With the project now back at the drawing boards of the Research Division, it was still to be 9 months before the final 'freezing' of the design. During this time, a major choice was made between the integral-strike box design (Exhibit 20(b)) and a compromise strike which was quickly modeled by the Research Division and was essentially the Product Design version but with the substitution of a plastic slider for the bendable tab (Exhibit 20(a) right). The former version was chosen for several reasons; for one, the immediate demands of the customers could be most quickly satisfied by producing an interim strike which would be made from a non-adjustable circular strike reworked to include a movable slide. The expensive non-adjustable strike dies could be used to perform the first manufacturing step. The main reason, however, was probably the higher quality appearance of the integral strike box version.

Decisions on details of the strike were being made both before and after the major choice concerning the integral strike box. Many months were spent on selection of the slider material; after a great deal of pricing and testing, black delrin was chosen. This plastic performed very well in the tests, and would give a slight reduction in contact noise as compared to metal. Design and testing time was also spent determining the best location for the two fastening screws. This work was completed before the pilot run was started in February. Many design modifications were made to the strike

EXHIBIT 20(a)   **Product design strike and compromise strike**

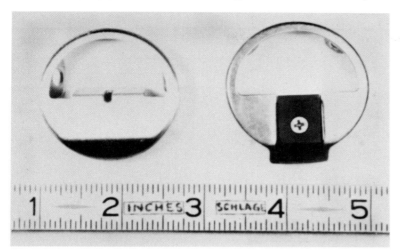

EXHIBIT 20(b)   **Pilot run sample of movable slide strike**

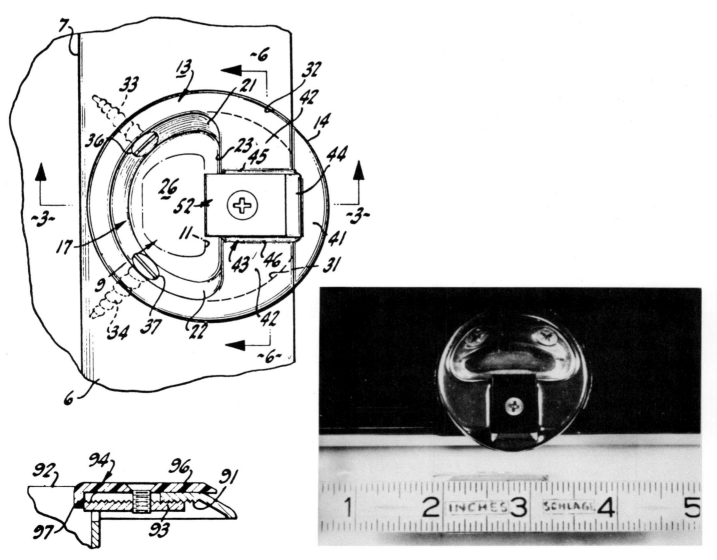

EXHIBIT 21   Final design

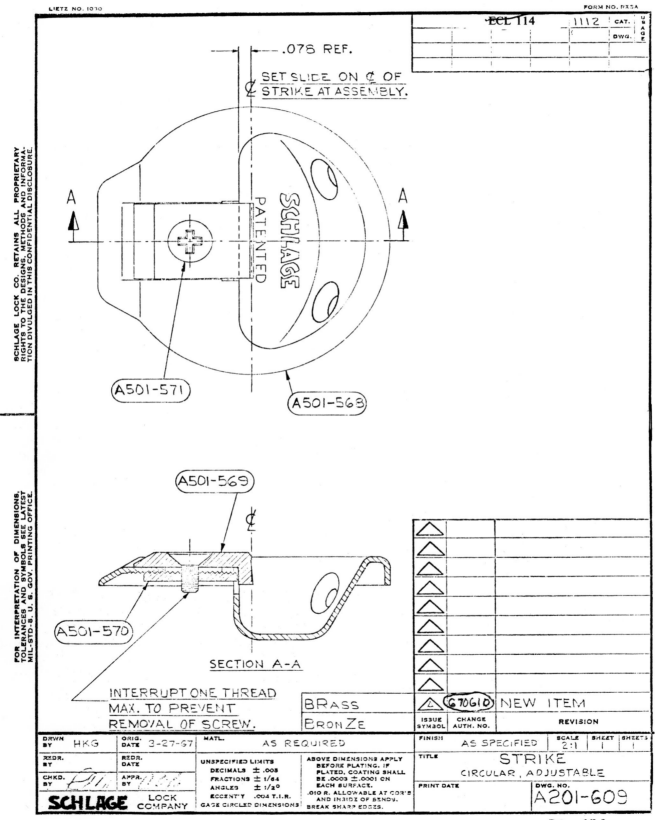

body before the freezing of the final production model in June 1967; these included an enlarging of the strike box and the addition of a short lip (the strike would no longer be completely circular) to prevent the latchbolt from striking the wood trim strip before the strike. The drawing of the strike body and the slider are included as Exhibit 21.

An evaluation of the pilot run indicated the adjustable circular strike would be a very popular item. After completion of the design in July 1967, work began on the production dies, including a new 14-station progressive die for the strike body. The plastic slides were being purchased from a vendor. As individual punch and die sets were completed, sample parts were made from them and checked against the specifications. If some dimensions were outside the tolerances, either the dies or the drawings had to be modified, depending on the location and seriousness of the misfit. Tooling for production was a long, slow process; in August of 1968, the main die was just being completed; full production was scheduled to begin in late 1968. A second patent application was filed covering the improvements made to the circular strike.

On looking back over the project, Mr Schlage commented, 'To most of my non-engineer colleagues, the amount of time, effort, talent and knowledge required to develop such an apparently simple piece of bent metal was greatly underestimated. Actually many disciplines were involved in its design. First, there was a familiarity with the product line, its performance in the field and its relationship with competitive lines. Salesmen's reports and customers' letters were analysed to obtain knowledge of changing conditions in the field which required product modifications or new items. The ability to invent and solve problems was present as well as a thorough understanding of issued patents relating to the product. The engineer took into account the capabilities of the production machinery and processes in his factory and knew from experience and training the physical properties of the materials which he selected for the product. Accurate drafting was prerequisite to the graphical determination of forces and the establishment of dimensions, clearances, and tolerances necessary for interchangeable manufacture. An aquaintance with labour, material and overhead costs as well as amortization of equipment and tooling costs was necessary when preparing product proposals. Also required was a familiarity with testing, aesthetics or industrial design, standards, purchasing practice, and field surveys. Finally, since the engineer is a member of a team, he had to demonstrate the ability to communicate, cooperate and get along well with others in his company.

'The delays, misunderstandings, failures, obstacles, frustrations and the duplication of effort in other departments were taken philosophically, and who knows but what they contribute in some mysterious way to the ultimate success of the final product.

'An unexpected benefit emerged from this project which promises to be very profitable for the company. The strike plate furnished with the vast majority of locks has the conventional Tee-shape and is stamped from heavy gauge brass. The strike box is a separate piece of zinc dichromated steel intricately folded by a multi-slide machine. This project has now taught us how to make this Tee strike of thin sheet brass. Sufficient strength is obtained by drawing down a peripheral skirt and incorporating an integral drawn strike box as was done on the circular strike. The savings in material costs are substantial and would even pay for the addition of the adjustable slide if desired.'

SUGGESTED STUDY

This case emphasizes the quality of engineering which can and must be applied to seemingly very simple familiar objects. Grandiose exotic projects are very useful in stimulating students' enthusiasm and imagination, but they need an antidote to arrive at a reasonably undistorted view of engineering. This case provides such an antidote. Mr Schlage's remarks quoted at the end of Part 4 summarize the technical and human content of the case.

Each part of the case suggests various questions and exercises. We mention a few:

Part 1:

1.1. Discuss the economic and engineering implications of pre-hung doors and prefabricated building construction as contrasted to the traditional on-site construction. *An exercise in engineering economic thinking suitable for class discussion.*

1.2. Give several examples of industries which are based on or which were revolutionized by inventions and name the inventions. *This exercise is intended to put engineering in perspective.*

1.3. Define terms: mortise lock, cylindrical lock, spring latch, deadlatch, dead bolt, strike plate, doorjamb, and doorstop. *Required for understanding of the case.*

1.4. Discuss the subject of patents, particularly their advantages, term in years, infringement, searches, drawings, specifications, claims, elements, and cited references.

1.5. Discuss the advisability of an engineer entering his thoughts, ideas, designs, and inventions in a notebook. *The extension to student projects is obvious.*

1.6. Invent six methods of latching doors which avoid the need of strike plates. These methods need not be practical, but should be feasible. *An exercise in invention. The latches of garden gates, automobile doors, suitcases, etc., can be used to provide starting points. Sketching or describing the inventions is a good exercise in communication.*

1.7. Sketch six strikes which meet the criteria set up by Mr Schlage. Pick the most promising two of the six. *An exercise in invention and communication similar to 1.6, but less freewheeling. The differences between some of the six methods are probably going to consist of details.*

1.8. Estimate the rebound force on the strike. Explain your method of estimation. *This requires fairly sophisticated reasoning, based on knowledge of dynamics, elasticity, and properties of materials. The estimate can be checked roughly by the knowledge that strikes and doors do withstand the rebound forces.*

Part 2:

2.1. Sketch and dimension the punch for the strike A501-883 shown in Exhibit 11. *A fairly straightforward exercise in mechanical drawing. Note that the 0·168 in diameter screw hole and 82 degree countersink will require a separate tool, which could be the subject of a separate exercise.*

2.2. What are the forces applied to the door by the kick test shown in Exhibit 9. *An exercise in applied mechanics. The forces depend mainly on the stiffness of the point at which the impacter leans on the door. This must be estimated.*

2.3. Propose a test to determine wear of strike plates. How many cycles do you think they should withstand? What shall determine 'failure'. *This is a wide-open design problem.*

2.4. What further improvements should be made on the strike plate? How would you accomplish them? *This is the main question of the case. It can be asked only if the students have not yet seen Parts 3 or 4.*

2.5. Propose standards, equipment, and procedures for testing a door lock to determine its resistance to corrosion, or to kicking, pulling, twisting, wearing, drilling, prying, hammering, picking the cylinder, or shimming the latch bolt. *Each of these questions is a good exercise in itself.*

2.6. What are the advantages of producing a pilot run of a product prior to tooling up for mass production? *Note that the pilot run almost always brings out points which* *were overlooked during the design of the product or of the tooling. Hardly ever is foresight equal to hindsight. The pilot run is one method of attaining hindsight quickly and at relatively small expense.*

Part 3:

3.1. Specify desirable properties for materials for the adjustable strike plates shown in Exhibits 15 and 16. Select real materials which meet these requirements as closely as possible. *A combination of ductility and rigidity is desired. Use of heat to increase ductility might be considered. Cost is a major consideration. The question emphasizes the need for balancing conflicting requirements.*

3.2. Sketch several other adjustable strike plates. *Another exercise in invention and communication.*

3.3. Discuss the following activities of an engineering department: inventing, designing, modeling, testing, and costing. *Class discussion will lead to the consideration of career choices by engineering functions rather than by fields.*

3.4. What factors should be considered before substituting plastic for metal? *Modulus of elasticity, creep, and electric conductivity have at times been overlooked.*

3.5. When considering a new product for manufacture, how does the engineering department's viewpoint differ from that of other departments in the organization? *Another wide-open question for class discussion.*

Part 4:

4.1. The case refers several times to 'higher quality appearance'; explain how one of the designs 'appears' to be of higher quality. State whether this appearance corresponds to a functional advantage or not. If yes, how?

4.2. Does the design shown in Exhibit 21 meet the specifications listed at the end of Part 4? Explain.

4.3. Make a time table of the whole circular strike project. Indicate where the 'delays' and 'duplication of effort' mentioned on page 369 are in the time table. *Try a PERT chart.*

4.4. Sketch the T-strike with adjustable slide mentioned in the last paragraph.

4.5. Make a fully dimensioned drawing of the T-strike with adjustable slide.

4.6. Compare the design considerations involved in a small, apparently simple, mass production item like the Schlage strike plate with those involved in the design of one spacecraft.

4.7. If you had been assigned the design of an improved strike plate, how would you have proceeded with the benefit of the hindsight you now have? Consider both technical and the human aspects of the problem.

4.8. Discuss each of the following terms and their relevance to the design and introduction of a product: interchangeable manufacturing; dimensions; tolerances; clearances; drafting; drawing; changes; competitors' patents; owned patents; aesthetics; committee meetings; communications; customers; competitors; salesmen; and top management.

Case 10

Development of the Mariner C solar panel deployment system

by R. Kerr, R. Weitzmann, and C. Yokomizo

Prepared under the supervision of Professor R. F. Steidel, Jr. at the University of California at Berkeley with the aid of a grant from the Ford Foundation. Condensed by Sue Hays with support from the National Science Foundation for the Stanford Engineering Case Program. Assistance from the Jet Propulsion Laboratory of the National Aeronautics and Space Administration is gratefully acknowledged.

This case is available in pamphlet form as ECL 89 from the Engineering Case Library, Room 500, Stanford University, Stanford, California 94305.

PART 1   Background

In April 1962 the Jet Propulsion Laboratory (JPL)* completed the preliminary design of a spacecraft known as Mariner B. The National Aeronautics and Space Administration (NASA) planned to send the Mariner B to Mars in 1964 using a Centaur boost vehicle. Soon after JPL began detailed design work on the spacecraft, difficulties arose with the Centaur. NASA then asked JPL to investigate the feasibility of building a spacecraft which an Agena D could boost. A study showed that this was possible and JPL was told to proceed on the program. Mars would come within shooting range of the Agena D during November 1964, and would not be within range again for another 25 months. JPL would have to deliver three assembled spacecraft to Cape Kennedy in September 1964. There would be less than $2\frac{1}{2}$ years in which to develop the Mariner C. Since the Agena D had only about half the thrust of the Centaur, the new spacecraft (to be called Mariner C) would be severely weight constrained.

JPL, which is operated by the California Institute of Technology under a NASA contract, develops unmanned spacecraft to aid in increasing scientific knowledge of lunar and interplanetary space. The Laboratory has seven technical divisions: Systems, Space Science, Telecommunications, Guidance and Control, Engineering Mechanics, Engineering (test) Facilities, and Propulsion. In addition, each major project has its own organization. (Exhibit 1 shows part of the organization for the Mariner C project.) At the time NASA asked JPL to consider developing the Mariner C, previous JPL projects had included the Explorer program, which produced the first United States satellite, the Pioneer program, the Mariner II program, and the Ranger program, which resulted in a rough landing attempt on the moon. A Ranger program to photograph the lunar surface would proceed simultaneously with and have priority over the Mariner C project. This would complicate the scheduling of test facilities for the Mariner C.

The conceptual design of a spacecraft requires interaction between people representing many technical disciplines and interests. At JPL, such efforts are coordinated by a Spacecraft System Project Engineer. The man assigned to head up the conceptual design on Mariner C was Mr John Casani. The Mariner C design team adapted a unique approach for defining the mission in greater detail. Every morning, a list of tasks for Mariner C to perform on its mission would be outlined. During the day the team would discuss such aspects of the proposed mission as feasibility, power, and weight limitations. A new mission would then be outlined for consideration the next day.

Detailed study was limited to a 'fly-by' mission; that is, to a mission which would orbit the spacecraft relatively close to Mars. This type of mission was thought to be quite likely to succeed and to yield more desirable data than missions which would place the spacecraft on an impact course with Mars or place it in a very large orbit about Mars. Originally, the design team constrained itself to working with modifications of spacecraft such as Mariner II and Ranger. Such an approach is an economical one if the desired performance can be obtained. However, after a week of considering various fly-by requirements, it became apparent to the design team that more than simple modifications would be required to meet the stringent weight limitations.

The Mariner C Project Engineer for the Spacecraft Development Section of the Engineering Mechanics Division was Mr James Wilson. This Division had responsibility for spacecraft structures, mechanisms, temperature control, electronic packaging and cabling, a large part of the configurational design, ground handling equipment, and mechanical design support to other JPL technical divisions. After a

* The efforts described in this case study represent the results of one phase of research carried out at the Jet Propulsion Laboratory, California Institute of Technology, under Contract No. NAS 7-100, sponsored by the National Aeronautics and Space Administration.

EXHIBIT 1   Mariner C Solar panel development and support

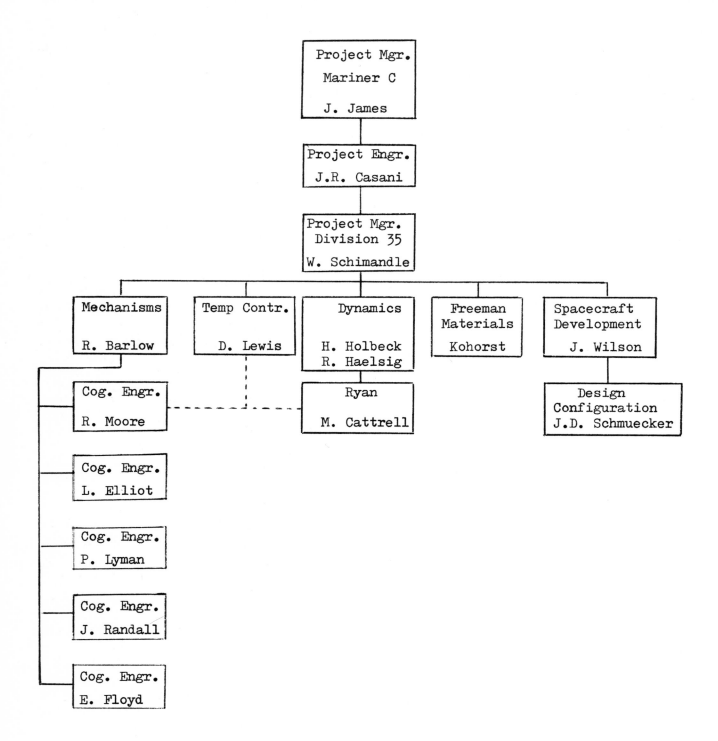

EXHIBIT 2   Mariner C sliding concentric tube damper

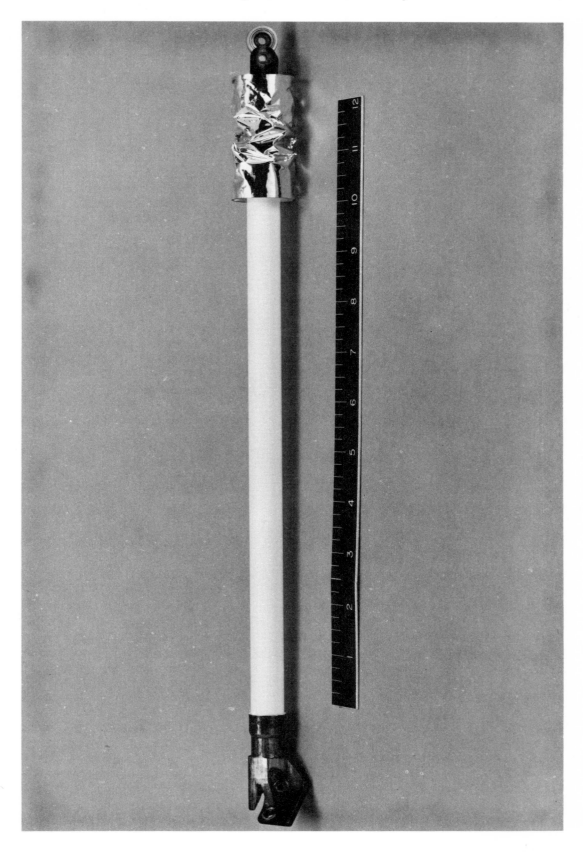

EXHIBIT 3   Mariner C spacecraft folded

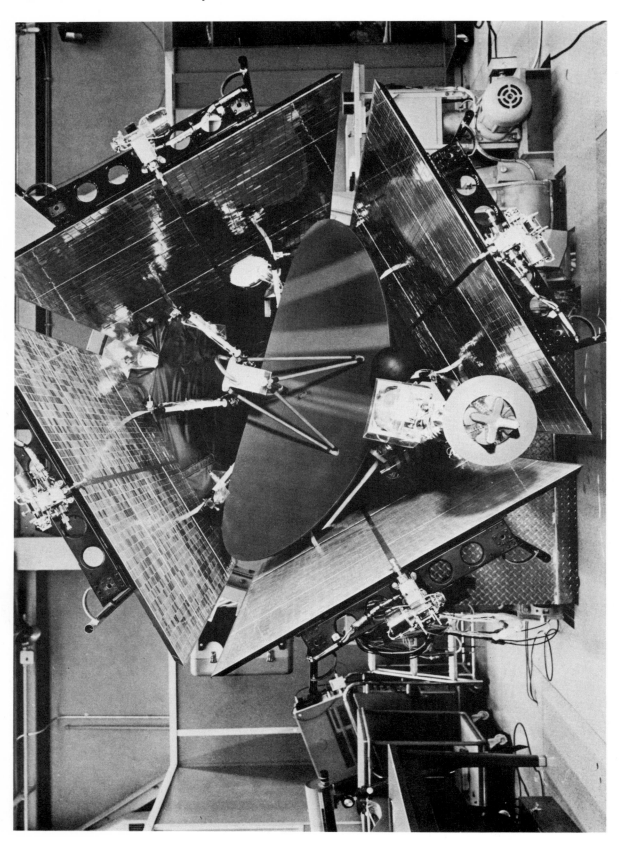

EXHIBIT 4   Mariner C spacecraft with solar panels deployed

week of preliminary design effort, Mr Wilson told Mr Casani that he felt that 50 lb of weight could be saved by going to a fresh configuration rather than attempting to retain Mariner II and Ranger structural concepts. Mr Casani gave Mr Wilson the go-ahead on a Friday afternoon, and Mr Wilson's section spent most of the week-end looking at new configurations. Quick layouts and weight calculations verified the estimated weight savings. The system weight could be reduced to a value within the capability of the Atlas Agena D by using some rather radical structural approaches.

Because of decreased solar intensity in the vicinity of Mars solar power sources for Mars spacecraft assume increasing importance from a weight standpoint. During the early phases of the Mariner C study, nuclear power was considered. However, because of the problems of nuclear integration, the relatively small amount of power required, and successful experience with solar power, the design team decided quite early to use solar panels on the Mariner C. Mr Wilson's section was responsible for the design of the geometrical configuration and structure of the solar panels. The electrical aspects of the solar panel design and fabrication were the responsibility of the Guidance and Control Division. The engineers involved took into consideration the performance of solar panels on previous spacecraft as well as advances in the state of the art. They decided to use four 3-ft by 6-ft rectangular panels. Attitude control jets would be mounted on the tips of the panels rather than on the body of the spacecraft. Since the jets would have greater moment arms, the required jet force would be less, and 15 lb of spacecraft weight would be saved.

As detailed design of the Mariner C progressed, the design began to 'gain weight'; most spacecraft designs do. When the first weight crisis arose in late 1962, Mr William Layman, an engineer in the Engineering Mechanics Division, had an idea about a way to reduce weight. He estimated that the weight of the solar panels could be halved if point-damped supports were used to control their fundamental resonance during the initial boost. Mr Layman had developed a sliding concentric tube damper (see Exhibit 2) to reduce the resonance displacements of an antenna on the Mariner II. Layman had later suggested that the damper be used to support the solar panels on the Ranger spacecraft. Although vibration tests verified that the dampers would reduce panel vibration, problems arose in developing the sliding tube dampers, and they had not been flown on the Ranger.

The structural engineers were the chief objectors to the idea of using damped solar panels on the Mariner C. They argued that their analytic capabilities were limited to lightly damped structures with linear characteristics. Layman's proposal involved heavy, nonlinear damping. Furthermore, no previous spacecraft system had had its success contingent upon the successful operation of dampers. Their use involved some risk and appeared to violate Mariner C project guidelines.

Mr Layman, however, was able to convince Mr Wilson that the solar panels should be damped. The two of them convinced Section Chief William Schimandle. Because Mr Schimandle was also Project Manager for the Division, he was able to override the protests of the structural engineers. When the use of damped solar panels was recommended to the Project Office, the Office decided to accept the risk involved. In December 1962, the Office modified the proposed plan for the solar panel system to include the use of dampers.

The spacecraft would be an octagonal structure with four damped solar panels extending from the structure to form a cross. Radio antennas would be located above the octagon. Optical equipment for taking television pictures of Mars and for tracking Canopus, the reference star, would be located below the octagon. The spacecraft would be about $9\frac{1}{2}$ ft in height, 5 ft in diameter and about $22\frac{1}{2}$ ft from panel tip to panel tip. See Exhibits 3 and 4 for photographs.

PART 2   The Deployment Systems

It was crucial to the success of the Mariner C mission that the solar panels deploy properly and that they not be damaged. Not only did the panels provide the main source of power for the spacecraft, but also the attitude control jets would be mounted to the tips of the panels. Unless all four panels deployed satisfactorily, it would be impossible to stabilize the spacecraft properly. Mechanisms such as those used to deploy solar panels are the proverbial nail in the horseshoe. If they fail, the entire mission may be lost. In the case of Mariner C, the penalty would have been $125 000 000 and two years' delay in the exploration of Mars. As a result, development of spacecraft mechanisms is never taken lightly. However, mechanism design and development is a very tricky business. No matter how conceptually simple, performance is seldom completely predictable.

The panel deployment system used by JPL on the Ranger and Mariner II had worked satisfactorily, and JPL engineers originally intended to use the same system, with a few minor changes in hardware, on the Mariner C. This deployment system was a set of hydraulically damped linear actuators – one for each solar panel. Each actuator consisted of a compression spring supplying force to a push rod with silicone oil as the damping fluid.

In addition to the damped actuators used for deployment, the solar panels would need 'boost dampers' and 'cruise dampers'. Two boost dampers would hold each solar panel in a vertical position on top of Mariner's octagonal base structure during the boost. These dampers would protect the delicate panels from excessive vibration. When the panels reached their full deployment positions, they would latch onto cruise dampers which would keep the panels from 'flapping' during the mid-course maneuver.

As development of the Mariner C system progressed, many engineers began to worry about the existing linear solar panel actuators. They were afraid that the silicone oil in the actuators might leak during the long flight (eight months) and contaminate the temperature control surfaces and the optical instruments. In addition, the engineers discovered that they would have to extend the stroke of the actuators to adapt them to the Mariner C. This meant adding an additional linkage to each actuator thereby increasing weight.

During the development of the Mariner II, an earlier JPL spacecraft, Mr Richard Moore of Mr Schimandle's section had tried to think of alternate ways of deploying the solar panels. One concept which occurred to him was to use a torsion bar actuator to supply the deployment torque and a centrifugal friction brake to damp the panel motion. Each torsion bar and each friction brake would have one end mounted to a solar panel and the other to the octagonal base structure. A planetary gear box would multiply the panel velocity, in order to produce a centrifugal force which would cause friction pads in the brake to contact a brake lining. This friction damping would hold the deployment velocity constant.

This scheme seemed potentially advantageous because of its rotary operation. Mr Moore was therefore told to design such an actuator to be used on the Mariner C.

In January 1963 Mr Moore became Cognizant Engineer for the entire panel system. At that time the status of the deployment system was indicated by the following definition:

> 'Four panels 3 by 6 ft, hinged at one end to the octagonal base structure of the spacecraft, are to be deployed simultaneously at a controlled rate through approximately 90°. Deployment shall be completed within 45 sec after the panels are released from the boost dampers by explosive squib pin pullers. The initial design of the torsion bar actuator is to be continued and, once developed, shall be used to deploy the panels.
>
> Dynamic coupling between the panels and the base structure shall be provided by means of a damper for each panel to specifications which will be given

by the Dynamics Group. Contact between the panel and the damper shall be established by latching the deploying panel to the damper.'

For the most part, the deployment performance requirements (45 sec) were based on experience with previous spacecraft. As Mr Barlow, Mr Moore's group leader, said, 'This is the way we had always done it, earlier missions had successfully been flown with these specifications under similar panel configurations. After a program has begun, it is almost always too late to try something which has never been done before. There are too many risks involved with radically new concepts, and these risks are difficult to accept on a limited time basis.'

Soon after Mr Barlow made Mr Moore Cognizant Engineer, he offered an engineering position to Mr Elliot. Mr Elliot had recently graduated from college with a degree in Mechanical Engineering and had been working at JPL on the Ranger program for about a year. Barlow gave Elliot some of the jobs Moore had had before he became Cognizant Engineer. Mr Elliot now became responsible for the solar panel deployment system.

By March 1963 a prototype of the torsion bar actuator had been made, and Mr Elliot began testing it. The actuator is shown in Exhibit 5. These actuators were rather heavy – four of them weighed a total of 4·5 lb. As is usually the case with newly developed mechanisms, several problems occurred in test. The planetary gear drive was improperly designed so that actuation torque was inadequate. The high stress level in the torsion bar caused unacceptable plastic deformation. The design obviously required additional development and was nowhere near as satisfying in hardware form as it had been in concept. At this stage there was one year remaining before the completed spacecraft was to begin final flight acceptance testing. The hardware freeze date was rapidly approaching. Quite a bit of design and development effort had been invested in the torsional actuator. However, since more effort was required, Mr Elliot and Mr Moore decided to consider the deployment problem anew.

After some thought, Mr Elliot suggested that the torsional actuators be abandoned. He suggested that the panels be individually spring loaded as before, but that all of them be retarded by one central retarder. He hoped his idea would increase the reliability of the system and decrease its weight. The central retarder which he proposed would be attached by cables to each of the panels and would allow them to open simultaneously in about 45 sec. The retarder assembly itself would be a hydraulic vane torque brake which would damp motion by forcing fluid from one side of a vane to the other through a tapered orifice. Mr Elliot had tried to minimize the possibility of fluid leakage in his design. Furthermore, he wanted to use alcohol instead of silicone oil as the damping fluid because alcohol, should it leak out, would not leave a film on the optical lenses.

In late June 1963 Mr Elliot presented his central retarder idea to a Design Review Board which consisted of certain JPL engineering staff members. The only difficulty Elliot foresaw with his idea was that air bubbles might form in the liquid as the retarder was filled. Air bubbles would prevent smooth retardation of the panels. In general, the Design Review Board liked the central retarder idea. The Board members agreed that the new design would be lighter, cheaper, and more reliable than the torsion bar actuator design. The total weight of the new system would be 3·5 lb – 1 lb less than the weight of the torsion bar system. The Board decided that the filling of the retarder could be done in a vacuum to prevent formation of air bubbles. The Board members could not agree as to whether alcohol or silicone oil should be used, however. One group of members wanted to pursue the alcohol idea. Silicone oil could be used if trouble developed, they claimed. The other members argued that all development time should be spent on the oil system. They pointed out that the actuator parts would be better lubricated and better damped. Eventually, the Board decided to try using alcohol as the damping fluid. This decided, the Board examined the mechanics of the retarder, the connecting cables, and the deployment springs.

EXHIBIT 5   Torsion Bar actuator

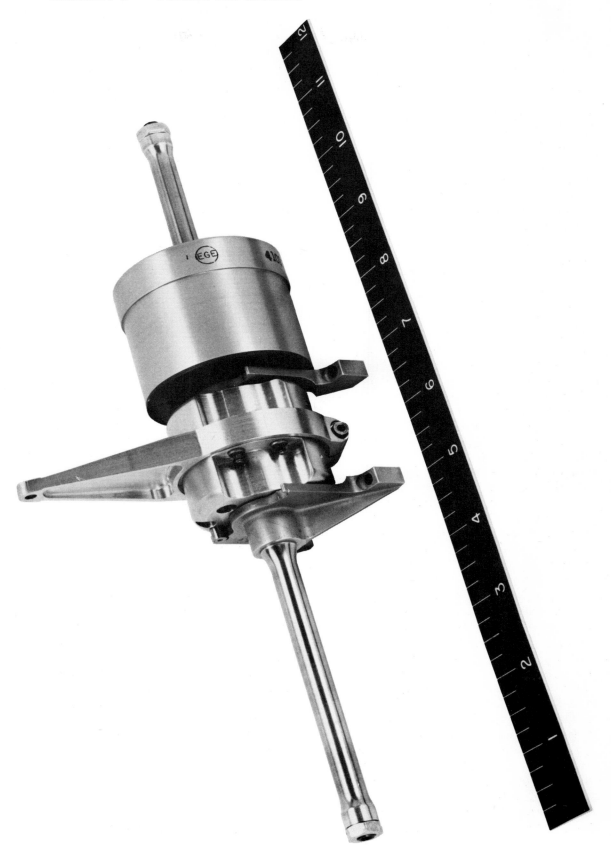

EXHIBIT 6   Central retarder

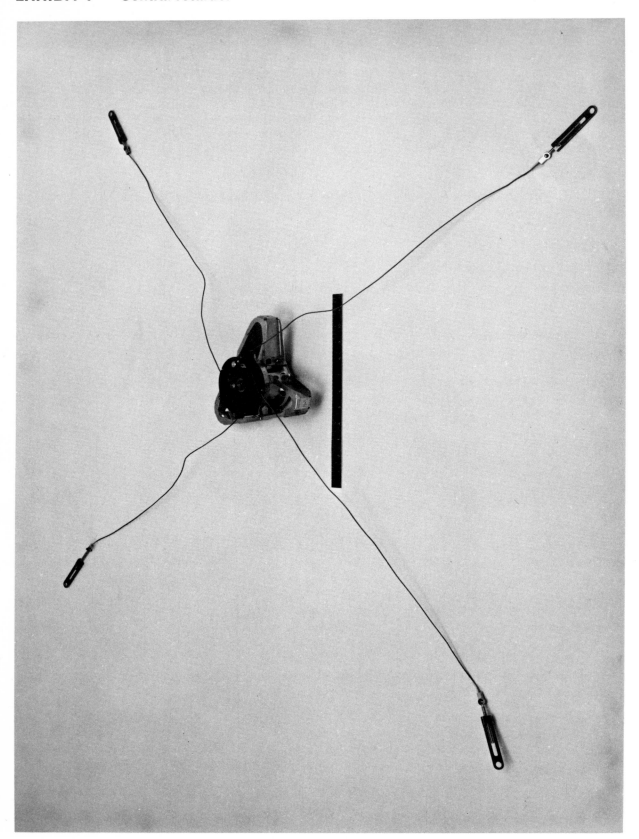

A total of sixteen recommendations were made on the recommended design. These recommendations ranged from the use of better seals to the generation of test plans to uncover any additional problems which might develop.

Mr Barlow and Mr Wilson decided to do a complete failure mode analysis of the central retarder concept. The purpose of the analysis would be to examine critically the failure possibilities in the system, to assess the impact these failures would have on the spacecraft, and to recommend corrective measures. Exhibit 7 is a summary of the failure mode analysis which Mr Barlow and Mr Wilson conducted.

The Design Review Board met again in August 1963 to discuss the failure analysis. The Board felt that the most serious, and most probable failure mode would be unretarded motion of a solar panel. (This is referred to in the summary as 'free fall' although the panels are deployed by springs, not by gravity.)

At this meeting, the Board was less impressed by the central retarder design. An excerpt from a summary of the Board's views says:

> 'The general consensus of the Board was that the current design left a bad taste in its mouth. Whether these feelings are the result of poor engineering or poor aesthetics may never be ascertained. At any rate, the feeling was that the basic system should be reexamined, simplified, cleaned up aesthetically, thoroughly tested, and thrown away.'

The Board made several specific recommendations which included the following:

1. Latch cables to panel after panel is folded up. This should minimize the possibility of cable snarling.
2. Check case of panel free fall to determine if panel is damaged.
3. Look for alternative deployment approaches (one likely contender is to let the panels free fall, stopping them with the damper which is to provide the dynamic coupling and a piece of balsa wood to absorb excess energy).

Since June, Mr Elliot had been following the recommendations made in the first Review Board meeting. Tests made on prototype models demonstrated that using alcohol would require extremely tight fits and tolerances. He therefore had to discard the alcohol idea and return to silicone oil. Other tests revealed other difficulties with the central retarder. Oil leakage was still a problem. The temperature compensation scheme was inadequate. There were also difficulties with the negator leaf springs which were to be used to deploy the panels.

In January 1964, Mr Elliot left JPL. The Mariner C time schedule dictated that the complete spacecraft enter the space simulation chamber for a flight acceptance test in July 1964 – six months remained in which to complete the panel deployment system. Considering the short time left before testing was to begin, Mr Barlow was now faced not only with a very tight time schedule in which to ready the deployment system, but also with the task of redistributing the workload of the central retarder development. The problem had become a crisis. The hiring and training of a new staff member would have required many months and was obviously not feasible at this time. As a result, he reassigned the deployment system as an additional responsibility to other members of his group. The central retarder was given to Mr E. Floyd for further work and the deployment springs to Mr J. Randall.

Mr Floyd, a mechanical engineer on Mr Barlow's staff who had been with JPL for some years, was Cognizant Engineer on the Mariner C scan actuator drive. He had a large amount of experience in solving unexpected design problems. Mr J. Randall was Cognizant Engineer for the solar vane system which was part of the attitude control subsystem.

Mr Floyd continued to carry out the various recommendations of the Design Review Board. Working on an overtime basis, he gradually developed a workable unit. Mr Randall decided to work with clock springs rather than the negator leaf springs

EXHIBIT 7   Division 35 subsystem failure mode analysis review

Cog. Engineer L. Elliott   Division 35 Subsystem Failure Mode Analysis Review

Assembly Part	Description	Description of Assumed Failure	Item Failure Mode	Influence on Mission	Possibility Factor	Possible Methods to Eliminate Failure Mode
Retarder	Empty	Little or no retard force	Panels open very fast - break on impact	Probable failure if panels break completely	1	1. Proper assembly and filling
	Air bubbles	Uneven retardation at higher air % No effect at very low %	1. Possible panel damage 2. Strap(s) break-panel uncontrolled	Failure if panels break-up	Probability is inversely to % air	1. Proper filling technique 2. Leak observation 3. Actuation check
	Foreign particles	1. Seize up of unit 2. Partial orifice clogging	Dependent on particle size (30 micron estimated critical point.) 2. Slower opening time	No problem unless total seize-up	3 to 10 mic 1 over 10 mic	1. Filter Fluid to 5 microns 2. Clean room assembly 3. Clean filling equipment 4. Assembly, actuate, disassemble and clean, reassemble
	Broken rewind spring	1. Cables slack, retarder opens partially to panel deployed position	1. Panels move rapidly until slack takes up	1. Possible panel damage - depends on amount free travel	1	1. Spring is low stressed 2. Proper tension on cables at button-up
	Kinked strap	Bind up (dependent on location of kink)	One panel either stops completely or slows up temporarily, then catches up - rapidly	1. Partially open panel is failure 2. Slow down and catch up is like air bubbles	2	1. Proper installation and care in assembly 2. Visual check during panel installation and button-up

EXHIBIT 7 *(contd.)*

Cog. Engineer L. Elliott

Division 35 Subsystem Failure Mode Analysis Review

Assembly Part	Description	Description of Assumed Failure	Item Failure Mode	Influence on Mission	Possibility Factor	Possible Methods to Eliminate Failure Mode
	Twisted strap	Extra friction if twisted area is in pulley or guides	Slow opening of all panels	None	3	Visual inspection
	Notched or nicked strap	Stress concentration	Strap may break allowing one cable to free fall	Free falling panel will probably break up	3	1. Visual inspection 2. Proof load of each strap
	Improper strap adjustment	Slack in one or more straps	Panels "jump" to take up slack	None	1	1. Proper installation and adjustment during panel installation
	Strap out of pulley groove between guards	Probably nothing, but in severe case, loop could form causing kink and jamming retarder	Nothing unless jam up, then panels don't open	Failure if panels do not open, otherwise, none	1	1. Proper installation and adjustment of cables during panel installation 2. Visual inspection
	Retarder temp. outside expected limits	Cold-slows down. Hot-speeds up	Nothing	None	1	1. Preflight calculations and tests
	Thermal gradient	Would take about 400F gradient across retarder to cause binding	None	None	0 - in order of magnitude to cause failure	1. Retarder was designed to avoid this problem
One panel does not release with others	1. Panel does not release at all 2. Panel releases later	1. Panel stays folded / Panel comes down uncontrolled	1. Low power A/C fouled up 2. Panel breaks up on impact	1. Failure 2. Failure	1. 2 2. 1	Pray
Power spring	One spring breaks	Total torque on retarder is lower than normal	Panels all come down, but slower	None	1	1. Design has two springs per panel Either spring will open panel but at reduced rate 2. Design is such that broken springs ends curl inward with no other damage

Mr Elliot had used because there was a better established technology for clock springs.

In March 1964 tests were begun on the revised central retarder deployment system together with the cruise dampers and latches which had been developed. The retarder is shown in Exhibit 6. The purpose of these tests was to observe the entire solar panel release, opening, and latching sequence. Tests were conducted by positioning the spacecraft on its 'side'. The two panels with vertical hinge axes were then released and allowed to move to the latched position. When the tests were carried out, the panels stopped short of latching onto the cruise dampers. After some analysis, the engineers came to the conclusion that each of the deployment system components was performing to its requirements, but that the component requirements were inconsistent. Another result of the tests was that the panels were not damaged when they were allowed to deploy unretarded in air. The deployment time was $5\frac{1}{2}$ sec, and the cruise dampers and latches were considerably overloaded, but without any serious effects.

Encouraged by the test results, Mr Barlow decided to further investigate the possibility of letting the panels deploy unretarded. He knew that the terminal velocity of the panels in space would be considerably greater than that in air, but if this free-fall scheme worked, it would increase the reliability of the system and decrease its weight. Mr Barlow assigned the detailed investigation of the idea to Dr P. Lyman who up to now had been engaged in the development of the cruise damper. The general approach was to calculate the capacity of the solar panel, all of its associated hardware and the octagon structure to survive the opening shock. This would yield the allowable deceleration rate. The cruise dampers would then be redesigned to provide the desirable deceleration characteristics, and a test programme would be conducted to verify the approach. After a high confidence had been achieved a final test would be run in a vacuum chamber using all flight type equipment. In addition, Mr Barlow established the following ground rules to be followed in the free-fall development program:

1. Each solar panel shall have two deployment springs, each of which is capable of accomplishing panel deployment.
2. All hardware must be capable of handling the load imposed with both deployment springs operating at the maximum specified torque.
3. There shall be no changes in the solar panel with the possible exception of an extra bolted-on doubler in the latch area. It must be remembered that the panel will be fully equipped with solar cells in place at the time this modification is made (if it is made).
4. There shall be no changes to the electronic chassis or its mounting with the exception of longer bolts for installation of the cruise damper.
5. If at all possible, the damping factor and the spring rate of the cruise damper, when acting as a cruise damper shall not be changed.

Meanwhile, Mr Floyd and Mr Randall worked on the latching problem revealed by the tests. In order to get the panels to latch to the cruise dampers, they increased the deployment torque on the panels by increasing the number of deployment springs used. This, along with other features added to solve various problems, increased the weight and complexity of the central retarder system. One member of the Design Review Board compared the central retarder with the original linear actuators used on the Ranger/Mariner II and came to the conclusion that the linear actuator scheme was slightly better than the central retarder design.

In April 1964 Randall and Floyd finished their revisions on the central retarder system and began testing a prototype. The tests revealed no serious difficulties or additional problems, so they initiated the fabrication of flight hardware.

After completing a dynamic analysis of free-all deployment, Dr Lyman decided that the free-fall approach was feasible. Exhibit 8 shows his analysis. During the

EXHIBIT 8   Dynamic analysis of free-fall deployment

Title _PANEL DEPLOYMENT SYSTEM_

Prepared by _R. T. Harling_   Date _4/24/64_

Checked by _____   Date _____

**JET PROPULSION
LABORATORY
CALIFORNIA INSTITUTE
OF TECHNOLOGY**

Report No. _ST 5.02.04_ Page _____

Project _MARINER C_

Classification _UNCLASS._

PROBLEM IDEALIZATION

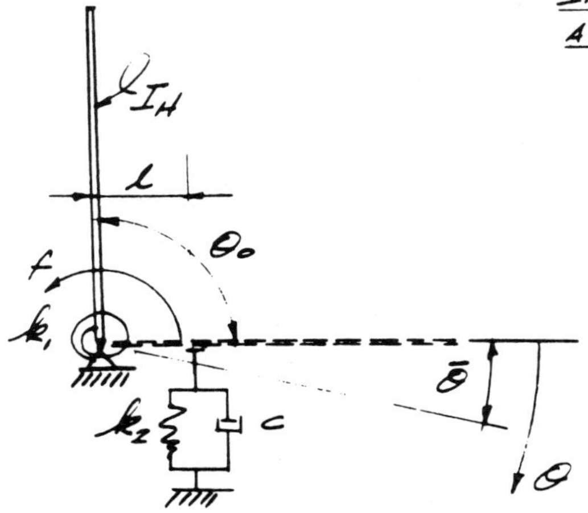

IN THIS SKETCH THE FOLLOWING ABBREVIATIONS ARE EMPLOYED:

I_H = ROTARY INERTIA OF PANEL ABOUT THE HINGE, IN·LB·SEC².

k_1 = TORSIONAL ACTUATOR SPRING RATE

f = CONSTANT TORQUE FRICTION

θ_0 = ANGLE BETWEEN PANEL BOOST POSITION AND CRUISE LATCH POSITION

l = MOMENT ARM FOR CRUISE DAMPER.

$\bar{\theta}$ = ANGULAR PRELOAD ON ACTUATOR

k_2 = CRUISE DAMPER SPRING RATE

c = " " DAMPING COEFFICIENT

θ = ANGULAR COORDINATE

PRIOR TO ENGAGEMENT OF THE CRUISE DAMPER, THE BELOW FREE BODY DIAGRAM IS APPLICABLE:

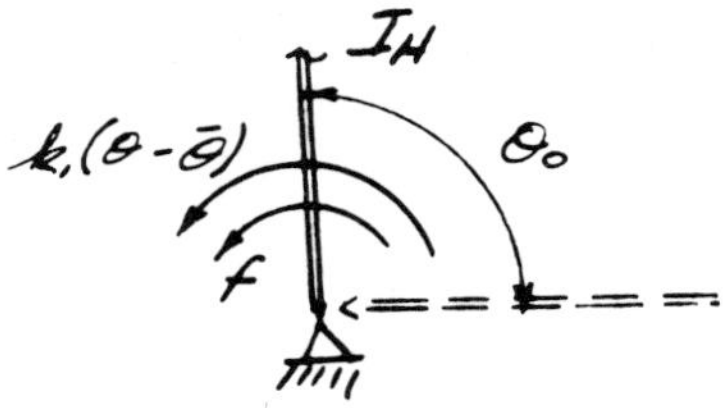

$$I_H \ddot{\theta} = -f - k_1(\theta - \bar{\theta})$$

$$I_H \dot{\theta}\frac{d\dot{\theta}}{d\theta} = (k_1\bar{\theta} - f) - k_1\theta ; \; LET : q = (k_1\bar{\theta} - f)$$

$$I_H \int_0^{\dot{\theta}_0} \dot{\theta}\,d\dot{\theta} = \int_{-\theta_0}^{0} \left[q - k_1\theta \right] d\theta$$

$$\frac{I_H}{2}\dot{\theta}_0^2 = \left[q\theta - \frac{k_1\theta^2}{2} \right]_{-\theta_0}^{0}$$

$$\dot{\theta}_0^2 = \left[q\theta_0 + \frac{k_1\theta_0^2}{2} \right]\left(\frac{2}{I_H}\right)$$

$$\boxed{\dot{\theta}_0 = \left[\frac{2q\theta_0 + k_1\theta_0^2}{I_H} \right]^{1/2}} \qquad Eq (1)$$

Eq (1) EXPRESSES THE VELOCITY AT PANEL-TO-CRUISE DAMPER ENGAGEMENT.

(Page 1 of 7)

JPL 0999-1-5   MAR 63

EXHIBIT 8 *(contd.)*

Title _BASIC EQUATIONS_

Prepared by _R. T. Hailey_ Date _4/24/64_

Checked by _____ Date _____

**JET PROPULSION
LABORATORY
CALIFORNIA INSTITUTE
OF TECHNOLOGY**

Report No. _55 50304_ Page _____

Project _MARINER C_

Classification _UNCLASS._

FOLLOWING ENGAGEMENT WITH THE CRUISE DAMPER, THE DIFFERENTIAL
EQUATION BECOMES:

$$I_H \ddot{\theta} = -f - k_1(\theta - \bar{\theta}) - k_2 l^2 \theta - c l^2 \dot{\theta}$$

$$= \underbrace{(k_1 \bar{\theta} - f)}_{q} - \theta(k_1 + k_2 l^2) - c l^2 \dot{\theta}$$

$$\ddot{\theta} + \frac{c l^2}{I_H} \dot{\theta} + \frac{(k_1 + k_2 l^2)}{I_H} \theta = \frac{q}{I_H}$$

TAKING THE LAPLACE TRANSFORM OF THIS D.E.:

$$s^2 F(s) - s(0) - \dot{\theta}_0 + \frac{c l^2}{I_H} s F(s) - \frac{c l^2}{I_H}(0)$$

$$+ \frac{(k_1 + k_2 l^2)}{I_H} F(s) = \frac{q}{I_H s}$$

$$\left[s^2 + \left(\frac{c l^2}{I_H}\right) s + \frac{(k_1 + k_2 l^2)}{I_H} \right] F(s) = \frac{q}{I_H s} + \dot{\theta}_0$$

FACTORING THE TERMS IN THE SQUARE BRACES:

$$s_{1,2} = \frac{-c l^2}{2 I_H} \pm \frac{1}{2} \sqrt{\left(\frac{c l^2}{I_H}\right)^2 - 4\left(\frac{k_1 + k_2 l^2}{I_H}\right)}$$

$$s_{1,2} = -S_A \pm S_B \; ; \; S_A = \frac{c l^2}{2 I_H} \qquad\qquad Eq\ (2)$$

$$S_B = \frac{1}{2 I_H} \sqrt{(c l^2)^2 - 4 I_H (k_1 + k_2 l^2)}$$

DEPENDING UPON THE VALUE OF S_B, THREE SEPARATE
FORMS OF THE SOLUTION EXIST:

$$(c l^2)^2 > 4 I_H (k_1 + k_2 l^2) \qquad s_{1,2} \text{ ARE REAL}$$

$$(c l^2)^2 = 4 I_H (k_1 + k_2 l^2) \qquad s_{1,2} \text{ ARE IDENTICAL}$$

$$(c l^2)^2 < 4 I_H (k_1 + k_2 l^2) \qquad s_{1,2} \text{ ARE COMPLEX}$$

EACH CASE WILL NOW BE SEPARATELY SOLVED.

(Page 2 of 7)

JPL 0999-1-S MAR 63

EXHIBIT 8 (contd.)

Title _REAL ROOT CASE_

Prepared by _R. T. Harling_ Date _4/24/64_

Checked by _____ Date _____

JET PROPULSION LABORATORY CALIFORNIA INSTITUTE OF TECHNOLOGY

Report No. _503.04_ Page ___ ___

Project _MARINER C_

Classification _UNCLASS._

$$F(s) = \frac{\dot{\theta}_0}{(s-s_1)(s-s_2)} + \frac{q}{I_H s(s-s_1)(s-s_2)}$$

USING THE METHOD OF PARTIAL FRACTIONS

$$F(s) = \frac{\frac{q}{I_H s_1 s_2}}{s} + \frac{\left[\frac{q + \dot{\theta}_0 I_H s_1}{I_H s_1 (s_1 - s_2)}\right]}{(s-s_1)} - \frac{\left[\frac{q + \dot{\theta}_0 I_H s_2}{I_H s_2 (s_1 - s_2)}\right]}{(s-s_2)} \qquad Eq\,(3)$$

PERFORMING THE INVERSE LAPLACE TRANSFORM:

$$\theta = \frac{q}{I_H s_1 s_2} + \left[\frac{q + \dot{\theta}_0 I_H s_1}{I_H s_1 (s_1 - s_2)}\right]e^{s_1 t} - \left[\frac{q + \dot{\theta}_0 I_H s_2}{I_H s_2 (s_1 - s_2)}\right]e^{s_2 t} \qquad Eq\,(4)$$

FOR $\theta = \theta_{MAX}$ SET $\dot{\theta} = 0$:

$$\dot{\theta} = \left[\frac{q + \dot{\theta}_0 I_H s_1}{I_H (s_1 - s_2)}\right]e^{s_1 t} - \left[\frac{q + \dot{\theta}_0 I_H s_2}{I_H (s_1 - s_2)}\right]e^{s_2 t} = 0$$

$$t_1 = \ln\left[\frac{q + \dot{\theta}_0 I_H s_2}{q + \dot{\theta}_0 I_H s_1}\right]\left(\frac{1}{s_1 - s_2}\right) \qquad Eq\,(5)$$

SUBSTITUTING Eq(5) IN Eq(4):

$$\theta_{MAX} = \frac{q}{I_H s_1 s_2} + \left[\frac{q + \dot{\theta}_0 I_H s_1}{I_H s_1 (s_1 - s_2)}\right]\left[\frac{q + \dot{\theta}_0 I_H s_2}{q + \dot{\theta}_0 I_H s_1}\right]^{\left(\frac{s_1}{s_1 - s_2}\right)}$$

$$- \left[\frac{q + \dot{\theta}_0 I_H s_2}{I_H s_2 (s_1 - s_2)}\right]\left[\frac{q + \dot{\theta}_0 I_H s_2}{q + \dot{\theta}_0 I_H s_1}\right]^{\left(\frac{s_2}{s_1 - s_2}\right)}$$

REAL ROOT CASE   Eq(6)

FOR MAXIMUM ACCELERATION: SET $\dddot{\theta} = 0$

$$\ddot{\theta} = s_1\left[\frac{q + \dot{\theta}_0 I_H s_1}{I_H (s_1 - s_2)}\right]e^{s_1 t} - s_2\left[\frac{q + \dot{\theta}_0 I_H s_2}{I_H (s_1 - s_2)}\right]e^{s_2 t}$$

$$\dddot{\theta} = s_1^2[\;]e^{s_1 t} - s_2^2[\;]e^{s_2 t} = 0 \;;\; t_2 = \ln\left\{\left[\frac{q + \dot{\theta}_0 I_H s_2}{q + \dot{\theta}_0 I_H s_1}\right]\left(\frac{s_2}{s_1}\right)^2\right\}$$

$$\ddot{\theta}_{MAX} = s_1\left[\frac{q + \dot{\theta}_0 I_H s_1}{I_H (s_1 - s_2)}\right]\left\{\left[\frac{q + \dot{\theta}_0 I_H s_2}{q + \dot{\theta}_0 I_H s_1}\right]\left(\frac{s_2}{s_1}\right)^2\right\}^{\left(\frac{s_1}{s_1 - s_2}\right)}$$

$$- s_2\left[\frac{q + \dot{\theta}_0 I_H s_2}{I_H (s_1 - s_2)}\right]\left\{\left[\frac{q + \dot{\theta}_0 I_H s_2}{q + \dot{\theta}_0 I_H s_1}\right]\left(\frac{s_2}{s_1}\right)^2\right\}^{\left(\frac{s_2}{s_1 - s_2}\right)} \;\; \text{REAL ROOT CASE} \;\; Eq\,(7)$$

EXHIBIT 8 *(contd.)*

Title __Complex Root Case__

Prepared by __R.J. Hailey__ Date __4/24/64__

Checked by _____ Date _____

JET PROPULSION LABORATORY
CALIFORNIA INSTITUTE OF TECHNOLOGY

Report No. __ST 5.0204__ Page ____

Project __Mariner C__

Classification __UNCLASS__

$$F(s) = \frac{\dot{\theta}_0}{(s+S_A)^2 + S_B{}^2} + \frac{\frac{q}{I_H s}}{I_H s\left[(s+S_A)^2 + S_B{}^2\right]} \quad ; \quad \begin{array}{l} S_1 = -S_A + iS_B \\ S_2 = -S_A - iS_B \end{array}$$

TAKING THE INVERSE LAPLACE TRANSFORM:

$$\theta = \left(\frac{\dot{\theta}_0}{S_B}\right)\left[e^{-S_A t}\sin S_B t\right] + \left(\frac{q}{I_H S_B}\right)\int_0^t \left[e^{-S_A t}\sin S_B t\right]dt$$

$$= \left(\frac{\dot{\theta}_0}{S_B}\right)\left[e^{-S_A t}\sin S_B t\right] + \frac{q}{I_H S_B (S_A{}^2 + S_B{}^2)}\left[S_B - e^{-S_A t}(S_A \sin S_B t + S_B \cos S_B t)\right]$$

$$\boxed{\theta = q'\left(1 - e^{-S_A t_3}\cos S_B t_3\right) + \left(\frac{\dot{\theta}_0 - q' S_A}{S_B}\right)e^{-S_A t}\sin S_B t_3}$$

WHERE: $q' = \frac{q}{I_H (S_A{}^2 + S_B{}^2)}$ COMPLEX ROOT CASE   Eq (8)

FOR θ_{MAX}, FIND t_3 @ $\dot{\theta} = 0$:

$$\dot{\theta} = -q' e^{-S_A t}(-S_B \sin S_B t) - q'(-S_A e^{-S_A t})\cos S_A t$$

$$+ \left(\frac{\dot{\theta}_0 - q' S_A}{S_B}\right)\left[e^{-S_A t}(S_B \cos S_A t) + (-S_A e^{-S_A t}\sin S_A t)\right]$$

$$\dot{\theta} = e^{-S_A t}\left[a \sin S_A t + b \cos S_A t\right] = 0$$

$$a = q' S_B - S_A \left(\frac{\dot{\theta}_0 - q' S_A}{S_B}\right)$$

$$b = q' S_A + S_B \left(\frac{\dot{\theta}_0 - q' S_A}{S_B}\right)$$

$$t_3 = \frac{1}{S_B}\tan^{-1}\left(\frac{b}{a}\right) = \boxed{\frac{1}{S_B}\tan^{-1}\left[\frac{S_B \dot{\theta}_0}{S_A \dot{\theta}_0 - q/I_H}\right] = t_3} \quad Eq (9)$$

COMPLEX ROOT CASE

FOR MAXIMUM ACCELERATION: $\dddot{\theta} = 0$

$$\ddot{\theta} = e^{-S_A t}\left\{\underbrace{(a S_B - b S_A)}_{b'}\cos S_A t + \underbrace{(-b S_B - a S_A)}_{a'}\sin S_A t\right\}$$

$$\dddot{\theta} = e^{-S_A t}\left\{(a' S_B - b' S_A)\cos S_A t + (-b' S_B - a' S_A)\sin S_A t\right\} = 0$$

EXHIBIT 8 *(contd.)*

Title _COMPLEX ROOT CASE (CONT'D)_

Prepared by _R.T. Hurley_   Date _4/24/64_

Checked by _____   Date _____

**JET PROPULSION
LABORATORY
CALIFORNIA INSTITUTE
OF TECHNOLOGY**

Report No. _JT 502.04_ Page _____

Project _MARINER C_

Classification _UNCLASS._

EQUATING THE TERM IN THE CURLY BRACKETS TO ZERO:

$$t_4 = \frac{1}{s_B} \tan^{-1}\left[\frac{a's_B - b's_A}{b's_B + a's_A}\right]$$

$$t_4 = \frac{1}{s_B}\tan^{-1}\left\{\frac{\dot\theta_0(3s_A^2 - s_B^2) - 2s_A \frac{q}{I_H}}{\frac{\dot\theta_0 s_A}{s_B}(s_A^2 - 3s_B^2) + \frac{q}{I_H s_B}(s_B^2 - s_A^2)}\right\} \qquad Eq(10)$$

$$\ddot\theta_{MAX} = e^{-s_A t_4}\left\{\left(\frac{q}{I_H} - 2s_A\dot\theta_0\right)\cos s_B t_4 + \left[\frac{\dot\theta_0}{s_B}(s_A^2 - s_B^2) - \frac{q s_A}{I_H s_B}\right]\sin s_B t_4\right.$$

COMPLEX ROOT CASE   Eq (11)

IDENTICAL ROOT CASE: $s_1 = s_2 = -s_A$

$$F(s) = \frac{\dot\theta_0}{(s + s_A)^2} + \frac{q/I_H}{s(s + s_A)^2}$$

TAKING THE INVERSE LAPLACE TRANSFORM:

$$\theta = \dot\theta_0 t_5 e^{-s_A t_5} + \frac{q}{I_H s_A}\left[\frac{1}{s_A} - e^{-s_A t_5}\left(t_5 + \frac{1}{s_A}\right)\right]$$

$$\theta_{MAX} = \frac{1}{I_H s_A^2}\left\{q + e^{-s_A t_5}\left[t_5 \dot\theta_0 I_H s_A^2 - q(s_A t_5 + 1)\right]\right\}$$

IDENTICAL ROOT CASE   Eq (12)

FOR θ_{MAX}, $\dot\theta = 0$

$$\dot\theta = \frac{e^{-s_A t_5}}{I_H s_A^2}\left\{-s_A^3 t_5 \dot\theta_0 I_H + q s_A^2 t_5 + q s_A + \dot\theta_0 I_H s_A^2 - q s_A\right\} = 0$$

$$t_5 = \frac{\dot\theta_0 I_H}{s_A \dot\theta_0 I_H - q} \qquad Eq (13)$$

EXHIBIT 8 *(contd.)*

Title _IDENTICAL ROOT CASE_

Prepared by _R.T. Hadley_   Date _4/24/64_

Checked by _____   Date _____

**JET PROPULSION
LABORATORY
CALIFORNIA INSTITUTE
OF TECHNOLOGY**

Report No. _ST 5.02.04_ Page _____

Project _MARINER C_

Classification _UNCLASS._

FOR MAXIMUM ACCELERATION:

$$\ddot{\theta} = \frac{-sa\, e^{-sa\, t_6}}{I_H}\left[-sa\, t_6\, \dot{\theta}_0\, I_H + g\, t_6 + \dot{\theta}_0\, I_H\right] + \frac{e^{-sa\, t_6}}{I_H}\left[-sa\, \dot{\theta}_0\, I_H + g\right]$$

$$\ddot{\theta}_{MAX} = e^{-sa\, t_6}\left\{(t_6 sa - 1)(sa\, \dot{\theta}_0 - g/I_H) - sa\, \dot{\theta}_0\right\}$$

$$\underline{\text{IDENTICAL ROOT CASE}} \qquad Eg(14)$$

$$\dddot{\theta} = e^{-sa\, t_6}\left\{sa(sa\, \dot{\theta}_0 - g/I_H) - sa\left[(t\, sa - 1)(sa\, \dot{\theta}_0 - g/I_H) - sa\, \dot{\theta}_0\right]\right\} = 0$$

$$t_6 = \frac{3 sa\, \dot{\theta}_0\, I_H - 2g}{sa(sa\, \dot{\theta}_0\, I_H - g)} \qquad Eg(15)$$

<u>INITIAL DECELERATION</u>:   RETURNING TO THE D.E. ON ρ B2

$$\ddot{\theta} = \frac{g}{I_H} - \frac{c\ell^2}{I_H}\dot{\theta} - \frac{(k_1 + k_2\ell^2)}{I_H}\theta$$

$$A_T \quad t = 0 \qquad \theta = 0$$

$$\ddot{\theta}_I = \frac{g}{I_H} - \frac{c\ell^2}{I_H}\dot{\theta}_0 \qquad Eg(16)$$

<u>DEPLOYMENT TIME</u>

$$I_H \ddot{\theta} + k\theta = T_0 \qquad ; \; k = \begin{cases} 6/\pi & \text{FOR 2 SPRING} \\ 3/\pi & \text{FOR 1 SPRING} \end{cases} \; ; \; \omega^2 = \frac{k}{I_H}$$

$$\theta_0 = \dot{\theta}_0 = 0$$

$$\theta = \frac{T_0}{k}\left[1 - \cos\omega t\right]$$

$$A_T \; \theta = \pi/2 \; : \quad t = \sqrt{\frac{I_H}{k}}\, \cos^{-1}\left[1 - \frac{\pi k}{2 T_0}\right] \qquad Eg(17)$$

EXHIBIT 8 *(contd.)*

```
HINGE INERTIA= 138.250INLB-SEC2   FRICTION=   .000
TORQUE= 15.000IN-LB   DEGREES PRELOAD= 540.000

DEPLOY VELOCITY=  .5589RD/SEC    TIME=  5.458SEC
```

DAMPING RATIO	DISPLACEMENTS FOR ACTUAL AND NOMINAL CRUISE DAMPER SPRINGS=509/530LB/IN			
	PANEL ANGULAR DISPL (DEGREES)		DAMPER DISPL (INCHES)	
	ACTUAL	NOMINAL	ACTUAL	NOMINAL
.00	3.0814	3.0189	.2958	.2897
.05	2.8546	2.7967	.2740	.2684
.10	2.6564	2.6025	.2550	.2498
.15	2.4819	2.4315	.2382	.2334
.20	2.3272	2.2800	.2234	.2188
.25	2.1894	2.1450	.2101	.2059
.30	2.0659	2.0240	.1983	.1942
.35	1.9546	1.9150	.1876	.1838
.40	1.8539	1.8163	.1779	.1743
.45	1.7625	1.7267	.1691	.1657
.50	1.6790	1.6450	.1611	.1579
.55	1.6027	1.5702	.1538	.1507
.60	1.5326	1.5015	.1471	.1441
.65	1.4680	1.4383	.1409	.1380
.70	1.4083	1.3798	.1351	.1324
.75	1.3531	1.3257	.1298	.1272
.80	1.3018	1.2754	.1249	.1224
.85	1.2540	1.2286	.1203	.1179
.90	1.2094	1.1850	.1161	.1137
.95	1.1678	1.1441	.1121	.1098
1.00	1.1287	1.1059	.1083	.1061

months of April and early May of 1964, Dr Lyman continued to investigate the required changes necessary to use the available Mark I cruise damper with the free-fall system. The changes allowed at this late date had been specified by Mr Barlow and have been listed above. The mandatory change was to provide sufficient stroke, i.e., energy dissipation, to the damper so that it would not 'bottom-out' thus preventing shock loads from being transmitted into the bus or the solar panel. The biggest question that had to be answered was whether or not the spacecraft midcourse firing damping requirements were compatible with the retardation requirements for panel deployment. The energy which must be removed from the moving solar panels is just the change in potential energy of the panel deploy springs. The maximum stroke of the damper was limited by geometrical interface constraints between the solar panel and the bus. The spring rate of the damper springs was determined by the panel frequency requirements. With these parameters defined, or at least bounded, the analysis of the panel deploy dynamics mentioned earlier was performed with the gratifying result that the system requirements were not completely inconsistent. To implement the system, it was necessary to ask the spacecraft System Engineer for a change in the panel damping requirements from a damping ratio of 0·6 to 1·0 matched to 5%, to a ratio of 0·15 to 0·70 matched to within 10% which was, indeed, compatible with the spacecraft autopilot control. (Exhibit 9 shows the installed cruise damper.)

In May 1964 the Design Review Board met to discuss the progress of the free-fall concept. The members agreed that this concept complied with essentially all the original ground rules for the project. The net weight saved would be 2·41 lb. The Board outlined an extensive and tightly scheduled testing programme which included vacuum chamber tests. At this time, the central retarder was already a working system and flight hardware was being manufactured. However, the most probable failure mode of the central retarder was still the free-fall mode. The central retarder system was still called 'primary' because another system could not be chosen until it had been proved.

According to the Design Review Board, a final comparison of the central retarder system with the free-fall system revealed:

1. With the free-fall system, far fewer failure modes exist. As a matter of fact, it was not possible to conceive of a single failure mode, assuming that the pin pullers would release the panels, where the free-fall system could abort the mission. With the retarder, there were such failure modes.
2. Reliability of the free-fall system is inherently greater.
3. A source of oil contamination due to the central retarder is eliminated by using the free-fall system.
4. Use of the free-fall system would reduce the spacecraft weight by 2·4 lb.
5. The decision to utilize the free-fall system is reversible. In fact, should the vacuum deployment test result in surprise damage to the solar panels or structure, the retarder and four spring actuators could be reinstated without reinstating the old cruise damper. Thus the improved damper with its added ability to prevent many retarder failure modes could be used in any event.

In June 1964, the Design Review Board recommended that the free-fall system be made 'primary' and made an Engineering Change Request. The central retarder would be the back-up system. Tests of the free-fall system revealed no further problems, and witnesses to these tests described them as 'rather dull to watch'.

The first launch of the Mariner C on 5 November 1964 was a failure because of a launch vehicle shroud separation problem. The second launch, also in November of 1964, was successful and the spacecraft (now designated Mariner IV) performed almost perfectly. It passed within 6 120 miles of the Martian surface on 15 July 1965. It returned 325·5 million pieces of scientific data including photographs such as that

EXHIBIT 9   Installed cruise damper

EXHIBIT 10   Photograph Martian surface

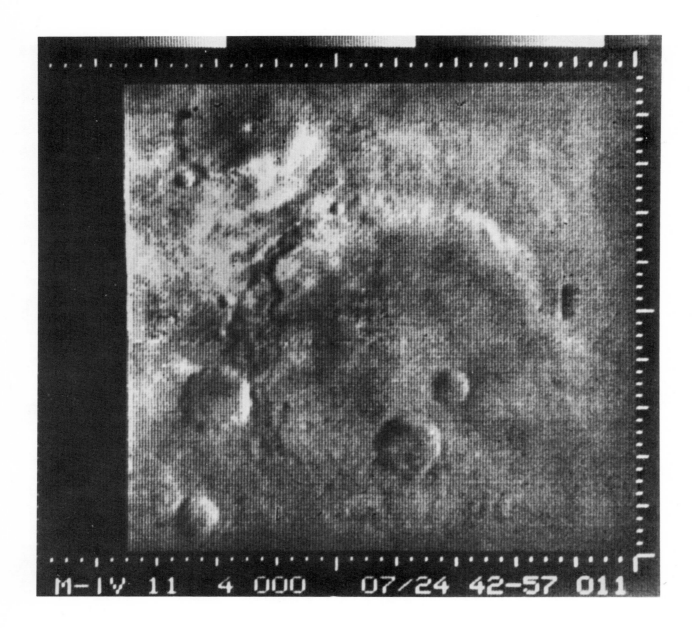

shown in Exhibit 10. The spacecraft functioned normally until December 1967 at which time it expended its supply of attitude control gas.

SUGGESTED STUDY

1. What is the problem?
2. What is the philosophy of design? What are the constraints?
3. What groups of engineers worked on the problem?
4. What projects were initiated in order to define the design? This could be posed after reading the preliminary design and the responses compared to what was actually done.
5. List the possible configurations with the advantages and disadvantages of each configuration.
6. Propose a concept of your own, including an estimate of the weight, time for delivery and cost.
7. Make a failure mode analysis (of a familiar device such as bicycle, brake system, a step-ladder, etc.).
8. Why would the hiring and training of a new staff member require many months?
9. How can one make progress if it is too late to try something after a program has begun?

ML

Professor Chengi Kuo

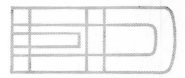

ENGINEERING DESIGN SERIES | EDITOR R K PENNY

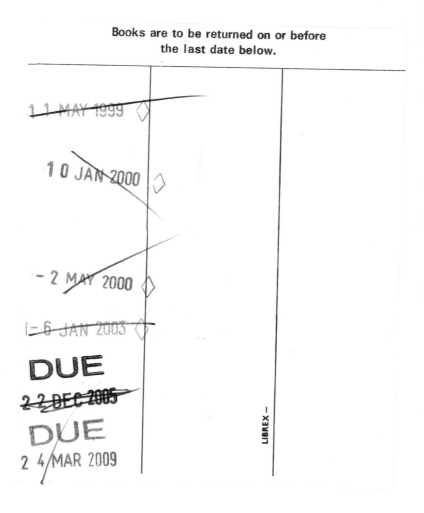